Multisensor Data Fusion

For a complete list of the *Artech House Radar Library*,
turn to the back of this book . . .

Multisensor Data Fusion

Edward Waltz
James Llinas

Artech House
Boston • London

Library of Congress Cataloging-in-Publication Data

Waltz, Edward
   Multisensor data fusion / Edward Waltz and James Llinas.
     p. cm.
   Includes bibliographical references and index.
   ISBN 0-89006-277-3
   1. Command and control systems. 2. Multisensor data fusion.
I. Llinas, James. II. Title.
UB212.W35   1990   90-41622
623--dc20   CIP

British Library Cataloguing in Publication Data

Waltz, Edward
   Multisensor data fusion.
   1. Military operations. Applications of computer systems.
   I. Title II. Llinas, James
   355.40285

   ISBN 0-89006-277-3

© 1990 Artech House, Inc.
685 Canton Street
Norwood, MA 02062

All rights reserved. Printed and bound in the United States of America. No part of this publication may be reproduced or utilized in any form or by any means, electronic or mechanical, including photocopying, recording, or by any information storage and retrieval system, without permission in writing from the publisher.

International Standard Book Number: 0-89006-277-3
Library of Congress Catalog Card Number: 90-41622

10   9   8   7   6   5   4   3   2   1

CONTENTS

Foreword	xi
Preface	xv
Chapter 1 Introduction	1
1.1 Motivations and Applications for Data Fusion	2
1.2 A Functional Model of the Data Fusion Process	5
1.3 Methods for Implementing Data Fusion	8
1.4 The Current State of Data Fusion	11
References	13
Chapter 2 Taxonomy of Functional Architectures	15
2.1 Generalized Processing Models for Data Fusion	15
2.2 Generic Level 1 Processing Architectures	20
2.3 Real-World Architectures	23
2.4 Functional Requirements for Situation and Threat Assessment	28
2.5 Blackboard Architectures and Opportunistic Reasoning	33
2.5.1 Problem-Solving Techniques for Situation and Threat Assessment	33
2.5.2 Concepts of Blackboard Processing	35
2.5.3 Pilot Aiding Applications	38
2.5.4 Some Advantages and Disadvantages of Blackboard Architecture	39
2.6 Achieving Real-Time Performance: Parallel Architectures	40
2.6.1 Parallelism in Rule-Based Systems	41
2.6.2 Parallelism in Semantic Networks	43
2.6.3 Parallelism in Logic Programming	44
2.7 Comments	44
References	46
Chapter 3 Defense Applications of Data Fusion	49
3.1 Representative Military Applications	50
3.1.1 Antisubmarine Warfare	51

		3.1.2	Tactical Air Warfare	54

	3.1.2	Tactical Air Warfare	54
	3.1.3	Land Battle Battlefield Warfare	57
3.2	Use of Sensed Information		58
3.3	Military Data Fusion-Decision Support Architecture		60
	3.3.1	Data Fusion Subsystem	61
	3.3.2	Decision Support Subsystem	66
	3.3.3	Human Command and Control	67
3.4	Characteristics of Fused Information		67
3.5	Special Considerations for Military Systems		73
	3.5.1	Information Security	73
	3.5.2	Multisensor Countermeasures	74
	3.5.3	Data Confidence Identification for Lethal Use	75
	3.5.4	Fusion System Classification	75
3.6	Comments		75
References			76
Chapter 4	Sensors, Sources, and Communication Links		77
4.1	Detection and Estimation Functions of Sensors		78
	4.1.1	A General Sensor Model	78
	4.1.2	Classical Detection and Estimation	80
	4.1.3	Sensor and System Performance Measures	84
	4.1.4	Detection and Estimation in Data Fusion Systems	86
4.2	Alternative Sensor Implementations		88
4.3	Comparison of Hard- and Soft-Decision Sensors		92
	4.3.1	Numerical Results for a Bayesian Example	93
	4.3.2	Criteria for Beneficial Combination	95
	4.3.3	Operational Benefits of Soft-Decision Sensors and Fusion	95
4.4	Soft-Decision Sensor Processing		97
	4.4.1	Target Signatures	97
	4.4.2	Features for Imaging and Nonimaging Applications	99
	4.4.3	Classification Database Requirements	100
	4.4.4	ATR Processes	100
4.5	Military Sensors		104
	4.5.1	Radar Surveillance and Fire Control Sensors	106
	4.5.2	Infrared Search-Track and Imaging Sensors	110
	4.5.3	Electro-Optical Sensors	113
	4.5.4	Identification Friend or Foe Sensors	113
	4.5.5	Electronic Support Measures Sensors	114
	4.5.6	Acoustic Sensors	118
4.6	Military Source and Data Link Characteristics		120
	4.6.1	Source Data	121
	4.6.2	Communication Links for Sensor and Source Data	122

4.7	A Simple Fusion Network Example	124
References		126

Chapter 5 Sensor Management — 129
- 5.1 Sensor Management Functions — 130
- 5.2 Sensor Interfaces — 135
- 5.3 Establishing Target Priority — 136
  - 5.3.1 Establishing Priority Autonomously — 138
  - 5.3.2 Establishing Priority Cooperatively — 141
- 5.4 Sensor-to-Target Assignment Methods — 143
- 5.5 Sensor Cueing and Hand-off — 147
- 5.6 Sensor Management Applications — 151
  - 5.6.1 An Air-Combat Sensor Management Example — 151
  - 5.6.2 A Distributed Sensor Network Example — 154

References — 157

Chapter 6 Data Fusion for State Estimation — 159
- 6.1 Association of Data and Tracking of Dynamic Targets — 160
  - 6.1.1 Static Data Association for Target Localization — 160
  - 6.1.2 Dynamic Data Association and Target Tracking — 163
  - 6.1.3 The Roles of Association and Estimation — 165
  - 6.1.4 Issues in Association and Tracking — 167
- 6.2 Static Data Association and Target Position Location — 168
  - 6.2.1 Multiple Sensor, Common Dimensionality — 168
  - 6.2.2 Multiple Sensor, Different Dimensionality — 170
  - 6.2.3 Multiple-Site, Multiple-Sensor — 172
- 6.3 Taxonomy of Dynamic Data Association and Tracking Algorithms — 175
  - 6.3.1 A General Association and Tracking Loop — 175
  - 6.3.2 Taxonomy of Design Approaches — 180
  - 6.3.3 Key Design Parameters — 185
- 6.4 Report-Track Data Association for Target Tracking — 187
- 6.5 Track-Track Data Association — 193
  - 6.5.1 Recursive Track-Track Association — 193
  - 6.5.2 Batch Track-Track Association — 197
- 6.6 State Estimators for Association and Tracking — 198
  - 6.6.1 State Estimation — 198
  - 6.6.2 Least Squares Estimator — 200
  - 6.6.3 Weighted Least Squares — 201
  - 6.6.4 Maximum Likelihood Estimator — 201
  - 6.6.5 Maximum Likelihood Estimation for Static Target Localization — 201
  - 6.6.6 Minimum Variance Estimators for Recursive Tracking — 203

6.6.7	The Kalman Filter	204
6.6.8	Tracking Filters for Maneuvering Targets	205
6.6.9	Tracking Filter Implementation Issues	207
6.6.10	Multiple Sensor Estimation Considerations	208

References 210

Chapter 7 Data Fusion for Object Identification — 213
- 7.1 Overview of Data Fusion Algorithms for Identity Estimation — 214
  - 7.1.1 Introduction — 214
  - 7.1.2 Taxonomy of Identity Fusion Algorithms — 214
  - 7.1.3 Algorithm Descriptions — 216
- 7.2 Comparisons Between Bayesian and Dempster-Shafer Techniques — 237
  - 7.2.1 Perspectives on the Formalisms — 238
  - 7.2.2 Bayesian Probability Theory — 239
  - 7.2.3 Dempster-Shafer Evidence Theory — 245
  - 7.2.4 Representative Comparison between Bayes and Dempster-Shafer — 252

References 260

Chapter 8 Military Concepts of Situation and Threat Assessment — 263
- 8.1 The Military Problem-Solving Process — 265
  - 8.1.1 The SHOR Model of Military Decision Processing — 269
  - 8.1.2 Data Processing Tasks in the SHOR Model — 272
  - 8.1.3 Information Processing Tasks in the SHOR Model — 274
- 8.2 Defining Situation Assessment — 277
- 8.3 Dealing with Concealment, Cover, and Deception — 278
- 8.4 Summary Comments on Situation Assessment — 283
- 8.5 The Character and Composition of the Threat — 284
  - 8.5.1 Elements of the Threat — 286
  - 8.5.2 Threat Assessment Functions — 288
- 8.6 Multiple Perspectives of Threat Elements — 288
- 8.7 Summary Comments on Threat Assessment — 289

References 290

Chapter 9 Implementation Approaches for Situation and Threat Assessment — 293
- 9.1 Issues and Problems of Methodological Selection — 295
- 9.2 Expectation Template-Based Techniques — 300
  - 9.2.1 Intelligence Preparation of the Battlefield — 301
  - 9.2.2 Event-Activity Profiling — 308
  - 9.2.3 Figure-of-Merit Techniques — 316
  - 9.2.4 Knowledge-Based and Expert-System Techniques — 327
- 9.3 Cooperative and Adaptive Approaches — 337
  - 9.3.1 ALLIES, STARS/PRM, and TAES: Cooperating and Adaptive Approaches — 340

9.4	The Role of Performance Models in STA	343
9.5	Comments	345
References		346

Chapter 10 Data Fusion System Architecture Design — 349
- 10.1 The Data Fusion Systems Engineering Process — 350
  - 10.1.1 Definition of Mission Requirements — 350
  - 10.1.2 Definition of Functional Requirements — 353
  - 10.1.3 Sensor Requirements Analysis — 353
  - 10.1.4 Subsystem Design Process — 354
  - 10.1.5 Design Synthesis — 358
- 10.2 Database Management for Data Fusion — 358
  - 10.2.1 Level 1: Association and Attribute Refinement — 358
  - 10.2.2 Levels 2 and 3: Situation Assessment and Threat Refinement — 359
  - 10.2.3 DBMS Considerations for Data Fusion Systems — 360
- 10.3 Data Processing for Data Fusion — 360
  - 10.3.1 Processing System Architecture — 361
  - 10.3.2 Methods of Applying Parallelism to Data Fusion — 363
  - 10.3.3 Parallel Computing Architectures Applied to Data Fusion — 368
  - 10.3.4 Connectionist Architectures Applied to Data Fusion — 370
- 10.4 Centralized System Architectures — 373
- 10.5 Integrated System Architectures — 376
  - 10.5.1 Integrated Sensors — 377
  - 10.5.2 Integrated Avionics Architecture — 378
- 10.6 Database and Processing Parametric Requirements — 382
- References — 386

Chapter 11 System Modeling and Performance Evaluation — 389
- 11.1 The Basic Theory of C^3 Systems — 390
  - 11.1.1 Lanchester Models of Combat — 390
  - 11.1.2 Command, Control, and Communication Models — 393
- 11.2 Formal Models of the Data Fusion Process — 397
- 11.3 Analysis of Data Fusion System Performance — 399
  - 11.3.1 Definition of Objective — 399
  - 11.3.2 Construct Alternatives — 400
  - 11.3.3 Establish Evaluation Criteria — 400
  - 11.3.4 Develop Modeling Approach — 400
  - 11.3.5 Analysis and Results — 402
- 11.4 Relating Fusion Performance to Military Effectiveness — 403
- 11.5 Data Fusion System Modeling Considerations — 406
  - 11.5.1 Scenario Definition — 409
  - 11.5.2 Model Fidelity — 409
  - 11.5.3 Sensor Modeling — 409

		11.5.4 Fusion Process Modeling	413

        11.5.4 Fusion Process Modeling   413
        11.5.5 Simulation Architecture   413
        11.5.6 Hierarchical Models   415
  11.6 Testbeds and Simulations   416
  11.7 Evaluating Military Worth   419
References   423

Chapter 12 The Emerging Role of Artificial Intelligence Techniques   425
  12.1 Broad Benefits of AI Technology   427
  12.2 Representative Prescriptive Solutions   433
        12.2.1 Applying Planning Theory   434
        12.2.2 Applying Knowledge-Based Approaches   437
  12.3 Technical Issues and Design Factors in Using KBS for Data Fusion   440
        12.3.1 Difficulties in the Application of AI Components to Data Fusion   441
        12.3.2 Real-Time Processing Requirements and Temporal Variance   443
        12.3.3 Combined Symbolic and Numeric Processing   445
        12.3.4 Large Data-Knowledge Requirements   446
        12.3.5 Uncertainty in Data and Knowledge   447
        12.3.6 Human Factors   451
  12.4 Special Aspects of Testing and Evaluating KBS   452
  12.5 The Range of Applications of AI to Fusion Problems   454
References   456

Index   461

The Authors   465

FOREWORD

During the 1980s, rapid advances in sensor technology and large increases in funding for sensor systems have resulted in vastly increased volumes of sensor data in a broad range of defense systems. The volume will continue to increase for the foreseeable future. This fact, coupled with the emergence and deployment of over-the-horizon weapon systems has fundamentally altered the information management problem in command, control, and intelligence (C^3I). Command and intelligence centers have been flooded by a mix of raw data, processed information, and the information demand of commanders. The field and afloat commanders, their view necessarily broadened by recognition of the over-the-horizon capabilities of their own and opposing weapons, have demanded (and continue to demand) more data and information, with heavy emphasis on speed and timeliness. In this environment, funding, research, and systems development in information processing have not kept pace with sensor improvements. Consequently, the critical intelligence analysis and the command and control (C^2) decision and action planning processes have been overwhelmed by the information volume and dependence on a preponderance of manual techniques. Obviously, a key to this information management problem is the ability to combine or "fuse" data, not only as a volume-reducing strategy, but also as a means to exploit the unique combinations of data that may be available. Despite the awareness of the need, provisions of this capability in fielded systems has remained elusive.

Recognition of this problem led the Joint Directors of Laboratories (JDL) Technical Panel for C^3 (TPC3) to establish the Data Fusion Sub-Panel. In late 1984, the subpanel began meeting to begin coordination and establish a joint data fusion project. In this endeavor the subpanel gradually recognized that the problem was greater than anyone had imagined. The subpanel discovered that, while the term data fusion is widely used, its meaning is subject to varying interpretations and is often misunderstood. Similarly, although there were many practitioners, there was no sense of common endeavor or community, no dedicated organization, no mechanism for technical exchange, nor a refereed journal. At best, "data fusion" was a topic relegated to the last afternoon of an AI or C^3 conference. An ever greater concern was the recognition that even the practitioners of the "art"—the analysis and decision makers—did not understand, or at least did not uniformly define, the data fusion process; were unaware of how automated fusion processing systems helped and sometimes hindered them; and, worse, there was no educational process in academia or the defense department that could help them.

As the subpanel searched for the right project, what became apparent was that data fusion research and development was being conducted under a wide variety of systems, names, and research endeavors. Rather than adding one more small drop to this ocean, the critical need was to bring a sense of order to the chaos of conflicting and overlapping concepts of data fusion from fundamental theory to practical application.

In coming to this conclusion the subpanel realized that (1) data fusion is a broadly applicable, fundamental human process of central importance to C^3I and (2) data fusion has a common basis in theory, which is independent of application and therefore is a discipline in its own right. Both of these realizations are important foundations for initiating and establishing a sense of order. The subpanel focused its efforts on identifying a data fusion community and providing a common frame of reference by developing a data fusion model and corresponding taxonomy and lexicon. The subpanel next began organizing the community by establishing coordinating mechanisms for communication and technical exchange through annual symposia, newsletters, and surveys. However, a major stumbling block to this effort has been the lack of a definitive, yet general textbook on data fusion that is approachable by all components of the community, from highly technical theoreticians to defense analysts and decision makers, and around which academic and defense schools could begin to structure curricula.

This book begins to fill that need. It clearly and lucidly identifies the full spectrum of data fusion processes and activities and, fortunately, provides the framework for understanding them in concept and application. The book is of interest to the full range of the community membership, for although the formal, detailed exposition of the theoretical is excluded (by design), the text provides a framework for the theoretician, a guideline for the systems developer, and a knowledge base of great importance to the field analyst and C^2 decision maker. It is very important to the JDL Sub-Panel's efforts to develop and nurture a community dealing with the tough issues of data fusion in a defense environment.

This book is quite important to US defense forces and our allies, particularly NATO, for it comes at a time when, potentially, we will have to do more with less personnel. As the numbers of the active defense and fighting forces diminish, the role of those who remain "ever-vigilant" increases in importance. As the new and improved sensor systems are already purchased or in the field, those who remain will be inundated by the available data, as well as having to deal with a less defined threat in a worldwide, low-intensity, limited-objective environment. The problem is daunting.

I have focused in this Foreword on the defense applications of data fusion. This is natural because this has been my focus throughout my career and that of the Data Fusion Sub-Panel. However, the book does not focus exclusively on this application, nor should it, for its importance transcends military boundaries. For example, the current national "War on Drugs" is totally dependent on effective data fusion in order to interdict the flow of narcotics into the United States. This effort represents a very complex data fusion environment with many legal and political ramifications. Similarly, decision makers in all aspects of our increasingly complex, information-rich society can benefit from a clear understanding of data fusion, its mechanisms, strengths, and limitations as presented here. For, although the application problems may differ, the fundamental techniques and approaches are widely applicable.

This book is long overdue, and in view of the nature of the problem, it is not the end or the final word, but, at last, a very good beginning. I thank the authors for their toil and commend them for their effort.

Franklin E. White, Jr.
Chairman, Data Fusion Sub-Panel
Joint Directors of Laboratories—Technology Panel for C^3
Naval Ocean Systems Center
San Diego, California

PREFACE

The objective of this text is to introduce researchers, engineers, and military operations personnel to one of the most rapidly emerging fields of technology in the area of military command and control: the automation of processes to combine diverse sets of sensed information. This field of technology has been appropriately termed *data fusion* because the objective of its processes is to combine elements of raw data from different sources into a single set of meaningful information that is of greater benefit than the sum of its contributing parts.

As a technology, data fusion is actually the integration and application of many traditional disciplines and new areas of engineering to achieve the fusion of data. These areas include communication and decision theory, epistemology and uncertainty management, estimation theory, digital signal processing, computer science, and artificial intelligence. Methods for representing and processing data (signals) are adapted from each of these disciplines to perform data fusion.

As a specialized area of military *command, control, communication, and intelligence* (C^3I), data fusion describes the techniques or systems that process sensor and intelligence data from diverse sources to provide the military commander with a complete and coherent picture of situations of interest. In this case, data fusion may include a combination of automated processes, procedures, and coordinated military operations to collect, fuse, and assess the military significance of sensed data.

This book presents a system-level introduction to data fusion, both as a technology and as a specialized area of C^3I. A general background in the previously mentioned engineering disciplines is beneficial, but not required for an understanding of the concepts presented. An attempt has been made to provide a thorough overview of the many disciplines involved, while moderating the presentation of theoretical details and mathematical formalisms provided in texts and references cited throughout the book. This system-level presentation has been taken to enhance the text's usefulness to the researcher and system developer as well as military operations personnel.

The book is suitable as a graduate-level text, and is applicable to industrial robotics, commercial air traffic control, and operations research courses as well as military disciplines. Standard introductory courses in linear systems theory, probability and statistics, and computer science may be helpful, but are not necessary.

Chapter 1 follows the general sequence of the book, providing a rapid overview of the motivation and applications for data fusion, the general approaches to implement fusion and the current state of data fusion research, development, and deployment. Important terminology and helpful background references are provided to aid the reader in the study of subsequent chapters.

Chapter 2 categorizes the elements of fusion processing using the standard model adopted by the U.S. Department of Defense Tri-service Data Fusion Subpanel. Functional architectural concepts and options are also introduced in this chapter, although physical system architectures are reserved for Chapter 10 after processing methods are described.

Chapter 3 introduces military applications and unique operational considerations for implementing data fusion systems and Chapter 4 introduces the characteristics of sensed data and the methods of representing the data quantitatively for mathematical processing. Categories of sensors and data sources are reviewed and a generic sensor model is introduced to describe the characteristics required to specify technical performance of sensors for data fusion. The processes by which individual sensors are controlled and multiple sensors are coordinated to enhance the effectiveness of data collection are then described in Chapter 5.

Chapters 6 and 7 describe the processing methodologies for fusing both state and attribute information, respectively. Traditional direction-finding, position-locating, and tracking methods using multiple sensors are described first, emphasizing the procedures for associating measurements. Next, the procedures for combining associated data are described. In each chapter, a taxonomy of approaches is provided to relate the many techniques that are available to implement these functions.

Chapters 8 and 9 detail the requirements and processes for higher-level analysis and assessment of the meaning of the combined data base of target and event state and attribute information. Situation and threat assessment techniques are introduced and related to position or attribute fusion processes.

Chapter 10 describes the systems engineering approach necessary to define requirements, analyze performance or effectiveness, and synthesize physical architectures (designs) for the hardware and software components that make up real fusion systems. Processing, database, and other architectural considerations are described and example architectures are introduced. Chapter 11 defines the methods of modeling and quantitatively assessing the performance and military effectiveness of data fusion processes. Chapter 12 identifies the emerging application of artificial intelligence technologies to the fusion of symbolic and highly uncertain data.

We conceived this book to be a thorough introduction to the field and as the first text to focus on the many diverse disciplines that have been applied to data fusion problems. Our hope is that this book will introduce many to this emerging field and provide an excellent foundation for those who will apply tomorrow's technology to the very difficult problems of fusing data for a wide range of applications.

JAMES LLINAS
BUFFALO, NEW YORK

EDWARD L. WALTZ
BALTIMORE, MARYLAND

ACKNOWLEDGEMENTS

The authors wish to acknowledge the invaluable encouragement and assistance of many colleagues in the data fusion community. Special thanks are due to Franklin E. White, Jr., of the Naval Ocean Systems Center for his Foreward contribution. Sprinkled throughout this book are many thoughts undoubtedly attributable to numerous helpful discussions with Frank, in which his advice and vision about data fusion were solicited.

Chapter 8 of this book is special in its inclusion of the writings of Dr. David L. Hall and of Dr. Dennis M. Buede, whose specific contributions are gratefully acknowledged. The authors and Drs. Hall and Buede are all mutual colleagues but over the years Drs. Hall and Llinas and Dr. Buede and Mr. Edward Waltz have worked in close, pair-wise relationships which have enriched the authors' views of the data fusion process. The explicit and implicit contributions resulting from these relationships are also appreciated and recognized.

James Llinas also wishes to thank the members of the Joint Directors of Laboratories Data Fusion Subpanel, of which he is a member, conversations with whom have also been valuable sources of ideas. He also wishes to thank his employer, Arvin/Calspan Corporation, for support and encouragement in his work in data fusion. The contributions of Ms. Bernadette Fitzgerald and of Ms. Joan Lus, Ms. Donna Weir, and Ms. Alice Miller for drafting of some of the manuscript, and of Mr. Ernest L. Hoefner and the graphics staff for some of the artwork, are especially appreciated.

Edward Waltz wishes to express his appreciation to his employer, the Allied-Signal Aerospace Company, Bendix Communication Division, for support and encouragement in this project. He is very appreciative of the encouragement and support provided by Ralph Ormsby, Jerry Woodall, E.J. Raimondi, Jack Shaul and Al Sinsky. He is also deeply indepted to the insight and understanding provided by his colleagues at Bendix in the development of data fusion systems: Dr. Richard Burne, Richard Clarke, Tom Haskins, John Kean, Robert Kelley, Jim Lyons, Marge Martin, and Jon Sipos.

*To Our Wives,
Barbara and Sarah*

Chapter 1
INTRODUCTION

This book is about the synergistic use of sensory data from multiple sensors to extract the greatest amount of information possible about the sensed environment. This *fusion* of multisensory data is a fundamental function observed in human and other biological systems and we are highly motivated to understand and automate this function in intelligent systems. Humans naturally apply this ability to combine data (sight, sound, scent, touch) from the body's sensors (eyes, ears, nose, fingers) with prior knowledge to assess the world and the events occurring about them. Because the human senses measure different physical phenomena over different spatial volumes with different measurement characteristics, this process is both complex and adaptive. The conversion of the data (images, sounds, smells, and physical shapes or textures) into a meaningful perception of the environment requires a large number of distinct intelligence processes and a base of knowledge sufficient to interpret the meaning of the properly combined data.

In recent years, many of these human functions have been emulated in automated systems that intelligently combine data from multiple sensors to derive meaningful information not available from any individual sensor. This process of combining data has been variously referred to as multisensor or multisource correlation, multisource integration, sensor blending, or data fusion. This text adopts the latter terminology because *fusion* is a nonmathematical term meaning "the process of combining or blending into a whole." It avoids the mathematical terms correlation and integration, both of which are distinct mathematical operations performed within data fusion.

Perhaps the most comprehensive definition of data fusion in the military context is "A multilevel, multifaceted process dealing with the detection, association, correlation, estimation and combination of data and information from multiple sources to achieve refined state and identity estimation, and complete and timely assessments of situation and threat" [1]. This definition focuses on three central aspects of fusion:

- Data fusion is a *process* performed on multisource data at several levels, each of which represents a different level of abstraction of data.

- The process of data fusion includes detection, association, correlation, estimation, and combination of data.
- The results (products) of data fusion include state and identity estimates at the lower levels and assessments of overall tactical situations at the higher levels.

The basic objective of data fusion, then, is simply to derive more information, through combining, than is present in any individual element of input data. This is the effect of synergy: the enhancement of the sensor system's effectiveness by taking advantage of the cooperative and joint operation of the multiple sensors.

1.1 MOTIVATIONS AND APPLICATIONS FOR DATA FUSION

The fusion of remotely sensed data (e.g., radar, sonar, voice reports) has long been performed by humans without the aid of tools to automate the process. Military intelligence gathering, *command and control* (C^2), weather prediction, air traffic control, and navigation are typical traditional applications using manual data fusion techniques. Mental reasoning methods with manual aids (e.g., "grease pencil" plot charts, cross-check lists, overlayed displays) and without aids have successfully combined sensor and source data because of the powerful associative reasoning powers of the human brain.

As the pace of modern warfare has increased, a number of factors have provided the motivation for developing *automated* data fusion systems:

- Increases in target mobility and weapon lethality demand earlier detection-identification time, shorter reaction time.
- More complex threats (variety and density of platforms, low observability, countermeasure sophistication) require improved detection-discrimination capability.
- Increased cost of personnel and risk to operators in some missions have dictated remotely controlled or autonomous weapon systems that require data fusion.

These demanding requirements in many multisensor data systems have exceeded the ability of human operators to cope with the volume and rate of incoming information to be merged [2]. In addition, the complexity of the data (e.g., multidimensional measurements) and ambiguity between measurements has exceeded the human ability to associate and classify data without decision aids. These factors have demanded the automation of all or part of the individual processes of fusion to provide timely, accurate, and easily understood information, rather than masses of raw data. In the case of robotic applications, automated data fusion will allow removing the human from the loop entirely. As a result, much of the motivation for data fusion is based on the need to emulate the human processes of fusion to increase speed, capacity, or improve accuracy of the process.

Military application for data fusion include a wide range of *command, control, communication and intelligence* (C³I) missions for both tactical and strategic warfare. The extent of applications can be illustrated by the following list:

- Autonomous weaponry employing multiple sensors
- Broad area surveillance systems employing single weapons platforms (e.g., ships, airborne surveillance, ground sites, surveillance spacecraft) or distributed sensor networks
- Fire control systems employing multiple sensors for acquisition, tracking, and command guidance
- Intelligence collection systems
- *Indications and warning* (I&W) systems, the mission of which is to assess threat and hostile intent
- Command and control nodes for military forces

We illustrate three representative military applications in detail (see Chapter 3), however, the principles and implementations are applicable to virtually all these application areas. Industrial, robotic, and commercial remote sensing applications of data fusion exist, and, although this text does not deal explicitly with these applications *per se*, the principles also are directly applicable to these areas.

The application of multiple sensors (and the fusion of their data) to the problems of detection, tracking, and identification offer numerous potential performance benefits over traditional single-sensor approaches. These performance benefits must, of course, be weighed against the additional cost, complexity, and interface requirements introduced for any given application. Characteristics of multisensor systems that provide operational benefits in specific applications include the following:

1. *Robust operational performance* is provided because any one sensor has the potential to contribute information while others are unavailable, denied (jammed), or lacking coverage of an event or target.
2. *Extended spatial coverage* is provided because one sensor can look where another sensor cannot.
3. *Extended temporal coverage* is also provided because one sensor can detect or measure an event at times that others cannot.
4. *Increased confidence* (a relative measure of uncertainty in the measured information) is accrued when multiple independent measurements are made on the same event or target.
5. *Reduced ambiguity* in measured information is achieved when the information provided by multiple sensors reduces the set of hypotheses about the target or event.

6. *Improved detection performance* results from the effective integration of multiple, separate measurements of the same event or target.
7. *Enhanced spatial resolution* is provided when multiple sensors can geometrically form a synthetic sensor aperture capable of greater resolution than any single sensor.
8. *Improved system operational reliability* may result from the inherent redundancy of a multisensor suite.
9. *Increased dimensionality* of the measurement space (i.e., different sensors measuring various portions of the electromagnetic spectrum) reduces vulnerability to denial (countermeasures, jamming, weather, noise, *et cetera*) of any single portion of the measurement space.

Table 1.1 compares the tactical benefits that can be derived in military systems by using multiple sensors with appropriate data fusion. All of these benefits are accrued, of course, at the expense of increased system complexity. This complexity, however, has its drawbacks over the single-sensor system: increased cost, the potential for reduced total system availability, increased equipment physical factors (e.g., size, weight, power), and increased system observability due to emissions. The performance benefits derived from multiple sensors must be weighed

Table 1.1
Benefits Provided by Multisensor Systems over Single-Sensor Systems

Improvement	*Specific Characteristics*	*Tactical Benefits*
Improved Spatial & Temporal Coverage	• Extended spatial coverage provided by multiple, overlapping sensor fields of regard • Increased detection probability due to synergy of multiple sensors	• Increased probability of threat detection • Enhanced situation awareness and mutual support
Improved Measurement Performance	• Reduced ambiguity in data with multiple measurements • Improved detection, tracking, & identification when data is integrated from multiple looks at the target • Reduced uncertainty in data when multiple independent measurements are integrated	• Target acquisition, track, & ID at longer ranges • Accurate discrimination of target type and military capability (threat)
Enhanced Operational Robustness	• Possible increase in mission reliability due to redundancy • Degraded (sensor subset) modes available • Robust performance provided by multispectral sensors	• Reduced vulnerability to denial of a single sensor's data • Availability of alternate sensor modes to reduce observability

against these drawbacks in each specific mission application. The achievement of these benefits can be evaluated by the system designer *prior to* the commitment to implementation using methodologies that perform quantitative assessments of the effectiveness of fusion systems (Chapter 12).

1.2 A FUNCTIONAL MODEL OF THE DATA FUSION PROCESS

To provide an overview of the data fusion process discussed in this book, we first introduce a simple multisensor model to illustrate the functional elements of a general fusion system and the relationship among these functions. The model assumes three sensors that view the same surveillance volume, which contains multiple moving targets of different types (classes). The data processing flow is discussed and information management issues (e.g., data base and sensor management) are not discussed for the sake of simplicity. Although there are a number of architectural alternatives, we present here the most straightforward implementation for purposes of illustration.

The primary elements of the model are depicted in Figure 1.1. Note that the model distinguishes two *levels* of processing functions as suggested by the definition of data fusion presented at the opening of this chapter. At the first level, the processing is primarily numerical in nature, resulting in numerical results (e.g., locations, velocities, target types). At the second level the processing is primarily symbolic, resulting in more abstract results (e.g., threat, intent, goals). This functional model will be more fully developed (Chapter 2) to refine the specific functions that constitute the general fusion process.

The model also illustrates how the functions of detection, association estimation, and classification form the kernel operations of data fusion.

Sensors scan the surveillance volume and report all detected targets within the volume once per scan. Each sensor makes independent measurements and applies a binary decision process to detect the presence of targets on the basis of signal characteristics. Once detected, the measurement parameters (target signature parameters and target state parameters) are reported to the fusion process. Some sensors may apply a sequential decision process, requiring a sequence of detections over several sample measurements prior to declaring the presence of a target and issuing a report.

Data association is performed on the reports collected at the completion of each scan. Each new report is correlated with each other report *and* the predicted locations of already detected targets from previous scans. On the basis of state (position and velocity) and attribute (parameter measurements that infer target class) correlation, an m-ary decision is made, with each new report being assigned to one of the following possible hypotheses:

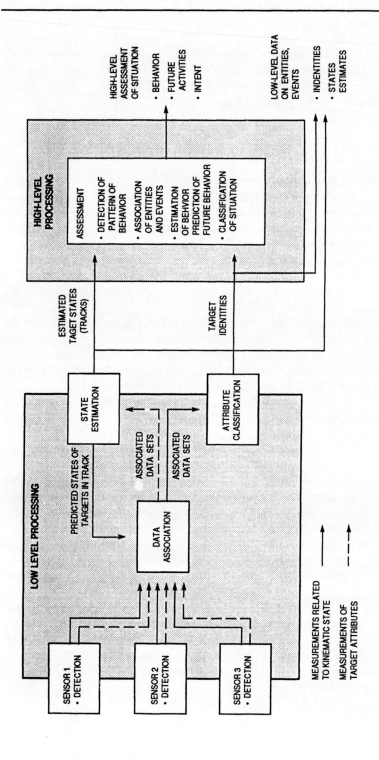

Figure 1.1 Elements of a basic data fusion system.

1. A new target detection set, establishing the report to be a new target not yet detected. The set may include multiple sensor reports, mutually correlated.
2. An existing target set, identifying the source of the report as a previously detected target.
3. A false alarm, assuming the detection did not result from a real target; the report is dropped from further consideration.

State estimation (target tracking) is made at the end of each scan from the sets of new and existing (previous scan) target reports. The estimator assumes a dynamic model of behavior, and estimates the parameters (e.g., position coordinates, velocity components) based upon the sensor measurements. These estimates are used to *predict* the location of targets at the point at which sensors will observe them in the next scan. These predictions are fed back for association on subsequent scans. The output of this function is the estimate of state of each target.

Attribute classification (target identification) is based on target signature parameters reported by different sensors to form a composite n-dimensional measurement, with each dimension being an independent attribute of the target. The attributes for all known target types have been measured previously and are the basis for an m-ary decision process that attempts to classify each composite measurement into one of the m possible target classes. The output of this function is the classification decision for each target.

Assessment is when the set of data for all targets (target states and classification) is compared (correlated) with previously identified behavioral patterns for all possible situations to describe the status of activities in the surveillance volume. Spatial and temporal characteristics that discriminate between abstract behavior patterns (e.g., normal activity, threatening conditions, suspicious activity, erroneous behavior) are the factors used in the correlation process, and an m-ary decision is made to determine which behavior pattern most closely matches the state of all targets.

This simple model illustrates how the basic detection, classification, association, and estimation processes are used repeatedly throughout the fusion process. It should be noted that the sequence in which these processes are applied has a great influence on the architecture, processing characteristics, and performance of fusion systems. Table 1.2 illustrates how the functions just described may be centralized at the fusion process or distributed among the sensors and the fusion process. Throughout this book, the tradeoff considerations and characteristics of these alternatives are discussed in detail because they are the principal functional architecture decisions facing the fusion system designer. In Chapter 4, the issue of centralized or distributed detection is discussed; and Chapter 6 contrasts centralized and distributed estimation and tracking. The implementation issues involved in designing distributed, centralized, and integrated system architectures are detailed in Chapter 10.

Table 1.2
Alternatives for Centralization-Distribution of Data Fusion Functions

Function	Distributed between Sensors and Fusion	Centralized at Fusion Node
Detection: Binary decision to determine the presence of a signal (target), based upon sensor measurements	• Perform local detection at each sensor • Collect sensor detections at fusion node, apply global detection decision role (AND, OR, etc.)	• All sensor data is passed to fusion node, where a composite detection decision is based upon all measurements
Association: Cross-correlation of measurements and m-ary decision to partition all measurements into sets of common origin	• Sequential measurements at each sensor are associated and used to derive sensor-level estimates (tracks)	
State Estimation: Estimation of state, based on multiple, associated measurements that can be directly related to the equations of state	• Sensor tracks (not measurements) are passed to fusion node for track-track association • Fusion-level state estimate is based on multiple, associated sensor tracks	• All sensor measurements are passed to fusion node for composite association and state estimation.
Classification: M-ary decision, using associated sensor measurements to assign the measurements to one of m classes	• Perform m-ary local classification at each sensor, using an n-dimension subspace • Collect sensor classifications at fusion node, perform global classification	• All sensor data is passed to fusion node, where all measurements are placed in an n-dimension space for m-ary classification

1.3 METHODS FOR IMPLEMENTING DATA FUSION

The methods available for fusing multiple measurements have been the subject of numerous mathematical developments, each motivated by specific applications that constitute a subset of the overall fusion problem. These developments have a rich historical background, and we briefly cite some of the most significant developments in the fundamental areas of decision-detection, estimation, association, and uncertainty management theories that have led to more recent applications to data fusion.

Decision or detection theory, in its basic application, bases detection and classification on decision theory, in which measurements are compared with alternative hypotheses to decide which hypothesis "best" describes the measurement. Reverend Thomas Bayes (1702–1761) introduced the foundational concepts of classical statistical decision theory in a paper published two years after his death. The theory assumes probabilistic descriptions of the measurement values and prior knowledge to compute a probability value for each hypothesis [3]. Bayes's method provided a means to update one's degree of belief (expressed as a probability) about a hypothesis, on the basis of prior knowledge and recent observations. For the first time, this provided a tool for quantitative inferences or learning. Bayes's theory also provided a means of associating a "cost" with implementing each possible decision to provide further control over the decision process. Bayesian and related statistical decision approaches are directly applied to the processing of multisensor data to perform binary decisions (the detection problem: signal present or absent) or m-ary decisions (the classification problem: identifying into which of m class hypotheses the measurement best fits). Although originally applied to single sensors, extensions [4] to multiple sensors have developed optimal decision criteria for distributed decision making when the decisions are distributed between sensors and the central node where decisions are fused.

Estimation theory is an extension of the decision process whereby the estimation of a parameter is made by using multiple observed measurements of variables that can be directly related to the parameter in question. The use of astronomical observations to estimate the location of asteroids and planets was an early application that produced the first statistical estimators. The development of the least squares method by Gauss in 1795 [5] introduced the concept of using multiple measurements, with an estimate of measurement errors, and a linearized model of the measurement to parameter relationships to minimize the sum of the squares of *residuals* (the differences between the observed measurements and the measurements computed from the estimated parameter). In 1912, Fisher [6] applied the probability density function of measurements to maximize the logarithm of the probability density function of the estimate in the maximum likelihood method of estimation. In the early 1940s, Kolmogorov and Weiner [7] extended these statistical estimation concepts to apply to continuous and time-discrete sequences of measurements. Their method was based upon the minimization of the mean square error and ultimately led to the development of a practical recursive implementation, the Kalman filter (estimator) [8]. Subsequent nonlinear and more efficient methods have been introduced and ongoing developments continue to refine the methods of estimation using multiple observations. In each of these developments, several general assumptions were made:

- A linearized model that relates measurements to the parameter in question can be developed.
- Knowledge exists regarding the statistics of the measurement errors and, in some cases, the behavior of the estimated parameter.

- The measurements are *known* to be from the same origin, as a result of a prior association process (i.e., they are from the same target or other entity).

This final assumption is often *not* the case in data fusion problems, where there are many sources of measurement, leading to the need for the next area of mathematical development.

Association occurs when many sources of measurements exist, and measurements from common sources must hence be associated with each other prior to, or at least in conjunction with, classification or estimation. To partition measurements into sets, with each set having a common origin, a correlation process can be performed to quantify a measure of the correlation (similarity) among all measurements. On the basis of the correlation measure, measurements can be separated (partitioned) into sets, with those in each set "associated" with a common source. The general concept of correlation is treated in communication theory, where the matched filter is developed for the optimal detection of signals in noise [9]. Blackman [10] and Goodman [11] have summarized the historical application of correlation to sequential sensor and then to multisensor data. The earliest applications to automate radar tracking in the 1960s applied correlation to estimate target tracks for batches of measurements. It was not until the 1970s that recursive estimators (the Kalman filter) were preceded by correlators to perform time-sequential association and estimation. In the 1980s, more complex multisensor, multitarget systems were developed to face the more complex association problems of closely spaced targets, with measurement time intervals large relative to target dynamics, and sensors that include both measurement and alignment errors.

Although these developments have provided the basic tools for fusing data from multiple sources, concurrent developments in the methods to represent the uncertainty in measurements (and decisions or estimates) have led to a variety of representations for uncertainty and calculi for combining uncertain data:

Uncertainty management stems from classical methods for representing uncertainty in measurements (evidence) that use the Bayesian probability model to express the degree of belief in each hypothesis (or proposition) with a probability. The hypotheses must be mutually exclusive and exhaustive. This requires that all hypotheses must form a complete set of possibilities and the probabilities must sum to 1. Because there is no means to represent ignorance or incompleteness in information (probabilities *must* be assigned to each hypothesis, even if not known), alternative models have been sought. Shafer [12] introduced the concept of probability intervals to provide a means to express ignorance and provide a calculus for combining evidence that reduces to the Bayesian model when complete knowledge is present. Other heuristic models [13] and fuzzy calculus [14] have also been applied to uncertainty representation for fusion applications. The debate between classical Bayesian and these other approaches still continues (for example, see [14] *versus* [15] and the debate in [16]) and is expected to continue as quantitative analyses continue to compare the performance of the alternatives in real-world applications.

1.4 THE CURRENT STATE OF DATA FUSION

At the writing of this book, for over a decade, reports in the open literature of progress on data fusion research and military applications are available to define current capabilities. These sources provide a rich and thorough history of the developments in recent years. In the late 1970s early efforts at automation of data fusion functions for military C^3I systems began to be reported [2, 17], emphasizing the need for increased computational capabilities, methods for efficiently combining data, and improved sensors. Perhaps most notable among the early developmental systems was the *battlefield exploitation and target acquisition* system (BETA), which experienced development problems [18], highlighting the complexity and difficulty of the new technology as well as the need for a disciplined systems engineering approach to future developments. The system ultimately achieved successful limited operational capability in Europe by the mid-1980s, demonstrating the feasibility while providing a greater appreciation for the problems associated with data fusion.

Throughout the 1980s the three U.S. military services pursued the development of tactical and strategic surveillance systems employing data fusion and supported extensive research in the areas of target tracking, target identification by noncooperative sensors, algorithm development for correlation (association) and classification, and the application of intelligent systems to situation assessment. The results of much of this work appeared in the open literature and is heavily referenced throughout this book. The large amount of fusion-related work in this period raised some concern over possible duplication of effort. As a result, the Joint Directors of U.S. Department of Defense (DoD) Laboratories convened a Data Fusion Subpanel to (1) survey the activities across all services, (2) establish a forum for the exchange of research and technology, and (3) develop models, terminology, and a taxonomy of the areas of research, development, and operational systems.

At the close of the 1980s, there existed a small number of deployed first-generation data fusion systems that effectively fused data from existing or modified military sensors. These systems range from large, strategic, ocean surveillance systems [19] to small, tactical systems [20]. In addition to these operational systems, a significant number of prototypes and testbeds [21] have been tested to evaluate the real contribution of data fusion to specific military applications. The next generation of systems, incorporating sensors and processors designed for fusion, remains under advanced and full-scale development as of this writing. Concurrently, a number of promising theoretical developments and technologies are emerging that will be directly applicable to the fusion developments of the 1990s. Among those are the following:

- Development of optimum decentralized detection, estimation, and tracking methods.
- Development of representation and management calculi for reasoning in the presence of uncertainty.
- Development of spatial and contextual reasoning processes suitable for assessment of activity and intent on the basis of temporal and spatial behavior.
- Development of integrated, smart sensors with soft-decision signal processing for use in systems that perform numeric and symbolic reasoning in uncertainty.
- Development of both distributed and integrated data processing architectures applicable to data fusion.
- Decomposition of processing algorithms into parallel processes for implementation on parallel machines.
- Application of neural network technology to detection, tracking, classification, and assessment problems.

For a continued update of the advanced research and application developments in this field, the following sources are suggested:

IEEE Transactions (general articles on data fusion, C^3I)

IEEE Proceedings on Man, Systems and Cybernetics (situation and threat assessment for C^3I, AI systems, systems analysis)

IEEE Proceedings on Aerospace and Electronic Systems (target detection and classification, sensors and signal processing)

IEEE Proceedings on Automatic Control (target tracking, sensor management)

Proceedings of International Society for Optical Engineering: SPIE, (in the areas of sensor fusion, remote sensing, signal processing, and related areas)

Proceedings of National Sensor Fusion Symposia: Society of Photo-Optical Instrumentation Engineers, SPIE (classified military technology and research in data fusion)

Proceedings of U.S. Department of Defense Tri-Service Data Fusion Conferences (classified military technology and research in data fusion; some proceedings are not classified or restricted, some are not classified with export restrictions, and others are classified)

Proceedings of the Military Operations Research Society, MORS (classified research in the quantitative operations analysis of military systems, including command and control systems, including data fusion)

Proceedings of IEE International Conferences on C^3MIS (command, control, communications, and management information systems)

REFERENCES

1. This is the definition developed by the Joint Directors of Laboratories Data Fusion Subpanel, with two revisions introduced by the authors: (1) The function of detection has been added, and (2) the estimation of *position* has been replaced by estimation of *state* to include the broader concept of kinematic state (e.g., higher-order derivatives, velocity) as well as other states of behavior (e.g., electronic state, fuel state).
2. "Intelligence Fusion Pushed," *Aviation Week and Space Technology*, January 29, 1979, pp. 205–211.
3. For a discussion of Bayesian statistics and a reprint of Bayes's original essay, see Press S. James, *Bayesian Statistics*, John Wiley and Sons, New York, 1989.
4. Tenney, R.R., and Sandall, N.R., Jr., "Detection with Distributed Sensors," *IEEE Trans. on AES*, Vol. AES-17, No. 4, July 1981, pp. 501–510.
5. Gauss, K.G., "Theory of Motion of the Heavenly Bodies," Dover Books, New York, 1963.
6. Fisher, R.A., "On an Absolute Criterion for Fitting Frequency Curves," *Messenger of Math*, Vol. 41, 1912, p. 155.
7. Weiner, N., *The Extrapolation, Interpolation and Smoothing of Stationary Time Series*, John Wiley and Sons, New York, 1949.
8. Kalman, R.E., "A New Approach to Linear Filtering and Prediction Problems," *J. Basic Eng.*, Vol. 82D, March 1960, pp. 34–35.
9. Wozencraft, J.M., and I.M. Jacobs, *Principles of Communication Engineering*, John Wiley and Sons, New York, 1965.
10. Blackman, Samuel S., *Multiple Target Tracking with Radar Applications*, Artech House, Norwood, MA, 1986, pp. 1–2.
11. Goodman, I.R., et al., *Naval Ocean-Surveillance Correlation Handbook, 1979*. NRL Report, Naval Research Laboratory, Washington, DC, pp. 5–17.
12. Shafer, G., *A Mathematical Theory of Evidence*, Princeton University Press, Princeton, NJ, 1976.
13. Shortliffe, E.H., and B.G. Buchanan, "A Model of Inexact Reasoning in Medicine," *Math. Biosci.*, Vol. 23, 1975, pp. 351–379.
14. Zadeh, L.A., *Fuzzy Sets and Systems*, North-Holland, Amsterdam, 1978.
15. Bogler, Philip L., "Shafer-Dempster Reasoning with Applications to Multisensor Target Identification Systems," *IEEE Trans. on Systems, Man and Cybernetics*, Vol. SMC-17, No. 6, November–December 1987, pp. 968–977.
16. Buede, Dennis M., "Shafer-Dempster and Bayesian Reasoning: A Response to 'Shafer-Dempster Reasoning with Applications to Multisensor Target Identification Systems,'" *IEEE Trans. on Systems, Man and Cybernetics*, Vol. SMC-18, No. 6, November–December 1988.
17. Waltz, E.L., "Computational Considerations for Fusion in Target Identification Systems," *IEEE Proc. NAECON*, May 1981, pp. 492–497.
18. "Evaluation of Defense Attempts to Manage Battlefield Intelligence Data," LCD-81-82, Comptroller General Rept. to Congress of the U.S., February 24, 1981.
19. Gravely, V. Adm. Samuel L., "The Ocean Surveillance Information System (OSIS)," *Signal*, October, 1982, pp. 30–36.
20. Rawles, James W., "Army IEW: New Structure for a New Century," *Defense Electron.*, June 1989, pp. 69–80.
21. Lippermeier, Col. G., and R. Vernon, "IFFN: Solving the Identification Riddle," *Defense Electron.*, January 1988, pp. 83–88.

Chapter 2
TAXONOMY OF FUNCTIONAL ARCHITECTURES

Chapter 1 described a rather general and abstract data fusion model (see Figure 1.1) and the associated functions and processes. This chapter will elaborate on this model and the various functional architectures employed in representative applications. The evolution of reasonably detailed process models and functional architectures is a sign of gradual maturation of the data fusion discipline. Such models serve as important communication vehicles through which design concepts, algorithms, and operational strategies can be discussed and evaluated. Models also aid in moving the data fusion community toward common functional modules, processing standards, and, importantly, interoperable multiservice systems.

2.1 GENERALIZED PROCESSING MODELS FOR DATA FUSION

What follows is paraphrased from White *et al.* [1], in which a generalized processing model for data fusion is presented. This model was developed by the Joint Directors of Laboratories (JDL) Data Fusion Subpanel (DFS) to provide a framework and common reference for addressing data fusion issues and problems.

In its efforts to establish common language and concepts, the DFS has been evolving models of the data fusion process. To date, it has evolved two versions (developed from different points of view) and still appreciates that further refinements to both models are warranted. The first model is shown in Figure 2.1; it identifies three levels of fusion processing products:

- Level 1—Fused position and identity estimates
- Level 2—Hostile or friendly military situation assessments
- Level 3—Hostile force threat assessments

These "levels" specify artificial logical separations in the overall data processing stream and were established in part to achieve common terminology regarding data

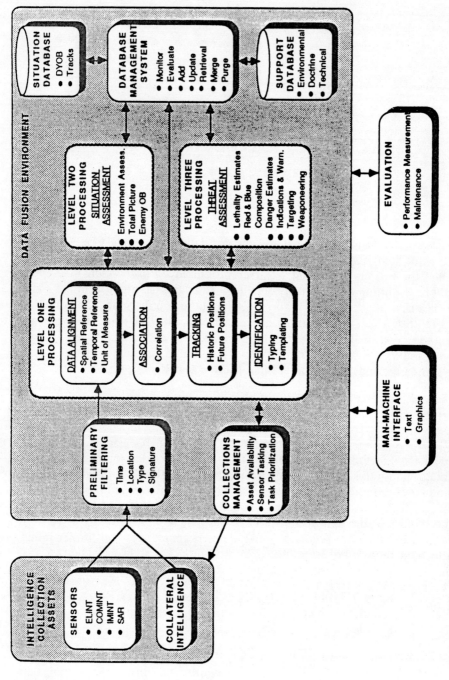

Figure 2.1 DFS product-oriented model of the data fusion process (from [1]).

processing activities. The focus in this model is on information "products," that is, distinctive steps, in the data fusion process and not on the architectural aspects of computing. This model also focuses on an inferential hierarchy as the process moves from Level 1 to Level 3. Across these steps, the generality of the results (products) increases from the very specific ("platform of type A at location B") to the more general ("ground-controlled intercept operations being conducted by air defense unit A"). Another distinction across these levels is in the nature of the processing. At Level 1, the processing operations are dominated by *numeric* procedures involving, for example, linear and nonlinear estimation techniques, pattern recognition processes, and various statistical operations. At Levels 2 and 3, the operations become dominated by *symbolic reasoning* processes involving various techniques from the field of artificial intelligence to support the formulation of higher levels of abstraction and inference. This model has been presented at some of the Tri-Service Symposia on Data Fusion (see [2, 3]) and has appeared to withstand the test of open discussion within the data fusion community.

Level 1 products result from single and multisource processing involving tracking, correlation, alignment, and association by sampling the external environment with multiple sensors and exploiting other available sources. The processing products are position and identity estimates for targets or platforms in the composite field of view. Level 2 processing involves situation abstraction and assessment. In Level 2, situation *abstraction* is defined as the construction of a generalized situational representation from incomplete sets of data to yield a contextual interpretation of the distribution of forces produced in Level 1. Situation *assessment* is defined as a multiperspective process of interpreting and expressing the environment based on contextual analyses and renditions based on the fusion of data from sensors and from technical and doctrinal data bases to yield indications and warnings, plans of action, and inferences about the distribution of forces and information. The third level of processing is threat assessment, which is also a multiperspective process of interpreting estimates of lethality and risk in terms of the ability of friendly forces to engage the enemy effectively. Threat assessment also involves estimating the vulnerability of one's "own forces" to engagement by the enemy and determining indications and warning of enemy intention by compiling products from technical and doctrinal data bases. The distinction between Levels 2 and 3 is that Level 3 products quantify the threat's *capability* and suggest the *intent* of hostile forces whereas Level 2 results are indicative of hostile *behavior patterns*. At each level, data processing, hypothesis testing, and decision making occurs; Goodman [4] has generalized and elaborated on such processes for fusion applications.

Another way to characterize the data fusion process is to examine it from the system architecture point of view. When this is done, various system features become evident:

- At Level 1, there are typically distinctions in processing data from multiple sensors of a given type (frequently called *commensurate sensors*) and those of different type *(noncommensurate sensors)*.
- The role of specialized, "level-dependent" data bases to support each of the processing steps is both crucial and distinctive at each point.
- The multilevel processes ideally are synergistic and work in combination with iterative feedback across multiple levels employed often in various systems.
- The cooperative data fusion process extends across different processing nodes that are frequently geographically distributed.
- Data fusion products from various processing levels must be distributed to the correct and multiple levels of operational forces.

The model created from this system architecture point of view is shown in Figure 2.2. The most significant characteristic of this alternative model is its distributed nature. "Bus"-type functional structures are now included, which accommodate broad notions of connectivity, both within a processing node (nodal intraconnectivity) and across processing nodes (nodal interconnectivity). Distinctions are also recognized in single and multisource fusion processing (labeled *similar* and *dissim*-

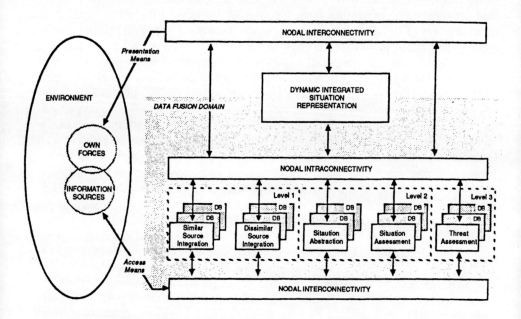

Figure 2.2 DFS system-architecture-oriented model of the data fusion process (from [1]).

ilar source integration in the figure). Acoustic arrays used in antisubmarine warfare are good examples of systems requiring single-source processing. The model also distinguishes the process of situation abstraction that emphasizes the symbolic nature of this step, a feature that characterizes it from the more analytic Level 1 processes, described earlier. Finally, the model acknowledges and emphasizes both the changeable nature of military environments and the fact that useful final products (as well as intermodule products) can result from a synthesis of the multilevel products; accordingly, it labels the "final" product a *dynamic, integrated situation representation* (DISR). Note that the relationship between the multilevel processing nodes and the DISR is similar to that between knowledge sources and a "blackboard" in knowledge-based systems (see Section 2.5 and also Chapter 12 on the role of artificial intelligence).

The dynamic situation (and threat) representation produces the type of information frequently required by the C^2 system decision maker. An aspect of both modeling and characterizing the data fusion process that must be remembered is that this overall process must satisfy both "push" requirements, driven from considerations of the data input flow, and "pull" requirements, driven by the needs of the operational application and the human analysts or commanders. Extensions of these models, reflecting in particular the interface to the human decision maker, are required. Richard Antony [5] has formulated a generic fusion processing paradigm that employs a biological metaphor. Since there are seven unique ways to compose the three primary classes of knowledge (short, medium, and long-term), Antony's fusion model reveals seven distinct data fusion classes as shown in Table 2.1.

Table 2.1
Examples of the Seven Data Fusion Classes (after Antony [5])

Fusion Class	Knowledge Classes			Composition Function	Product
	Short Term	Medium Term	Long Term		
1	Signals			Correlation	Signal correlation
2	Sensor report	Track file		Correlation	Track association
3	Signal		Linear filter	Convolution	Filtering
4		Target files		Correlation	Asynchronous multisensor correlation
5		Unit locations and status	Doctrinal templates	Correlation	Doctrinal templating
6			Elevation/soil-type maps	Intersection	Trafficability map
7	Sensor report	Track file	Road data base	Correlation	Context-sensitive target tracking

2.2 GENERIC LEVEL 1 PROCESSING ARCHITECTURES

One indicator of a gradual maturation of the fusion discipline is that its taxonomy is beginning to stabilize; this applies to the taxonomy of basic processing architectures as well. In various recent works (e.g., [6, 7]), researchers have offered concepts for generic processing architectures.

In the domain of fused position and identity estimation (Level 1 processing), Reiner [6] and Yannone [7] have offered the views shown in Figures 2.3 and 2.4. These arrangements generally reflect approaches that operate on raw data (called *centralized* or *measurement set approaches*), operate on preprocessed, "locally fused" data (called *autonomous* or *track file approaches*), or operate on both raw and preprocessed data (called *hybrid approaches*). Each approach has benefits and disadvantages, both in data processing and accuracy. For example, the centralized or measurement set architecture has the advantage of provably optimum position estimates for any sensor-specific measurement variances but requires the highest bandwidth buses (to pass the high-rate raw data) and a powerful central processor capability. The autonomous or track file approach has the advantage of tuning each sensor-specific estimation process to the nuances of that particular sensor's data and operating characteristics but at the expense of increased inaccuracy of the fused position estimate. Processing loads in this architecture are distributed according to specific data throughput requirements, another advantage of this approach. The hybrid approach permits selective transitions between these approaches according to the requirements of the operational problems.

Blackman [8] has also assessed a similar set of architectures (he calls them *central, sensor,* and *combined,* so our language requires a bit more standardization). He evaluates the architectures in terms of

- tracking continuity and accuracy (excellent, fair, and excellent, respectively)
- "survivability" via distributed tracking (low, high, high)
- invulnerability to degraded sensor data (low, moderate, high)
- computational time and complexity (moderate, moderate-high, high)
- data transfer load (high, moderate, very high)

Thus, it is clear that various trade-offs in performance will generally be required and still other hybrid architectures may be designed for specific applications.

In addition to the trade-offs required in selecting one of these (or other) processing architectures are the numerous actual algorithms to select from to perform the association and estimation processes (see Chapters 6 and 7 for specifics of position and identity estimation algorithms). Typically, each algorithm is optimized for some particular aspect of the problem environment; of course, it is difficult to anticipate exactly what precise problem the system may encounter and, in addition, military tracking problems usually change significantly over time. As one approach to overcome such restrictions (i.e., suboptimal performance), Reiner [6] has suggested

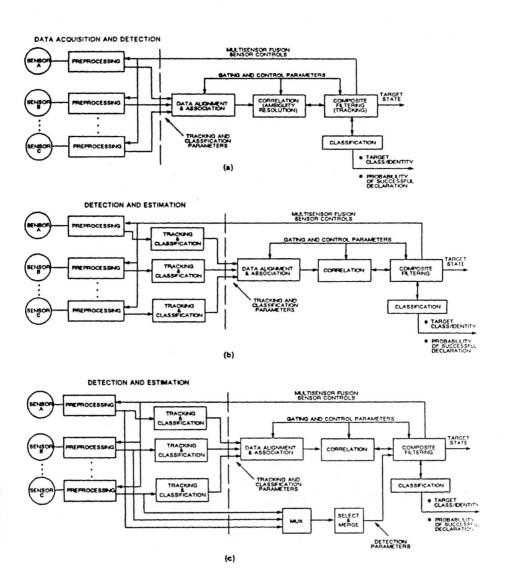

Figure 2.3 Taxonomy of tracker-correlator functional architectures: a = centralized; b = autonomous; c = hybrid (after [6]). © 1985 IEEE.

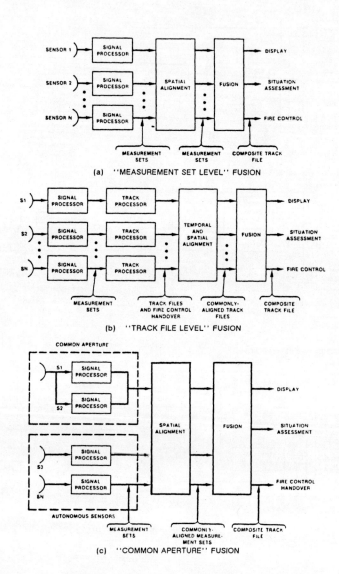

Figure 2.4 Tracker-correlator architectures: a = "measurement set level" fusion; b = "track file level" fusion; c = "common aperture" fusion (from [7]). © 1985 IEEE.

the concept shown in Figure 2.5. The general idea is that the system would have an inventory of tracker algorithms invoked by an expert system that monitors tracker performance in the changing environment. No details of implementation for such a concept have been seen in the literature, although the idea of an "algorithm manager" is not new to complex tracking problems.

Very similar taxonomy and architectural ideas can be found in readings on multisensor approaches to the *automatic target recognition* (ATR) problem, which involves estimating both target position and target classification or identification. Emphasis is usually on the classification approach because image-based tracking techniques are often variations of an approach based on target segment centroid tracking from frame to frame. Pemberton *et al.* [9] have suggested the architectural taxonomy for classification processing shown in Figure 2.6; note its great similarity to Figures 2.3, 2.4. The ATR problem is usually typified as an air-to-ground problem (although the requirement exists for air-to-air engagements as well) and most often involves imaging sensors. In such problem settings, even if the components of the image can be recognized, a context-dependent approach offers great advantages in selecting targets from the field of view (e.g., see King *et al.* [10]). Because knowledge-based approaches offer a way to understand the contextual makeup of an image, such techniques have also been suggested for ATR applications. Hence, we see such approaches as that in Figure 2.7 offered as a means to improve both the discrimination of targets and weapon autonomy.

2.3 REAL-WORLD ARCHITECTURES

For many real-world systems, the Level 1 architectural concepts shown in Section 2.2 represent only partial solutions to the overall requirement, as might be expected. For those involved in building real systems of relatively broad functionality (e.g., C^2 applications of fusion), it is important to realize that the specific fusion processing element of the system, although perhaps a challenging design problem, may represent a relatively small portion of the total system problem. In such applications in which the authors have been involved, fusion specific software has represented about 15–20 percent of the total software developed (based on line of code counts as a metric).

Figure 2.8 shows a representative real-world system functional architecture. In contrast with those of Section 2.2, this structure incorporates several additional functions:

- communication
- database management
- human-machine interfaces
- executive control

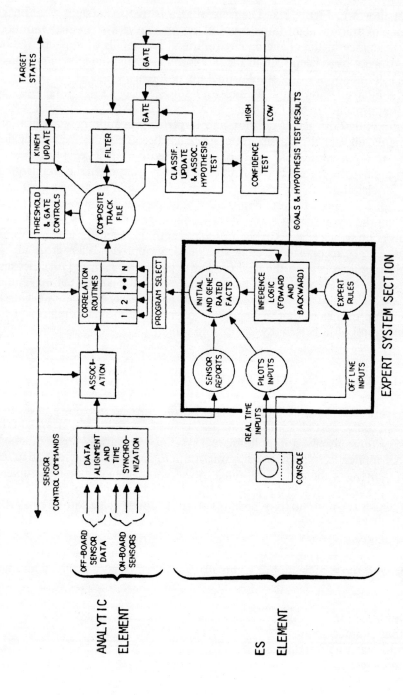

Figure 2.5 Expert system controlled tracker-correlator notional architecture (after [6]). © 1985 IEEE.

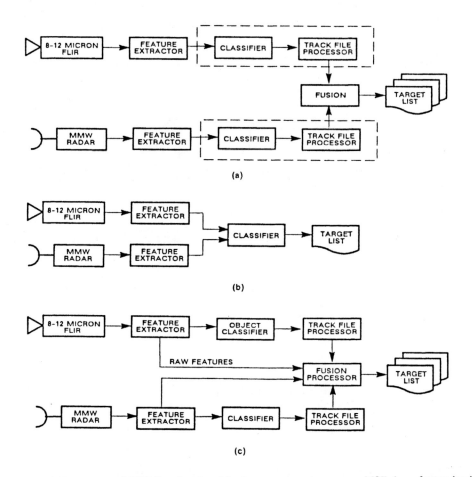

Figure 2.6 Taxonomy of ATR functional architectures: a = post-processor MSF; b = feature-level MSF; c = hybrid MSF (after [9]).

and still others could be added. This structure leads us into an important aspect of fusion processing—the computational aspect.

If the fusion function could be performed independent of the required interfaces to those other functions in real systems, computability would probably be a less critical issue. However, this is not usually the case, as the fusion system is part of the whole, not the reverse. Hence, the fusion system can be subject to constraints that, *a priori,* impose suboptimal performance. Nowhere is this clearer than when the sensor systems are outside of the control of the fusion process, making the fusion process open-looped, so that optimal sensor data collection patterns cannot be realized.

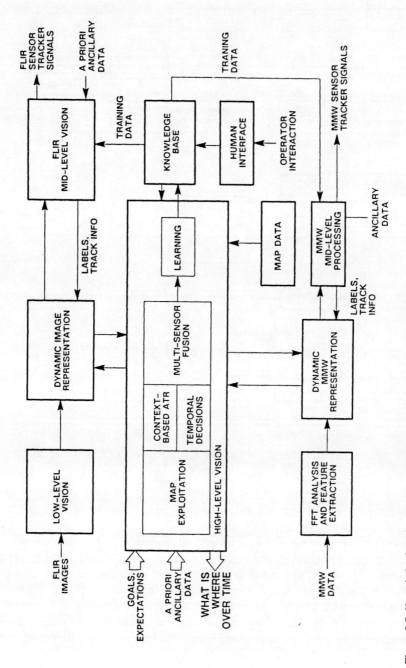

Figure 2.7 Knowledge-based approach to ATR with contextual-learning functions.

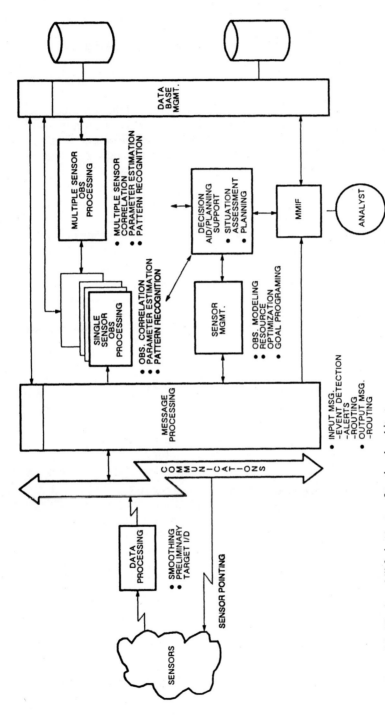

Figure 2.8 "Real-world" fusion system functional architecture.

Another driver to the design of real systems is the database management problem. It is clear that in fusion problems involving multiple targets in military scenarios, in which single platforms or events often associate with more than one possible intended hostile activity, the combinational aspects of the problem can grow quite rapidly. This situation leads to complex problem-solving logic, to requirements for fast computers, and especially to difficult database management problems. As an example of this requirement, on the *tactical command and control* (TCAC) system development, a predecessor to the *all source analysis system* (ASAS), about 30 percent of the code performed the database management, the single largest software function.

The important aspects of real-world systems architectures that the fusion system designer has to realize are the constraints and interfaces provided by the surrounding system elements. This may sound straightforward but in the real world these constraints can force dramatic alterations or complications in what were elegant research laboratory concepts. The combined requirements of complex combat environments and real-world system constraints have resulted in the consideration of parallel computing architectures for Level 1 systems (see, e.g., [11, 12]) so that calculation of fusion products do not bog down the overall system. So far, these architectures have been of an ad hoc, application-specific character, so we offer no extension of the taxonomy of Figures 2.3–2.7, but the reader should be aware that such architectures are evolving.

2.4 FUNCTIONAL REQUIREMENTS FOR SITUATION AND THREAT ASSESSMENT

Defining a class of architectures for *situation assessment* (SA) is a more difficult problem than developing a taxonomy of Level 1 functional architectures, because the SA process is relatively ill-defined (see Chapter 8) and many elements enter into a "situation." In general, however, we are attempting to perform a contextual analysis of Level 1 products (position and identity of individual entities). This next level of inference is thus concerned with force deployment and associated events and activities, knowledge derived from some type of pattern analysis applied to the Level 1 data. Moreover, this knowledge is coupled to a contextual background or setting that may involve both the physical and the sociopolitical environments.

Figure 2.9 shows a functional organization of major SA elements (for details see Chapter 8). Note that before the situation can be abstracted and assessed from the multisensor-knowledge source data, some assessment of countermeasure activity must be performed. Following this, the SA process usually attempts to determine *force disposition/deployment/location* (FDDL) in the context of an environmental or sociopolitical background or both. Note, too, that the SA process attempts to determine important intangibles for the forces of morale, level of training, and so on, as noted in Figure 2.9.

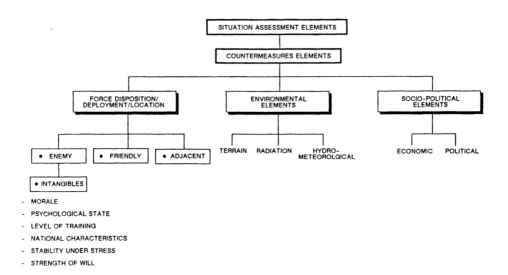

Figure 2.9 Elements of situation assessment.

Following this organizational description, we can assert a generic set of functional requirements for the SA process, as shown in Figure 2.10. Note that at the top we define a "struggle of motives" assessment function. This term follows the Soviet literature and reflects the notion that any given collection of SA data may have several interpretations and implications for decision making. Note also that ideally the SA analyst should be applying the concept of "shifting perspectives" to the data to develop an optimum viewpoint of the situation. This means examining the data from each of Red (as if he were the Red force commander), Blue (the traditional role as Blue force commander), and White viewpoints (studying the environmental aspects to determine how to exploit such aspects in his plans). Thus, a complete SA system includes data bases that support the formulation of these multiple views. Ideally, the SA process yields results that (a) reflect the true situation and (b) provide a basis for event-activity prediction and thereby a basis for optimal sensor management. The SA process therefore is concerned with what is happening and what events or activities are going to happen—as noted previously, it is focused on the *behavioral* aspects within the area of interest.

Most research and prototyping efforts include some or all of these functions, albeit at varying levels of detail. For example, SA analysis in the context of a pilot's estimate of the situation is totally different than that for a corps-level air-land battle. The time-criticality, complexity, and appropriateness of each function varies with the particular application.

Threat assessment (TA), although used liberally in the military literature, is equally ill-defined. The general character and some structural features of this pro-

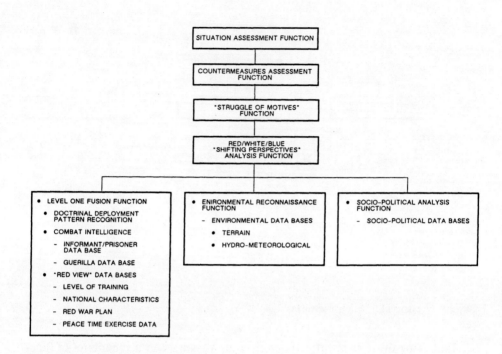

Figure 2.10 Functions of situation assessment.

cess are described in Chapter 8; there, it is asserted that such analyses generally attempt to define *force capability* (FC) and intent, and in the most quantitative way possible. The U.S. military literature is somewhat helpful in defining the elements constituting a TA; these are also discussed in Chapter 8.

Figure 2.11 depicts TA elements assembled from multiple literature sources in an organizational structure. Note that there are some similarities to the SA process (e.g., in countermeasure assessment) and the FDDL activity; it could be argued that critical node analysis could also be part of either process. As we can see, the focus of TA is to assess the likelihood of truly hostile actions and, if they were to occur, projected possible outcomes. Hence FC analysis includes an element labeled *Blue force loss estimates* as a quantitative indicator of potential danger.

As for SA, this elemental composition of a TA can be used to deduce a generic set of functional requirements for TA; this is shown in Figure 2.12. Comparison of Figures 2.9–2.12 reveals the characteristics of SA and TA. SA can be seen to focus on symbolic processing, pattern recognition, spatial, and contextually based reasoning. TA can be seen to include both numerical processing, using "counting" functions of different types, network assessment processes, as well as combat modeling techniques. Neither process is "pure" in employing these methods; that is,

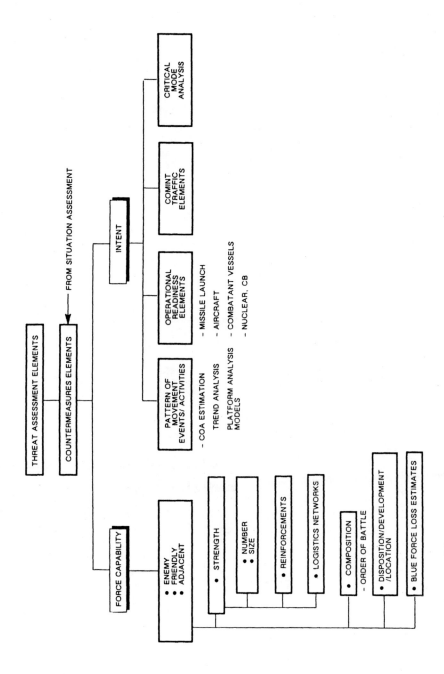

Figure 2.11 Elements of threat assessment.

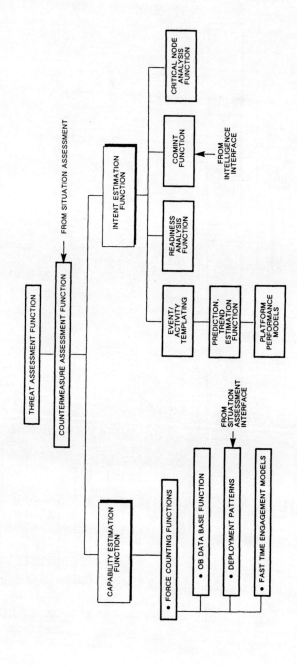

Figure 2.12 Functions of threat assessment.

both processes can and do employ symbolic and numeric processing, but each process usually emphasizes the types of techniques just cited.

2.5 BLACKBOARD ARCHITECTURES AND OPPORTUNISTIC REASONING

The SA and TA fusion processes are said to be more complex than those needed at Level 1 because numerous factors and parameters are involved as well as numerous viewpoints. As a result, the development of SA–Level 2 and TA–Level 3 products typically involves multiple types of expertise—a key characteristic of these processes in most problem settings (again, the single pilot is a special case but even the pilot would often prefer to have several experts in the cockpit). The requirement for multiple types of expertise is reflected in the functional requirements of Figures 2.10 and 2.12. These functional structures, although helpful in offering conceptual characterizations of necessary solutions, do not provide the computer scientist with a processing specification. What is needed is a problem-solving approach and a consequent processing approach.

2.5.1 Problem-Solving Techniques for Situation and Threat Assessment

For problem solutions involving multiple experts or knowledge sources, at least three problem-solving paradigms can be considered:

- *Opportunistic problem solving,* wherein multiple experts are (usually) collocated, monitor the evolving problem state and solution state, and contribute partial solutions asynchronously.
- *Communicative problem solving,* wherein multiple experts must employ some type of communication channel to share in problem and solution state awareness and contribute their partial solutions.
- *Cooperative problem solving,* wherein multiple experts caucus on interpretation of problem states and develop partial solutions.

Figure 2.13 diagrams the nature of each method. Of these methods, the first two have been the subject of most research and development; the cooperative approach (in which human experts are replaced with *knowledge-based systems,* or KBSs) is still the subject of basic research. Because the software implementation of the opportunistic approach is actually an implementation of the communicative approach (due to the implementation-based constraints that prevent truly opportunistic behavior), we focus here on this methodology. Readers interested in alternative concepts can refer to Newell [13] and Nii [14].

In the opportunistic paradigm pieces of knowledge are applied at the "most opportune" time. This paradigm has led to the so-called blackboard processing

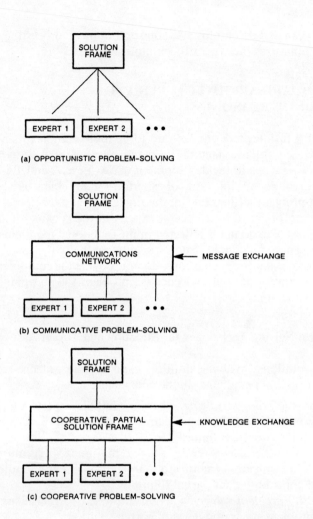

Figure 2.13 Representative problem-solving paradigms: a = opportunistic; b = communicative; c = cooperative.

scheme in numerous Level 2 and Level 3 data fusion applications, and it is probably the solution approach most often used. The concept of opportunistic problem solving notionally allows the experts (or knowledge sources) to participate in a completely ad hoc, asynchronous manner—the analogy to humans using a blackboard for problem solving is clear. This paradigm can employ *backward reasoning,* which reasons from a goal to be achieved toward a required initial state, examining the current data to see if the necessary state exists. Alternately, a *forward-reasoning* approach can be used, in which inference steps are applied from the dynamically changing state ("data fusion" state) to assess the existence of a goal. The key issue is, How and when should a particular piece of knowledge be applied?—that is, the problem of mediation or control. In most fusion applications, knowledge is applied sequentially, but recent efforts have explored the parallelism of fusion problems (see Section 2.6).

2.5.2 Concepts of Blackboard Processing

The fundamental blackboard processing approach was first proposed by Newell in 1962 [13]. However, the first practical realization of the approach is generally considered to have occurred in HEARSAY-II [14], a system intended to understand human speech.

This processing model is a highly structured, special case of opportunistic problem solving. The processing specification defines the organization of the domain knowledge (several knowledge sources), the dynamic input, and the flow of partial solutions to the blackboard. The blackboard is thus a global data base of hypothesized solutions to a problem; in real applications, the blackboard or solution space may be organized into one or more application-dependent hierarchies. The knowledge sources that each transform solution states at one level combine into extended solution states on the same or other levels. Opportunistic reasoning is applied so that the incremental generation of *the* solution results.

Most descriptions of this processing architecture or model consider three major components:

- the *knowledge source* (KS)
- the *blackboard data base* or data structure (BB)
- the monitoring or *control function* (CF)

Each knowledge source is designed to contribute information that will lead to a solution of the problem. They are typically represented by procedures, rule sets, or logical assertions. Only the KSs modify the BB; however, CFs may also be on the BB so that KSs can also modify CFs. Each KS has preconditions for application—when the precondition exists on the BB the KS is a candidate for activation. At any point, multiple KSs may be activation candidates and the CF must mediate their employment. Because of this requirement for KS mediation, the BB approach really reflects a communicative solution rather than an opportunistic solution.

The BB is fundamentally a storage mechanism; it stores solution state data needed and produced by the KSs. In this fashion, interaction among the KSs occurs solely through changes on the BB. Representational techniques for partial solutions in data fusion are usually either spatial or object oriented; Figure 2.14 shows such representations. Figure 2.15 shows how a particular military target object hierarchy may be formed. Figure 2.16 shows how these representations can be applied hier-

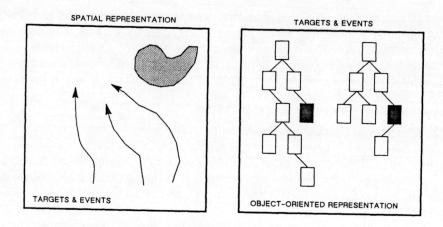

Figure 2.14 Spatial and object-oriented blackboard representation.

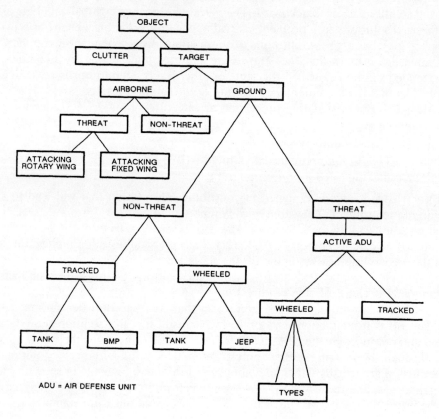

Figure 2.15 Representative object hierarchy.

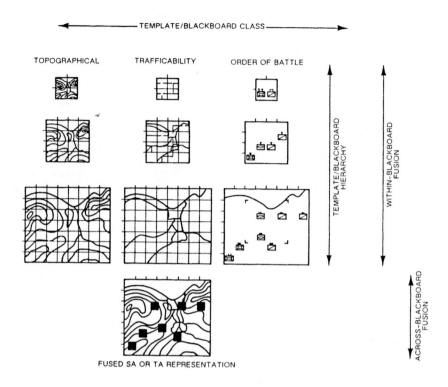

Figure 2.16 Multiple blackboard hierarchy and fusion.

archically in an air-land battle to yield a fused SA or TA. Thus, we see that the solution states are usually represented in a hierarchy or network structure according to distinct types and levels of data analysis. Each part of the solution space can have unique properties and a unique taxonomy-vocabulary through which the state is characterized. For most military problems, the BB usually has multiple "panels" of the type shown in Figure 2.16.

The frame is a common data structure used for a BB when the solution state is object oriented. A frame consists of slots, which in turn consists of facets. Named links between each element are used to define a network of objects and their properties and attributes. These links can reflect such relationships as "a level of," "part of," "in support of," "next to," "follows," *et cetera*. With this flexibility, the frame can also be used to define events and activities. In many applications the term *template* is used interchangeably with *frame;* Chapter 9 discusses several template- or frame-based approaches to data fusion for SA and TA purposes.

For either sequential or parallel processing of BB data by KSs, the primary purpose of the CF is to define the system's "focus of attention," which defines the next thing to be processed. A critical portion of BB system design is therefore the *criteria* used in the CF for determining the focus of attention. The state of these

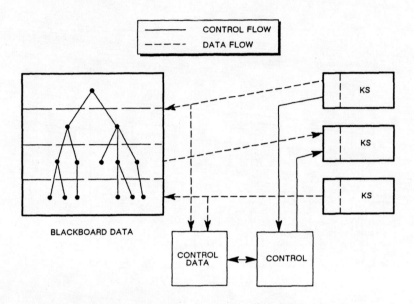

Figure 2.17 Nominal blackboard framework.

criteria may be held either on the BB itself or in a separate control data base. A control module employs its criteria to select (i.e., focus upon) either

- KS, in which case a blackboard object is defined by the CF to serve as the context of invocation for the KS
- BB object in which case a KS is defined by the CF to process that object
- termination condition

Figure 2.17 shows the general nature of the BB framework.

Note that the CF determines the knowledge-application strategy and thereby the problem-solving behavior of the system.

2.5.3 Pilot Aiding Applications

The *rapid expert assessment to counter threats* (REACT) system is a prototype artificial intelligence system described by Rosenking [15]. The purpose of REACT is to aid pilots in determining optimum response strategies during combat; so it is both an SA and TA system.

Figure 2.18 shows the REACT blackboard architecture, which employs four KSs and has a CF that also interfaces with aircraft systems and the pilot. The general purpose of each KS is as follows:

- Global planning: general route the aircraft will take
- Local planning: threat response planning (attack, maneuver, countermeasures, *et cetera.*)

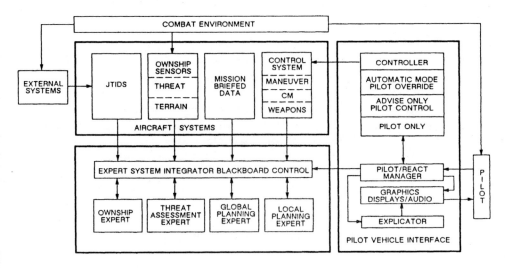

Figure 2.18 REACT architecture (from [15]).

- Threat assessment: fusion of aircraft systems and sensor data for TA
- Own-ship: monitors aircraft state

The generic CF, according to Rosenking [15], has been based on a numerical priority scheme, although some type of expertise is also being considered for the CF.

The specifics of the numerical priority scheme are not revealed by Rosenking [15], but a scheme of this type is discussed in Hiestand *et al.* [16]. The focus of this research was to develop a quantitative basis for evaluating the threat potential of hostile aircraft. Thus, the approach is consistent with our notion of TA as a quantitatively oriented process, and it reflects one particular methodology. The general technique used is that of *multiple attribute utility theory* (MAUT) to formulate the mathematical threat value model. The model employs solely those target and threat features available from radar processing: range, speed, heading, altitude. From these parameters, a *scalar scoring function* (SSF) is formed, called the *measurable threat value function*. The general form of the equations is given by either sums or products of weighted "component value functions," which are functions that provide a mapping from a measured attribute scale (e.g., range) onto the unit interval (see [16]). The weights and component value functions are determined from both analyses and expert judgement.

2.5.4 Some Advantages and Disadvantages of Blackboard Architecture

The BB architecture is certainly a flexible processing technique for effectively implementing the opportunistic approach to problem solving. As such, it provides a way to model the way a group of humans, such as analysts or officers, would attempt to solve an SA or TA problem. However, the disadvantages to this approach must be considered. Table 2.2, gleaned from both experience and a review of the literature, collects a set of such characteristics for these systems.

Frequently, contemporary reasoning systems that deal with complex problems employ *truth maintenance systems* (TMS). TMS is a problem-solving subsystem that records and maintains the reasons for program beliefs [17]; these data aid in guiding the course of action of the problem-solving method (e.g., a blackboard or optimistic approach) and in constructing explanations of program behavior. The key function of a TMS is to trace the reasons for a given set of beliefs (at a given time during program execution) to find and adjust to the consequences of changes in the set of assumptions upon which the beliefs are based. Such changes may happen "nonmonotonically" (see [17]). The TMS uses "dependency-directed backtracking" to locate assumptions associated with an argument, records any inconsistency in this set of assumptions, and changes one of the assumptions to reestablish consistency of the beliefs.

Table 2.2
Advantages and Disadvantages of BB Systems

Advantages	*Disadvantages*
• Flexible structure, inherent modularity for broad range of applications, especially ill-structured problems	• Scheduling or control can become very complex and is generally domain dependent
• Can be used for both static and dynamic aspects of a problem	• Robust generic development tools just now maturing
• With proper design can achieve graceful degradation in blackboard network hierarchy	• Limited (blackboard-based) communication among KSs.
• Easier overall knowledge base control and validation through KS partitioning	• Truth maintenance system often required as KSs operate autonomously
• Expansion and growth relatively easy by adding KSs	• All solution state knowledge on blackboard causes high blackboard-KS input-output
• Simple KS-KS communication protocol (no message-passing protocols required)	• Can be expensive to build and run
• Useful paradigm for exploratory research, rapid prototyping, and incremental development	• Incorrect problem decomposition, often discovered late, requires system restructuring

2.6 ACHIEVING REAL-TIME PERFORMANCE: PARALLEL ARCHITECTURES

For real-world SA or TA problems, it is highly probable that the amount of knowledge required for problem-solving will be quite large. Irrespective of the knowledge representation scheme, growth in the magnitude of the knowledge base will generally lead to slower execution of the system; in many military applications such reductions will be operationally unacceptable. Slow execution can be a problem in many other nonmilitary applications as well and, as a result, many research efforts have been directed to improving real-time performance of knowledge-based systems. These efforts have focused on various strategies for partitioning the solution method and exploitation of such partitioned solutions on the diverse parallel computers available today. Other efforts have examined the benefits, for example, of time-urgent search techniques [18], but in general the focus has been on exploring the benefits of parallelism.

2.6.1 Parallelism in Rule-Based Systems

Rule-based or production-based methods for knowledge representation in KBSs have dominated other approaches to knowledge representation in most expert systems developed over recent years. This method provides a unique, convenient formalism for knowledge representation and a programming environment that incorporates data-driven control and logic representation. These features have enabled production systems to be particularly appropriate for requirements-analysis tasks, rule-based expert systems, and models of human cognition.

In larger systems, these advantages can be offset by slow execution speed. As a result, many researchers have been looking into parallel algorithms and architectures for production system implementation. If we examine such systems, we see that the execution process involves the repeated execution of a "recognize-act" cycle, comprising a *match* between working memory and production memory, *conflict resolution* of the set of satisfied productions to select the rule to be executed, and *execution* of the action side of the rule. Most researchers agree that the match phase is the most involved computationally and consumes most of the execution time. To prevent this phase from being computationally explosive, nearly every implementation incorporates "state saving" and "indexing." *State-saving* is a strategy that limits the amount of working memory that must be examined in any match phrase; this strategy clearly is useful only when the working memory does not change drastically from cycle to cycle, although this is frequently true. *Indexing* is a strategy that maintains a pointing mechanism (an index) between each working memory element and the productions potentially affected by a change in that working memory; this avoids iterations over the entire production memory.

The match phase in the recognize-act cycle is frequently invoked by an assortment of match algorithms; two of the most popular are the RETE [19] and TREAT [20], algorithms with the RETE algorithm probably the most used. One approach to improved speed is therefore to examine the parallelism available in the RETE algorithm; this has been the subject of the studies in [21–23], which have examined shared-memory approaches. Another approach deals with the use of large-scale parallelism with many (thousands) of simple processing elements (e.g, [24–26]). Still another approach [27] involves an *n*-ary tree architecture, shown in Figure 2.19. This approach attempts to exploit production-level parallelism that evaluates all productions in each recognize-act cycle. Each production processor performs the entire match phase for a subset of all productions. Effective partitioning of the rule set thus is the key to this approach. The basic objective of rule partitioning is that, for each working memory change, the productions affected by the change should all be on different processors so that the speed-up factor over a sequential approach is equal to the number of affected productions processed simultaneously.

For fusion applications, Miles [28] has analyzed the parallelism issue in rule-based approaches. This work examines KS-level partitioning (similar to the KS subsetting described for [27]); rule-level partitioning, in which rules are partitioned according to class such as "correlation rules," "combination rules," *et cetera;* and subrule-level partitioning, which partitions rules according to processing requirements. Miles offers some possible architectures and strategies to effect such partitioning strategies, ranging from an architecture similar to that in Figure 2.13 (communicating or cooperative problem solving) to highly parallel object-oriented approaches.

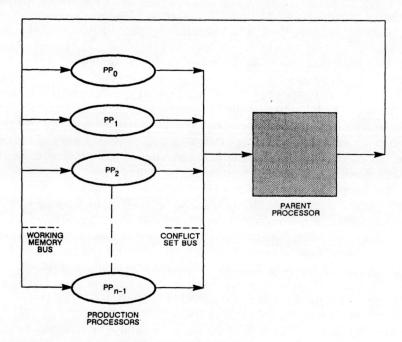

Figure 2.19 Production-level parallel architecture (from [27]).

2.6.2 Parallelism in Semantic Networks

A multiphased research program in the United Kingdom has been examining the multisensor, multiplatform data fusion problem for the Royal Navy [29, 30]. In one of the more recent pieces of work, MacRae and Byrne examined the use of semantic networks and "connectionist" or massively parallel architectures for real-world fusion applications. As suggested, these British researchers have found that execution speed for real-world situations (this work has grown to use NATO maritime exercise data) has become a problem. Initial research efforts used production-rule representations [29], and their analyses suggest that this paradigm may be inappropriate for real-time fusion processing [30]. As a result, they are exploring the use of semantic networks, in particular the "NETL" marker passing semantic network [31]. NETL is an implementation in hardware of a semantic network, in which the nodes represent objects (or concepts) and the links represent relationships between objects; an example is shown in Figure 2.20. Note the "inheritance" feature whereby particular species inherit properties of the genre.

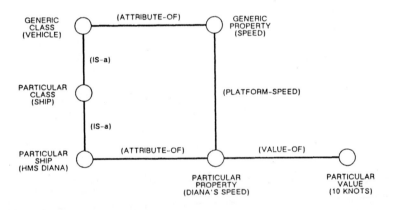

Figure 2.20 Semantic network example.

In semantic networks, the inferencing process is invoked in a different way than in production architectures. The time to complete search operations and carry out deductions is essentially constant regardless of the size of the knowledge base (i.e., network). Fundamentally, deductions are created in a semantic network by set intersection operations. Based on the incoming multisensor data, an initial active node set is identified by rapid marker propagation performed in the parallel architecture (each node having a processing element). As additional properties or attributes are identified by the sensor data, multiple properties are ascribed to (fused to) the different nodes by fast marker passing and inheritance. Nodes or objects having combined (fused) properties are rapidly identified by this intersection process, aiding in the SA or TA analysis. To investigate these issues further, researchers at ARE are building a NETL simulation. Comparisons of the rule-based and semantic network-connectionist approaches will then be undertaken.

2.6.3 Parallelism in Logic Programming

Another approach to building inference systems employs logic programming. As in other approaches described earlier, an analysis of potential parallelism can be carried out by considering the details of the process to discover methods for the introduction and exploitation of parallelism.

The dominant logic programming approach to inference systems is through the use of the PROLOG programming language. In the use of PROLOG, we formulate questions to interrogate the system knowledge (logic) base as a way to assess progress toward a goal state (answer to the question). Calling the question at each step a *goal*, the operation steps of PROLOG can be considered goal-reduction steps involving search and test to assess the existence of required conditions. In conventional sequential PROLOG, the search and test operations (called *unifications*) are executed one by one, but parallel search and test operations can be implemented through parallel architectures.

Parallelism of at least three types can be asserted for PROLOG applications [32]:

- to search for conditions for *all* literals in parallel
- to search and test all conditions for *each* literal in parallel
- to unify several arguments of a literal in parallel

Other operations in PROLOG can also be candidates for the introduction of parallelism.

An *n*-ary type architecture suggested for the exploitation of these parallelisms has been proposed by Tanaka [32]. The machine has a two-level organization, with a Level 1 cluster of inference units, a Level 2 that is a subset of Level 1 systems, and a system manager; see Figure 2.21. The system manager manages global activity through intercluster control of Level 1 systems. The rationale connecting the parallelism assertions to this proposed architecture is rather complex and the reader is referred to [32] for the details.

2.7 COMMENTS

Common language in any research area is important to the achievement of improved communications among researchers, interoperability of systems, and

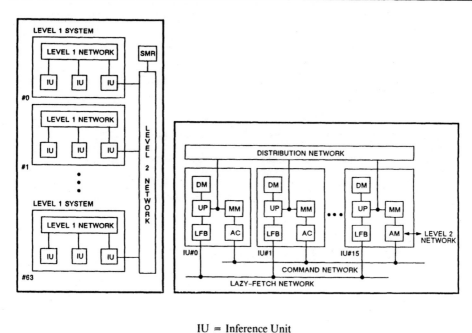

IU = Inference Unit
SMR = System Manager
DM = Definition Memory
UP = Unify Processor
LFB = "Lazy-Fetch" Buffer
MM = Memory Module
AC = Activity Controller
AM = Activity Manager
SM = Structure Memory

Figure 2.21 Overall organization of parallel basic architecture (from [32]). © 1986 IEEE.

standardization and the possible reuse of application modules. This is also true for an *architectural* taxonomy and its attendant language of terms. The data fusion community has begun to move toward such common architectural concepts, but much more work is needed in this area. Progress has been made for architectures of the type employed in laboratory and limited-scope prototypes, but little has been done in the area of larger systems, simply because so few have been built. The language and taxonomy being explored by the AI community for parallel inference architectures can be helpful in this regard and has been described here briefly because of this. Specialized applications of these parallel architectures to data fusion problems can be expected in the near future (e.g., [30]), and an improved architectural taxonomy should result.

REFERENCES

1. White, F., *et al.*, "A Model for Data Fusion," SPIE Conference on Sensor Fusion, Orlando, FL. April 1988.
2. Data Fusion Subpanel/JDL-TPC[3], *Proc. 1987 Tri-Service Data Fusion Symp.*, Johns Hopkins University, Baltimore, June 1987.
3. Data Fusion Subpanel/JDL-TPC[3], *Proc. 1988 Tri-Service Data Fusion Symp.*, Johns Hopkins University, Baltimore, May 1988.
4. Goodman, I., "A General Theory for the Fusion of Data," *Proc. 1987 Tri-Service Data Fusion Symp.*, Johns Hopkins University, Baltimore, June 1987.
5. Antony, R., "Sensor Fusion: Storage, Search, and Problem Solving Efficiency Issues Associated with Reasoning in Context," draft paper, CECOM Center for Signals Warfare, Warrenton, VA, 1989.
6. Reiner, J., "Application of Expert Systems to Sensor Fusion," *Proc. IEEE National Aerospace Electron. Conf.*, 1985.
7. Yannone, R.M. "The Role of Expert Systems in the Advanced Tactical Fighter of the 1990s," *Proc. IEEE National Aerospace Electron. Conf.*, 1985.
8. Blackman, S.S., *Multiple-Target Tracking with Radar Applications*, Artech House, Dedham, MA, 1986.
9. Pemberton, W., *et al.*, "An Overview of ATR Fusion Techniques," *Proc. 1987 Tri-Service Data Fusion Symp.*, Johns Hopkins University, Baltimore, June 1987.
10. King, J.H., *et al.*, "A Context Dependent Automatic Target Recognizer System," *Proc. SPIE*, Vol. 485, May 1984.
11. Rosen, J.A., and P.S. Schoenfeld, "Parallel Processing Applications to Multi-Source Correlation and Tracking," *Proc. 1st Tri-Service Data Fusion Symp.*, June 1987.
12. Opsah, I.T., *et al.*, "Target Detection and Track Formation on a Massively Parallel Computer for Space-Based Surveillance System," *Proc. 13th DARPA Strategic Systems Symp.*, October 1987.
13. Newell, A., "Some Problems of Basic Organization in Problem Solving Programs," in *Conference on Self-Organizing Systems*, ed. M.C. Yovits, Spartan Books, Washington, DC, 1962.
14. Nii, H.P., "Blackboard Systems: The Blackboard Model of Problem Solving and the Evolution of Blackboard Architectures," *AI Magazine*, Summer 1986.
15. Rosenking, J.P., "REACT: Cooperating Expert Systems via a Blackboard Architecture," *Applications of Artificial Intelligence VI*, SPIE Vol. 937, 1988.
16. Hiestand, D., *et al.*, "An Automated Threat Value Model," *Proc. 50th MORS Conf.*, March 1983.
17. Doyle, J., "A Truth Maintenance System," *Artificial Intelligence*, Vol. 12, 1979.
18. Slagle, J.R., *et al.*, "BATTLE: An Expert System for Fire Command and Control," Naval Research Lab Memo—Rep. 4847, July 1982.
19. Forgy, C.L., "RETE: A Fast Algorithm for Many Pattern/Many Object Pattern Match Problem," *Artificial Intelligence*, Vol. 19, 1982.
20. Miranker, D.P., "TREAT: A New and Efficient Match Algorithm for AI Production Systems," Technical Rep. TR 87-03, Department of Computer Science, University of Texas—Austin, 1987.
21. Forgy, C.L., *et al.*, "Initial Assessment for Architectures for Production Systems," *Proc. National Conf. Artificial Intelligence*, 1984.
22. Gupta, A., "Parallelism in Production Systems," PhD thesis, Carnegie-Mellon University, 1986.
23. Gupta, A., *et al.*, "Parallel Algorithms and Architectures for Rule-Based Systems," *Proc. 13th Int. Conf. Computer Architecture*, 1986.
24. Gupta, A., "Implementing OPS5 Production Systems on DADO," *Proc. IEEE Int. Conf. Parallel Processing*, 1984.

25. Stolfo, S.J., et al., "Architecture and Applications of DADO: A Large Scale Parallel Computer for Artificial Intelligence," *Proc. IJCAI*, 1983.
26. Stolfo, S.J., "Five Parallel Algorithms for Production System Execution on the DADO Machine," *Proc. National Conf. Artificial Intelligence*, 1984.
27. Sabharwal, A., et al., "Parallelism in Rule-Based Systems," *Applications of Artificial Intelligence VI*, SPIE Vol. 937, 1988.
28. Miles, J.A.H., "An Analysis of Parallelism in Rule-Based Data Fusion," Admirality Research Establishment, AXT3/JM Tech. Note 23, United Kingdom, February 1987.
29. Lakin, W.L., and J.A.H. Miles, "IKBS in Multi-Sensor Data Fusion," paper presented at the First Int. Conf. on Advances in C^3 Systems, Bournemouth, England, April 1985.
30. MacRae, J.R., and C.D. Byrne, "Connectionism Applied to a Real Time Expert System for Tactical Data Fusion," 3rd Annual Expert Systems in Government Conf., October 1987, Washington DC.
31. Fahlman, S.E., and G.E. Hinton, "Connectionist Architectures for Artificial Intelligence," *Computer*, January 1987.
32. Tanaka, H., "A Parallel Inference Machine," *IEEE Computer*, Vol. 19, No. 5, May 1986.

Chapter 3
DEFENSE APPLICATIONS OF DATA FUSION

Perhaps the most challenging information management problem in military *command, control, communication and intelligence* (C^3I) systems is the need for effective data fusion processes to merge diverse pieces of data into a single, coherent representation of the tactical or strategic situation. The command and control function must have an accurate and timely perception of both friendly and hostile forces for effective battle management decisions [1]. As tactical and strategic C^3I problems have increased in complexity and scope, the processes required to fuse data have surpassed the capabilities of traditional manual methods. This has brought about the need for improved methods that can handle larger volumes of input data, sustain high data rates, and produce accurate estimates. These new methods introduce various levels of automation to the data fusion process, as appropriate for the specific military applications.

Several trends have forced the move to automation of the data fusion process:

- Increases in target mobility or velocity and shorter reaction time lines demand faster information response times.
- Increases in the lethality of weapons require more responsive detection and countermeasure reactions.
- Reduction in target observability and control of signatures require more sensitive detection and identification processes.
- Where increased cost of personnel and risk to operators dictate remotely controlled or autonomous weapon systems, automation of the fusion process is usually required.
- More complex threats (variety, density of platforms, and sophistication of countermeasures) require improved discrimination between targets in the presence of increased levels of disruption, deception, and destructive countermeasures.

Concurrent with these needs, a number of technological developments and military employments have set the stage for the introduction of data fusion systems with increasing levels of automation:

- Employment of multisensor suites to acquire, detect, identify, and track targets using multispectral methods such as *radio frequency* (RF), *infrared* (IR), and *electro-optical* (EO) emissions and reflections.
- Availability of theatrewide tactical data links and worldwide strategic data networks to pass data between *command and control* (C^2) nodes, allowing the exchange of tracks and detections (events) for cross-correlation and sensor-to-sensor hand-off.
- Employment of distributed sensor networks with associated data links to provide coordinated surveillance and location capabilities with improved detection and countermeasure performance.
- Development of passive, low-observable weapon systems using multiple, complementary passive sensors (e.g., electronic support measures, infrared or electro-optical) in place or support of single, active sensor systems such as radar.

Several military missions that may employ data fusion are illustrated in this chapter. The generic functions described in Chapter 2 are applied to these military applications and the characteristics of the missions are contrasted.

3.1 REPRESENTATIVE MILITARY APPLICATIONS

Tactical and strategic military applications are excellent candidates for the implementation of multisensor suites with data fusion to provide the benefits described in Chapter 1. In Chapter 11, we will describe how potential military benefits can be quantified in terms of effectiveness measures to evaluate the relative improvements accrued from the addition of multiple sensors and automated fusion. In this section, applications are described for three tactical mission areas for which data fusion systems have been developed and achieved operational capability.

Table 3.1 [2] depicts the magnitude of data and decisions required to be processed by the C^3 function in the three tactical warfare areas. In each case, the volume of information to be collected, sorted, and acted upon imposes a formidable task for the area commander and staff. Automation of the fusion of data and quantitative evaluation of alternative actions are necessary tools for the decision makers, the human commanders.

Virtually every C^3I application employs a variety of sensors (e.g., radar, sonar, IR search-and-track or IR-EO imagers, seismometers, *et cetera*) and sources (e.g., humint, photoint, data-linked reports) to collect the information necessary to develop a perception of the military situation. In each C^3I application, the fusion requirements may vary widely due to the unique characteristics of sensors, collection systems, and target or event behavior. The primary mission-unique parameters that characterize each application include

1. Target quantities (number of targets to individually detect, identify, or track) and categories (levels of distinction between targets) for each application.

Table 3.1
Typical Information Requirements in Three Tactical C^3 Areas

Parameter	ASW	TACAIR	Land Battle
C^2 command level	Battle group commander	Air component commander	Division commander
Surveillance volume	2000 × 2000 km	800 km² × 20 km	500 × 200 km
Sensor systems	4 surveillance aircraft	6 airborne warning system aircraft	4 surveillance aircraft
	12 ASW ships	50 ground radars	20 outposts
	2 ASW subs	100 fighter A/C	6 ground radars
			1 RPV battalion
Targets in track (max)	100–200	500–1000	100–250
Reports/minute	1,000–5,000	50,000–100,000	100–500
C^2 Decisions/minute	1–5	25–50	10–25

These factors influence the size of the target data base and the processing requirements for updating the data base with new data.

2. Detection or decision rate requirements are a function of the relative temporal behavior of targets (speed, rate of fire, rate of activities, *et cetera*) and reaction time requirements of the C^2 system. These requirements dictate the necessary update rates for sensors, to acquire and track target behavior.
3. Quantity and variety of sensors greatly influence the architecture and processing requirements on the data fusion algorithm. Sensors may be collocated on a single weapon platform or they may be distributed. Differences in sensor characteristics (e.g., resolution, FOV, range, target revisit rates) influence the data association and sensor management functions.
4. Command and control processes although similar [2] in decision-making structure, impose unique requirements due to differences in the number of human operators (e.g., single pilot, hierarchy of personnel in a command center), the methods by which situation data is displayed to operators, and the needs for information to impose control.

Figures 3.1, 3.2, and 3.3 illustrate the three different C^2 applications showing the platforms, sensors, and operational scenarios that characterize the application. The diverse requirements and characteristics of each mission area are described in the following paragraphs.

3.1.1 Antisubmarine Warfare

The antisubmarine warfare (ASW) mission is crucial to the survival of the Navy task force. The coordination of numerous sensor systems is necessary to ensure that hostile attack submarines do not enter the outer defense zone undetected or

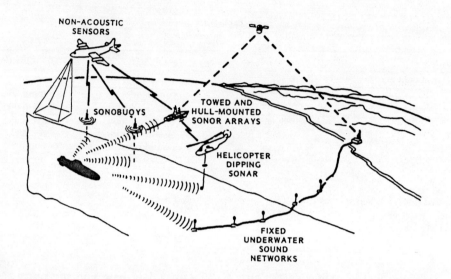

Figure 3.1 Antisubmarine warfare (ASW) C^2 mission (from [2]).

unidentified. Figure 3.1 illustrates the complement of air-, ship-, and land-based sensors that may be networked to acquire, track, and identify submarine threats. Aircraft (carrier-based and land-based) provide the over-the-horizon ASW sensing capability of the fleet, using air-dropped sonobuoys and nonacoustic sensors [3, 4, 5] to detect and track submarines. Towed array and hull-mounted sonars of ASW destroyers, deployed at the perimeter of the task force, form the outer surveillance barrier.

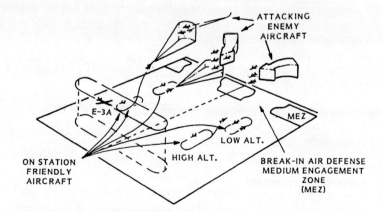

Figure 3.2 Tactical air warfare (TACAIR) C^2 mission (from [2]).

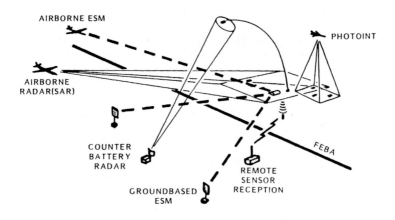

Figure 3.3 Land-battle battlefield C^2 mission (from [2]).

Several data fusion nodes process sensor data and exchange both sensor and target data with other nodes:

1. Land-based surveillance centers receive satellite and other intelligence source data to maintain a fleet-level assessment of which submarines are at port or at sea, and the anticipated missions and on-station periods. Fixed underwater sonar sensor systems (deployed at necessary passage points) also provide detection to confirm the passage of submarine traffic to and from ocean basins. The data are fused with every confirmed sighting and detection from all sources to maintain the highest-level situation assessment of the fleet under surveillance. The data are used to cue the at-sea elements of the sensor system.
2. An afloat fusion node is located on designated surface combatants that collect data from other local surface ships, as well as submarines and aircraft, to develop a local situation assessment for the battle group. Probable target tracks relayed by other naval forces and ground-based centers are used to predict threats approaching the outer perimeter. These reports are used to cue the local sensor platforms to search for the tracks. This requires the coordination of a wide variety of platforms to detect, track, and identify elusive targets hidden in the vast subsurface surveillance volume beneath the force. This node provides an integrated air-surface-subsurface picture of all targets in the area of interest for local commanders.
3. Individual ships, aircraft and submarines each require Level 1 fusion processing for the local sensors aboard each platform. Ships and attack submarines employ hull-mounted and towed array sonar as well as ESM sensors to listen for emissions. ASW aircraft sensors include *electronic support measures* (ESM) to detect emissions at long range or high resolution radar to detect

extended periscopes. These aircraft deploy a variety of active and passive sonobuoys that are dropped to the ocean surface in geometric patterns to relay acoustic detections back to the aircraft for locating the target.

3.1.2 Tactical Air Warfare

The defensive counterair mission depicted in Figure 3.2 shows fighter aircraft flying "lanecap" patrol with long range surveillance support from *airborne warning and control system* (AWACS) aircraft [6]. Two sources of data are available for data fusion on each fighter aircraft: (1) own-ship sensors may include fire control radar, IR search and track, radar-IR-EO warning sensors, *identification friend or foe* (IFF), and EO sensors; and (2) tactical data links provide the ability to pass target tracks and identities from AWACS to fighters or fighter to fighter. The fighter requires a moderate volume (10 to 30 tracks) data fusion and decision support capability to rapidly acquire, track, identify, and hand off targets as well as recommend attack options to the pilot. Such air-battle management systems require the ability to operate independently or cooperatively where cooperating fighters share sensor data and coordinate a multiple target attack. In this case, the data fusion-decision support function is effectively distributed among the air battle management algorithms, located in separate fighters and connected by the tactical data link.

Data fusion may be performed aboard individual fighters as well as on the AWACS aircraft. Wingmen exchange data on a subnet of the tactical data network to provide each with an integrated display of fused targets within an airspace well beyond the maximum envelope of beyond-visual-range weapons. The fighters also act as supporting sensors to pass fused track and identity information for fusion with the central target data base aboard the AWACS, which maintains the broad area surveillance data. The AWACS, in turn, passes track and identity information to cue fighters for their own volume of coverage. The network is also extended to ground-based sensors (surveillance radars, ESM systems, *et cetera* that contribute track information and receive cueing instructions from the broad surveillance data base.

Figure 3.4 illustrates a typical engagement time line in which blue fighters, B1 and B2, engage a hostile strike package (two medium bombers and four escort fighters) after being cued by the AWACS to locate the target complex based upon long-range detection. The sequence of events and contributions of fusion are described in steps 1 to 5 below.

Step 1

The AWACS long-range radar detects the strike package at the edge of the surveillance volume. The two escorts are detected and tracked first because the two bomb-

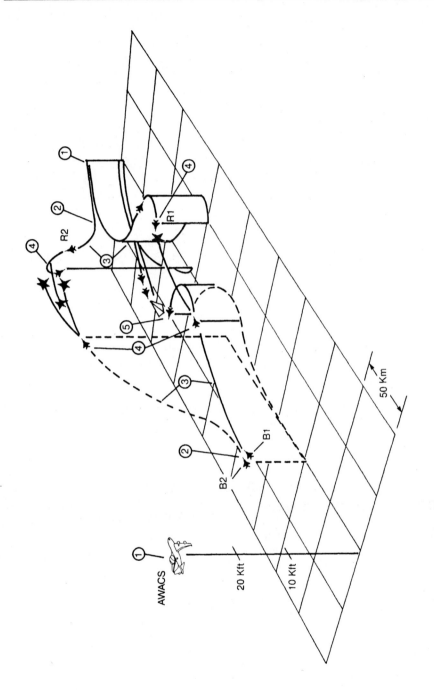

Figure 3.4 Beyond visual range TACAIR scenario.

ers are flying low and in trail to avoid detection by remaining in the clutter background. The ESM sensor detects the terrain-following radar emissions of the bombers and directs the radar to change pulse repetition frequency and processing over that sector for subsequent scans to improve detection of the targets.

Step 2

AWACS cues the two blue fighters to intercept the "probable hostile" targets for positive identification and, if hostile, to engage with medium (beyond visual) range missiles. The fighters fly to the predicted radar acquisition point at 10,000 feet and detect two red fighters, R1 and R2. The red fighters detect blues' radar illumination and break away from above the bomber, turning to beam at first to increase radar cross section and detectability (to ensure that blue will follow the decoy). R2 accelerates and climbs to gain altitude advantage and to attempt to pull the blue fighters further away from the bombers.

Step 3

B1's and B2's radars track the splitting fighters and exchange radar tracks to develop an optimal solution to assign the two fighters' sensors to the two diverging targets while concurrently searching for the, as yet, undetected terrain-following targets. Level 1 fusion processes aboard each blue fighter combine radar, *IR search and track* (IRST) and ESM data (now detecting the reds' fire control radars) to refine the tracks and attempt to associate the emitters with the targets to determine if the emitter is associated uniquely with a given aircraft type operated by a hostile nation. An IFF challenge is directed at each target to determine if the targets can reply to cryptographically secure queries. Neither target replies. The cooperating battle management algorithms aboard the blue fighters assign B1-to-R1 and B2-to-R2 to optimize sensor coverage and missile engagement opportunities. B1 turns to maintain R1 within its radar field of view; and B2 accelerates and climbs, switching the tracking assignment from radar to IRST to eliminate emissions as it can now track R2 against the cold sky background. Throughout the transition from intercept to potential engagement, and Level 2/3 situation and threat assessment displays aboard both fighters depict the same target pictures with symbols indicating identification of targets, blue-to-red assignments, potential threats to each own-ship, and missile launch envelope data against the assigned targets.

Step 4

As each fighter merges toward its assigned target, sensors are managed to acquire positive identification. Medium range noncooperative target recognition sensor modes are activated (at appropriate sensor signal levels) to attempt to identify the targets by sensor signatures. In the case of B1, a radar signature, combined with the nonunique ESM data permits discrimination of R1 as being a specific fighter type that is positively hostile. Knowing the airframe type, the Level 3 fusion process determines missile capabilities aboard the fighter and predicts the optimum engagement point. This information is data linked to B2, where the Level 3 fusion process infers that R2 is hostile due to its coordinated behavior with R1; its recent switch to a radar mode, which indicates that it may be about to engage B2; and the fact that it has begun to jam the radar of B2 (an act considered to be hostile). B1 and B2 launch missiles against R1 and R2, respectively, with the command-guidance phase of each missile supported by tracking data from the Level 1 fusion processors aboard each aircraft. Both fighters make their respective kills.

Step 5

Throughout the course of the engagement, ESM sensors aboard the AWACS and B1 have intermittently detected the bomber's terrain-following radar. The emitter track (bearing only) in B1's Level 1 fusion processor has been passed to AWACS, to permit passive localization of the bomber. As the baseline between AWACS and B1 increases and the distance between B1 and the bombers reduces, track estimates improve. This leads to a sufficiently accurate state estimate; and B1 is cued search for the bombers, who can be positively identified at short range by emitter or radar signature before engagement.

Although this scenario is a relatively simple, 2-on-4 head-on engagement, it does effectively illustrate the complexity of TACAIR fusion functions that must be performed in three dimensions, with maneuvering targets and in relatively short time intervals. All three levels of fusion must be performed aboard each aircraft, using data from the sensors of the other aircraft, as appropriate.

3.1.3 Land Battle Battlefield Warfare

Only the land battle portion of the total air-land battlefield scenario is depicted in Figure 3.3, emphasizing the sensors and fusion necessary to detect and locate ground-based targets. These targets include C^2 nodes, weapon systems (stationary and moving), troops, transportation systems, materiel, and other support forces. Detection and identification of these targets are achieved by a variety of phenomena of which the most significant are movement, RF-EO-IR emissions, and weapon engagement-related events (battlefield targets are characterized as *movers, emitters,* and *shooters,* respectively). The wide variety of sensors employed [7] include

1. Airborne radars detect and locate stationary and moving targets (MTI radar) at slant ranges extending well beyond the forward line of own troops (FLOT).

2. Airborne and ground-based *electronic intelligence* (ELINT) emitter detection-location systems provide the location of RF emitters (e.g., air defense and surveillance radars, communication nodes) beyond the FLOT.
3. Ground-based counterbattery radars track incoming rounds and compute the source weapon location.
4. Airborne *remotely piloted vehicles* (RPVs) as well as piloted reconnaissance aircraft provide targeting beyond the FLOT using TV/IR imaging *photo intelligence* (PHOTOINT) and ESM sensors.
5. Remotely deployed ground sensors (magnetics, seismic, acoustic) report the detection of ground events (e.g., troop movements, formation, and assembly activities) beyond the FLOT. These sensors can be placed on the ground by aircraft, artillery, or ground forces.

In addition, verbal reports from field commanders, outposts, patrols, and intelligence reports from interrogation of prisoners or exploitation of captured communication are used to develop a comprehensive picture of the battlefield situation.

Sensors and sources are fused at the platform level where appropriate, but are transferred to an all-source center that must coordinate the data collection and information fusion to develop battlefield-level assessment for area commanders. This all-source fusion center must perform all three levels of fusion, requiring the coordination of the assets of at least two services (Army and Air Force). Battlefield fusion requires an extensive data base of prior knowledge, including data such as

- Geographic terrain features (e.g., rivers, roads, *et cetera*) that restrict options for force movement, sensor line of sight, concealment, or resupply.
- Order of battle, including intelligence estimates of air and ground force weapons and electronic warfare assets.
- Logistics capabilities and supply support resources.
- Hydrological data and meterological predictions that provide estimates of conditions that may influence troop and air activities as well as the effectiveness of *nuclear-biological-chemical* (NBC) weapons.

Table 3.2 summarizes the differing characteristics of these three areas of warfare and highlights the similarity in the basic requirements imposed on the C^3 data fusion-decision support functions.

3.2 USE OF SENSED INFORMATION

Data fusion uses a combination of sensors and sources to collect information on the tactical situation. This information includes reports of events or target detections, target tracks (a model of state that generally maintains location plus derivatives to predict future behavior), factual statements and measures of uncertainty in

Table 3.2
Characteristics of Three C² Mission Areas

Characteristics	Antisubmarine Warfare	Offensive-Defensive Tactical Air Warfare	Battlefield
Targets	• Submarines • Submersibles • Mines	• Aircraft • Missiles	• Vehicles (movers) • C² centers (emitter) • Air defense units • Missile, artillery units (shooters)
Events	• Surface activity • Torpedo launch • Missile surface broach	• Aircraft maneuvers • Missile engagements • ECM (jamming)	• Weapon firing • Nuclear bursts • Engagements • Troop assembly, movement

Sensors	• Sonar signature
	  land based
	  towed array
	  hull mount
	  sonobuoy
	  helo dipping
	• Passive ESM
	• Satellite surveillance
	• Nonacoustic sensors
	  Magnetometers
	  IR/EO
	• Visual sightings
		• Radar
		  fighter fire control
		  airborne surveillance
		• IRST
		• IFF
		• Passive ESM
		
		• EO imager
		
		• Visual sightings
			• Airborne surveillance
			  SAR/SLAR
			  Passive ESM emitter location
			  photorecon
			• Recon teams
			• Counterbattery radar
			• Ground-based ESM (DF)
			• Remote ground sensor
			  acoustic
			  seismic
			• RPV
			  EO/FLIR
			  ESM

Sources	• Intelligence
	• National sources
		• Preflight intelligence
		• Tactical data link
		  Passed tracks
		  Friend locations
			• Intelligence
			• National sources

Command-Control decisions required	• Submarine
	  which tactical approach
	  when to initiate tactical approach
	  when to engage?
		• Counter air
		  air order of battle
		  aggregation
		  allocation of A/C
		  assignment of A/C to raids
		  weapons choice
		• Target prioritization
		• Route planning
		• C³CM planning
			• Central battle
			  which avenue of approach?
			• Second echelon attack
			  where? when?
			  which targets?
			  with what weapon?

the data. This data is used to detect, locate, and discriminate between tactical targets and events.

Discrimination between targets of different nationalities or types (i.e., identification) can be achieved by any one or a combination of sensed variables. The discrimination process can infer identity by measured target attributes, target behavior, or contextual clues provided by multiple sensors.

Directly measured features include attributes of the target (e.g., spectral signatures, spatial characteristics) or of phenomena that can be associated with the target (e.g., effects on the environment, secondary effects, or events linked to the target). These attributes are measured by the sensor directly or result from preprocessing operations (parameter estimation) that refine and combine the raw measurements into a single attribute. Level 1 fusion processing performs the detection association and classification of these measured features.

Behavior of the target includes temporal behavior (velocity, acceleration, maneuvering, direction of travel, *et cetera* and tactical activities (emitter status or mode; hostile acts such as jamming, deception, or engagement actions). Behavior of a target or emitter can be used for identification only when a unique relationship exists between the measured behavior and target type. Physical or doctrinal models must be developed and validated for uniqueness to confidently use behavior as a hostile identifier. Air corridors and restricted zones known only to friendly aircraft are examples of procedures used in tactical air warfare to provide discrimination of foes by restricting the behavior of friends. The security of such procedures must be maintained, of course, to prohibit foes from presenting themselves as friends. The processing of behavior data can be performed at Level 1 (simple discrimination) or at Levels 2 and 3 of fusion processing.

Context includes the total spatial, spectral, and temporal situation in which the target (or event) is found. Spatial context includes location and relationship to other targets or participants in the scenario. The spectral context may include sensed attributes in relationship to other activities in the same spectrum (e.g., communication exchanges, countermeasure activities, levels of noise). Temporal contextual information includes the relative timing of sensed events or target activities and their implications for inferring coordinated group behavior. The processing of contextual data is performed exclusively at Levels 2 and 3 of fusion processing.

Table 3.3 provides typical examples of measured features, behavioral characteristics, and contextual information that can be used to discriminate submarine and aircraft targets in the ASW and TACAIR missions, respectively.

3.3 MILITARY DATA FUSION-DECISION SUPPORT ARCHITECTURE

The elements of a representative C^3I system [8] are now described to illustrate a general system implementation in which the data fusion functions are partitioned into a Level 1 data fusion subsystem and a Level 2 and 3 decision support subsys-

Table 3.3
Examples of Measurable Discriminants for Antisubmarine and Tactical Air Warfare Mission

Discriminant Categories	ASW	TACAIR

Directly measured features	• Acoustic signatures
	• Magnetic signatures
	• Diesel fume spectra
	• Measurable phenomenologies
	  surface wave effects
	  biological effects
	  atmospheric effects
	  nuclear effects
		• Radar signatures
		• IR signature
		• Electronic emissions (PRF, pulse characteristics)
		• IFF replies
		• Acoustic signature
		• Imagery shape or size signature (IR-EO)

Behavioral characteristics	• Speed • Sustained speed • Depth • Maneuverability • Hostile act (cruise missile launch, torpedo launch) • Communication activities with hostile forces	• Velocity-altitude envelope • Maneuverability • Operation within an air corridor (friend) or a restricted volume (foe) • Hostile act (ECM, missile engagement against a friend, etc.) • Flight path tactics executed
Contextual information	• Origin of mission (home port) or ports of call • Location relative to friendly surface vessels	• Origin of flight from hostile or friendly base • Flight path • ECM activities • Relative formation with other knowns

tem. The system closely follows the functional architecture introduced in Chapter 2.

The C^3I system, depicted in Figure 3.5, uses a variety of sensors to collect data from the combat environment and then passes sensor reports on to the data fusion system. The sensors may provide different kinds of data: (1) detections (reports) that may include time-tagged events or targets with available location or identity information; (2) sensor-level tracks that model target state (behavior) and may include identity information; (3) statements of fact that are typically provided by intelligence sources and usually related to current or predicted (probable) behavior of hostile forces. The system performs the three levels of data fusion processing previously described in Chapter 2.

3.3.1 Data Fusion Subsystem

Level 1 fusion functions are performed in this subsystem to develop a common situation data base of associated, combined information from the raw sensor

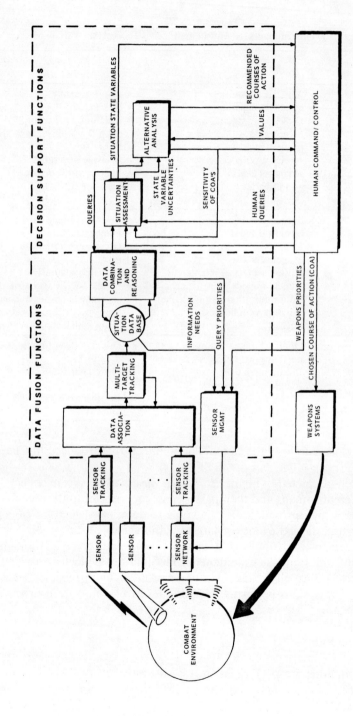

Figure 3.5 Data fusion and decision support architecture (from [2]).

reports. As sensor reports are accumulated in time, temporal tracks are developed to model the dynamic behavior of each target. These tracks of targets, histories of events and detections, and statements of fact are maintained in a situation data base. Numerical classification algorithms and symbolic processes combine the attribute data associated with individual targets to refine the estimate of target-event identities.

The three major functional elements of this subsystem are described in the following subsections, introducing the major implementation alternatives that will be described further in Chapters 5, 6, 7, and 8.

Data Association and Estimation for Mutiple Target Tracking

Detections (reports) and tracks (from sensors that maintain local track files of targets) must be associated in time and space to assign sensed data to individual target files. Because the reports and tracks arrive from sensors with differing geometries (relative to the targets), resolutions, fields of view, measurement spaces, and measurement times, it is necessary to transform all measurements to a common spatial reference (i.e., latitude-longitude, range-azimuth-elevation, X-Y-Z, *et cetera*) and temporal reference. Once the measurements are transformed to a common measurement space, association hypotheses can be developed and scored to determine probable report-report or report-track assignments. Figure 3.6 depicts the typical association and tracking functions, which include both tracking and nontracking (report-only) sensors. Each new report on track is first transformed to a common spatial coordinate system before hypothetical assignments (pairings) are formed to existing reports or tracks in the situation data base. Each hypothesis is scored on the basis of a metric that can include spatial distance measures between the sensed report and hypothetical assignment, and feature data measures that characterize the sensed report and hypothetical assignment. These hypotheses are then evaluated to select a single pairing or to minimize the set of acceptable pairs if multiple hypotheses are to be saved for further iterations, when more data is available.

When unique assignments of new report data are made to existing tracks, the dynamic tracking filter is updated to improve the estimate of the targets' behavior. Uncorrelated reports are passed to the situation data base for possible association with future detections. Because of the uncertainty in sensor measurements and density of targets, probabilistic methods are often required to score and evaluate the many hypotheses that can result in multiple sensor systems [9–11]. Hypothesis management over time and dynamic tracking of maneuvering targets are the most difficult challenges to the association-tracking function.

Attribute Fusion for Target Identification

The next function combines all the sensor data associated with each target in the data base to classify the target into a set or class, thereby inferring its identity. Con-

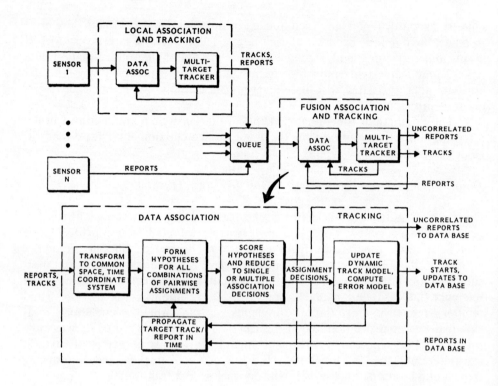

Figure 3.6 Data association and tracking functions (from [2]).

ventional, structured algorithms and symbolic reasoning processes are candidates to mechanize these combination-inference functions. Any combination of directly measured features, behavioral characteristics, and contextual information in the data base may be used to classify targets. Several categories of prior knowledge are stored to perform the combination process:

1. *Structural models* are maintained to define the relationship between measured target attributes (e.g., acoustic-IR-radar signatures, RF emissions, kinematic envelopes, and component parts) and targets or target classes. These relationships between attributes and entities are defined by the so-called IS-A and IS-A-PART-OF hierarchical connectives in a frame-like data base.
2. *Behavioral models* relate the temporal behavior (e.g., aircraft velocity-altitude flight envelopes, ground troop assembly-attack profiles) as well as spatial behavior (e.g., target formations, weapon ranges) of targets or events.
3. *Heuristic inference rules,* including simple IF-THEN and Boolean (AND, OR, COMPLEMENT) relationships, provide a general structure to construct inference templates. These rules include an antecedent (the IF conditions that

must be satisfied by the sensor data) and consequent (the THEN inference that can be concluded when the antecedent conditions are present).
4. *Numerical classification algorithms* combine incomplete, ambiguous multiple sensor data by computing composite hypothesis scores that use the independence between multiple measurements to reduce the ambiguity in the measurement data. Mathematical models are used for quantitative comparison of sensor data with prior information on known target types to classify the sensor data.

Because sensor measurements and intelligence data are often characterized by uncertainty, a number of methods may be employed to quantitatively represent the uncertainty in sensor and intelligence data. Probabilities [12], fuzzy variables [13], and probability intervals [14–17] are among the candidates described in Chapter 7 for representation of the uncertainty associated with sensor data. Numerical combination algorithms may be used to provide a more accurate composite estimate of the measurements, reducing uncertainty in target identity or location.

The inference process can be driven by the arrival of new sensor data or by queries fed back from the decision support subsystem. The process of combining new data with the existing situation data to determine if ambiguity is reduced or consequents can be achieved is referred to as *forward chaining*. The arrival of new associated report data (antecedents) on a given target will cause the combination algorithm to be iterated to refine the estimate of target state and determine if new inferences (consequents) can be drawn. The situation data base is then updated with any new consequents. In contrast, the queries from the decision support subsystem cause a *backward-chaining* process to determine if the antecedents in the data base support a hypothetical consequent supplied by the query. The query may request, for example, the degree of support for a consequent: What is the certainty that multiple hostile bombers with fighter support (a strike package) are en route to attack the airfield at location X within 10 minutes? Such a query may use each of the types of knowledge to verify consequent:

- Behavioral models define airfield attack profiles, and airspace regions in which potential strike packages must be located.
- Situation data base is then searched to locate hostile and unknown aircraft meeting the strike criteria or any related events that may indicate the presence of a strike.
- Structural models can then be compared with target identities or attributes, flight patterns, related electronic warfare events, *et cetera* from the data base to determine if a strike package can be located.
- Numerical combination algorithms compute the certainty for each hypothesis (each candidate strike package located).

The data fusion subsystem responds to the queries with single (hard decision) or multiple hypotheses, each represented by state variables (target locations, identities, *et cetera*) and associated certainties.

Sensor Management

The allocation of sensors to targets refers to the ability to control the pointing and dwell time of sensors on detected targets or regions of the surveillance volume. The management function must rank targets or regions and then allocate sensor resources to targets or regions on the basis of those priorities. Management of the sensors is based upon several driving functions:

- Information needs for targets in the data base are a function of the quality of state estimates and confidence in identity data. Additional data may be required to refine the track, resolve ambiguity in target location, or discriminate target identity. These needs may be ranked by the degree of threat of each target to friendly forces.
- Query priorities are developed by requests for additional data on targets to reinforce or disprove developing hypotheses.
- Weapons priorities from the C^2 function affect those sensors that directly support weapon systems, such as fire control sensors. These priorities result from C^2 decisions to engage targets with weapons that require direct sensor support.

The sensor management function merges these (possibly conflicting) requirements to allocate sensors to targets. The functions also include sensor cueing, hand-off, and human override.

The situation data base contains the following kinds of data as a result of Level 1 processing:

- Location of all targets and events (and the uncertainty associated with each).
- Estimated dynamic state of targets (tracks) from which future behavior can be predicted.
- Identification of targets and associated uncertainty or ambiguity between multiple possible identities.
- Situations of military importance resulting from individual target locations, behavior (e.g., threat weapon systems in attack positions), or aggregate behavior of multiple targets.
- Sensor allocation assignments, schedules, or priorities for each target.

3.3.2 Decision Support Subsystem

This data is passed to the decision support subsystem for Level 2 and 3 situation and threat analysis. The modeled situation includes uncertainty arising from limitations in the sensor measurements and coverage. As a result this uncertainty allows for multiple hypotheses concerning many of the elements of the situation model (e.g., target locations, identities, behaviors). The situation assessment function refines these hypotheses using doctrinal templates and probabilistic inference to select a finite set of important alternatives for analysis. The set of possible situations, plus uncertainties of the state variables are passed to the alternatives analysis function to create-analyze-rank alternate courses of military action.

The situation (and threat) assessment function performs those functions previously described for data combination and reasoning at a higher level to determine the *meaning or military significance* of the state of targets or events and their behavior as estimated in the situation data base. Human or automated queries search for behavior (spatial, temporal, or spectral) patterns from which can be inferred the situation or threat, as described by such items as

- The plans, orders of battle, available courses of action, intents, or activities of hostile forces.
- The opportunities available to friendly forces against hostile targets.
- The threats imposed by hostile targets to friendly forces and the countermeasures available to defeat those threats.

Alternative *courses of action* (COAs) are then developed and expected outcomes computed, with associated scores for each alternative. The most promising COAs are then recommended to the human operator or commander with scores for assessment and consideration. The operator may enter weighting values (expressing degrees of importance) for performance or effectiveness measures used in scoring to influence the means of recommending COAs.

3.3.3 Human Command and Control

The human commander (e.g., pilot, division commander, task force commander, tactical air control commander) views the situation display (range-azimuth display, map display, *et cetera*) with symbols overlaying target identities, velocity vectors, weapons, envelopes, and other information relevant to each target. In addition, the commander may view the alternative courses of action and expected results. The commander's action commits the weapon systems under command and completes the feedback loop by affecting the combat environment. Planned actions and procedures are also entered into the situation data base to aid the combination and reasoning process.

3.4 CHARACTERISTICS OF FUSED INFORMATION

The information provided by data fusion systems is characterized by multiple levels of information content and a representation of the degree of certainty (confidence) in the information provided. Figure 3.7 depicts an example of the levels of identification needed by a TACAIR decision support system. The hierarchy of aircraft identification includes four general levels of information content:

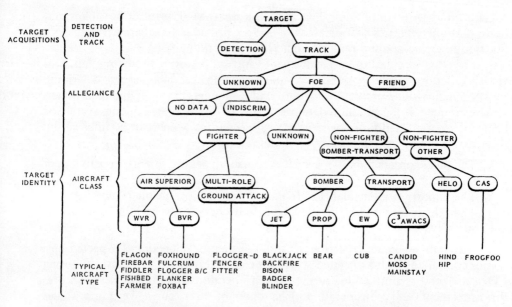

Figure 3.7 Hierarchy of data fusion information on a sensed aircraft target (from [2]).

1. Detection and tracking of a target is the first level of information provided. Detection information can include three-dimensional spatial coordinates (e.g., sensor coordinates in range-azimuth-elevation, or earth-centered coordinates) or only one-dimensional data (e.g., passive detection of an emitter with only azimuth or bearing information). When repeated detections are associated and a time-series representation of the target's dynamic behavior can be determined, a track can be established.
2. Aircraft allegiance is the traditional friend-foe-neutral identity that provides the first, and most important, level of situation awareness. If foe, the aircraft is a potential threat and a candidate to engage or evade.
3. Aircraft class provides a general identity of the aircraft mission, weapons (degree of threat), and performance capability. In the defensive counterair mission, for example, the discrimination of fighters from bombers permits the pilot to avoid threats (hostile fighters) and selectively engage strike aircraft (bombers). A knowledge of aircraft class permits direct inference to other attributes common to the class such as basing, weapons, communications, avionics, and kinematic performance capabilities.
4. The highest level of information is the discrimination of aircraft (airframe model) type. Soviet aircraft selected from *Jane's All the World's Aircraft, 1983–84* [18] are listed in the figure to illustrate the variety of aircraft types we want to discriminate. For an identified foe, this information will aid the

pilot to assess the degree of threat imposed by a fighter or to select the highest value bomber or fighter bomber to attack.

Additional levels of information that may be provided on targets include: weapons status (stores remaining), estimated fuel status (inferred from type and point-in-mission) and predicted mission intent (e.g., intended targets, next-action, *et cetera*). The various levels of information maintained in the situation data base are stored as hard or soft decisions:

1. Hard decisions (declarations) are statements of fact where the combined evidence exceeds a predefined threshold of certainty (e.g., greater than 0.99 probability of occurrence).
2. Soft decision data include a representation of uncertainty for each of multiple hypotheses using probabilities, measures of belief, fuzzy variables, or other metrics.

The information supplied to the decision support functions results from queries initiated by that function and by new sensor reports, which result in changes in the knowledge of the tactical situation. Such changes can include, for example,

- New targets detected as they enter the surveillance volume.
- Refinement or revision of target identities, unknown targets identified.
- Changes in target behavior or state.
- Loss of targets from track.
- Detection of events or aggregate target behavior that infers opponents' tactics or future behavior.

The information in the situation data base is reported to human operators in various display formats that depict targets, estimated state, associated identities, and other pertinent information (e.g., velocity vectors, relations to other targets). The information associated with each target can also be reported in a variety of methods. Hard decisions may be reported as single statements or ambiguity lists (lists of possible alternatives, each with an equal likelihood of occurrence). Soft decisions on the other hand, are reported as multiple hypotheses, each with a representation of the confidence or uncertainty associated with the hypothesis.

Military data displays for fusion information take a variety of forms, depending on the application. The most common display format to all applications is the spatial display of target information in a two-dimensional coordinate system; for example, the familiar radar *plan-position indicator* (PPI) in range-azimuth sensor coordinates or an x-y earth coordinate display with geographic map overlay. Targets and associated decision data are most often displayed using symbols and accompanying tabular-mnemonic alphanumeric text:

- Targets are depicted in their proper spatial location by symbols with shapes that are related to target identity (allegiance, class, type, or air-surface-subsurface classification).

- Targets or events that cannot be located with sufficient certainty to a single point display the region of probable location, which may be defined by a boundary with appropriate identification.
- Limitations to information the symbols can convey call for additional numerical or textual information provided by several common means: (1) display of the data for each target adjacent to the target symbol directly on the display, (2) display of the data for all targets in tabular form on a sidebar or pop-up windowed area on the display, (3) display of the data for any target designated by the operator (using a cursor or other means) on a sidebar or pop-up window on the target display or on a separate display.
- Information from Levels 2 and 3 processing and interactive queries to the decision support system by operator may be acquired via structured or natural language interfaces using keyboard terminals or speech synthesis-recognition equipment.

Conceptual situation displays for the tactical air and battlefield examples described earlier are presented in Figures 3.8 and 3.9, respectively. These concepts illustrate the formats that are appropriate for displaying Levels 1, 2, and 3 data fused information in an operationally useful format, but these are not specific formats for any operational system. Figure 3.8 illustrates the format of representative fighter aircraft displays that provide the pilot with three views of the tactical situation. The formats can be presented on individual displays, or the pilot can select one from a common integrated display. The formats shown are simplified versions of those on modern (1980s) fighters that have independent displays for the fire control radar and radar warning receiver sensors and a tactical data link, respectively.

A forward-looking sensor view (Figure 3.8a) displays targets detected by the primary radar or IRST sensors in a range-azimuth format. The location of the own-ship is at the center; at the bottom of the display associated targets are displayed with detections, tracks, and associated radar-IR targets distinguished by unique symbols. Tracked targets are characterized by velocity vector lines that indicate direction of travel. The figure illustrates two inbound hostiles (designated as triangles), a single outbound unidentified target (rectangular symbol), and a single friend (circular symbol) leaving the sensor's field of view to the left. A jam strobe is indicated at $+35°$ by vertical dashed lines. Additional altitude, velocity, multiple target, and ID alphanumeric data may be displayed adjacent to target symbols, if desired. When operating in a fire control mode, this display may also overlay graphical information indicating location of a designated target within the fighter's missile launch envelope.

An integrated ESM-IR-EO emitter warning view (Figure 3.8b) presents a $360°$ polar coordinate display of all emitters passively detected. The own-ship location is in the center, with emitters displayed in concentric range rings (if passive ranging methods provide range) or at the outer ring (if range is unknown). Symbol shapes

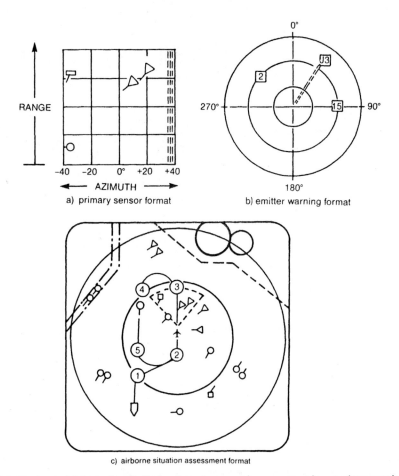

Figure 3.8 Conceptual fighter aircraft display formats: a = primary sensor; b = emitter warning; c = airborne situation assessment.

and alphanumerics indicate emitter type. Note that the radar jam strobe observed at +35° on the primary sensor display also appears here as emitter J3.

A situation display (Figure 3.8c) presents an overhead view of the fused local sensor data and global (data-linked) data in an earth-fixed x-y coordinate system. The own-ship is located at the center of the display, with the velocity vector pointing to the top of the format. Degree of threat, target multiplicity, target identity, or confidence in identity and other such information can be indicated by symbol shape or color. In this example, note the following items displayed:

1. One's own ship has flown a course from base (five-sided symbol) to waypoint 1 onto a lanecap racetrack oval described by the waypoints designated 2, 3, 4, and 5.

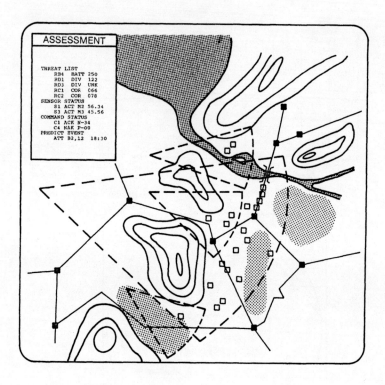

Figure 3.9 Conceptual battlefield situation display.

2. Forward line of troops is designated by the dashed line at the upper right corner of the display. *Surface-to-air missile* (SAM) launch envelopes are shown as circles beyond the FLOT.
3. An air corridor for safe passage of friendly strike aircraft is at the upper left, with two aircraft shown returning down the corridor.
4. Aircraft tracks are shown with velocity vectors and symbol shapes for identity (as on the primary sensor display). The four prime sensor targets can be seen within the angular sector that shows the combined radar-IRST fields of view. The wingman to one's own ship is seen on the return leg of the racetrack between waypoints 4 and 5. Other aircraft are also shown behind one's own ship and beyond its sensor's ranges that have been provided by other platforms over the data link.

The conceptual battlefield display format (Figure 3.9) illustrates the overlay of data fusion information on a map that depicts terrain features important to the battlefield commander. The simplified example depicts friendly forces at the left side of the display (the west side) with hostile forces being detected to the east of

the two terrain features at the center of the display. The following symbols characterize this display:
1. The underlying map shows terrain, rivers, major roads, bridges, towns or cities, and forested areas in formats as close to conventional maps as possible.
2. Sensor coverage sectors are shown (if selected by operator) for an airborne ELINT-MTI radar system and several counter battery sensors.
3. Hostile units detected are indicated by rectangular symbols in this example. Symbol shape, color, or interior alphanumeric may be used to indicate unit type (e.g., tank, mortar, artillery, air defense). Distinctions among company, battalion, regiment, and division levels of command are also distinguished, where possible.
4. Dynamic activities (movement detected by MTI sensor, firings by artillery, missile launches, *et cetera*) are indicated on appropriate targets by symbol change (shape or color).
5. Situation and threat assessment data from Level 2 or 3 processing is presented in tabular and textual form in the window in the upper left corner of the display. These assessments include measures of uncertainty in data, doctrinal templates that most closely match the battlefield situation, order of battle estimates, predicted threats, avenues of approach, and opponents objectives.

The density of battlefield targets and desired resolution of targets to be displayed (relative to area covered by the display) usually demands a high-resolution format to display all Level 1 fusion results. The use of color and careful selection of symbol shapes aids in the recognition of force patterns and behaviors. Level 2 or 3 assessments may be displayed as textual data in windowed blocks (as illustrated in the example) or as graphic overlays of predicted advancements or objectives of hostile forces.

3.5 SPECIAL CONSIDERATIONS FOR MILITARY SYSTEMS

The implementation of data fusion systems for military applications imposes several requirements that must be considered by the operational planner and designer. These requirements are not applicable to industrial and robotic applications and are therefore mentioned briefly here.

3.5.1 Information Security

As illustrated in the examples of this chapter, fusion processes bring together a variety of sensor and source data, often with different levels of security classification and access. Source intelligence data often discloses, by inference, the means of collection and has a higher classification than tactical sensors. The results of Level 2 and 3 fusion processes using such data must ensure that information revealing the collection means is clearly identified. The input and output data, therefore, must be carefully partitioned to maintain security in accordance with approved security procedures. The distribution of fused data products to various levels of users (e.g., national or local commanders and tactical forces) may require that the fusion process maintain an audit trail of data used to classify each element of output data and limit access to authorized users.

3.5.2 Multisensor Countermeasures

Hostile countermeasures will certainly be developed to deceive, disrupt, and exploit the functions of multisensor fusion systems. Coordinated, multisensor countermeasures may be very effective against data fusion systems that do not consider the means by which a countermeasure may present false information to their sensor suites. Several potential threats that must be considered include the following.

Coordinated Sensor Deception

The coordination of inputs to many sensors may permit an adversary to create a false detection or classification. Although more complex to perform than single-sensor deception, this countermeasure may achieve a higher degree of deception due to the high confidence level that may be assigned by the fusion of multiple (false) inputs.

Selective Disruption

Selective jamming of sensors in a multisensor suite may provide blind spots in temporal, spatial, or spectral coverage, degrading performance and providing an opportunity for undetected targets or activities.

Behavioral Deception

Knowledge-based systems that infer identity, intent, or plans based on behavior of adversaries must consider the elements of surprise, unlikely behavior, and human deception. The gullibility (a measure of deceptability) of early-generation Level 2 and 3 assessment systems may be difficult to determine and should not be expected to rival the performance of experienced military personnel.

3.5.3 Data Confidence Identification for Lethal Use

Military systems use fusion data for lethal (weapons commitment) as well as nonlethal (intelligence, reconnaissance, surveillance, target acquisition) purposes. The confidence in data (expressed as degree of belief, *a posteriori* probability, or other confidence factors) is operationally related to its usefulness for critical, lethal purposes. Target identifications made for commitment of weapons to a target, for example, require very low error rates to reduce fratricide (due to erroneous foe decisions) or threat to own forces (due to erroneous friend decisions). Low-confidence data is useful for warning, cueing, and planning; whereas only high-confidence data (e.g., "positive hostile identification") is used for engagement of targets under normal *rules of engagement* (ROE). For this reason, the specification of levels of confidence in data and the associated means of display are critical in military systems.

3.5.4 Fusion System Classification

In addition to the necessity for security in the protection of sensor, source, and processed information, the *methods* of data collection and processing may also require classification. Knowledge of these characteristics reveal the capabilities, performance limitations, and vulnerabilities of systems. The specifications for military sensors and data fusion systems are generally classified and the software code may also be classified to maintain such security.

3.6 COMMENTS

This chapter has presented representative military applications for data fusion, with illustrations from three tactical warfare areas. Additional strategic military applications not discussed here include indications and warnings of nuclear attack, strategic intelligence fusion, and the fusion of layered ground-air-space sensors for strategic defense against ballistic missiles. These and many more applications have been reported as data fusion systems have moved from R&D status to full scale development and into operational weapon and surveillance systems. References to open literature descriptions of such systems are cited throughout this book, where specific features of those systems are described.

Although the focus of this book is on military applications, data fusion is certainly not limited to this domain. Additional applications reported in the literature include the remote sensing of natural resources, nonmilitary surveillance, robotics, navigation, and industrial materials handling, inspection, and assembly [19]. These applications have similar functional, architectural and physical implementation properties, with the differences being in operational implementations. Special considerations that distinguish military systems are the need for information security, countermeasure protection, data confidence identification, and system classification.

REFERENCES

1. Edwards, Col. Albert J., USAF, "Information Challenges from Tactical Air Operations," *Signal*, November 1985, pp. 75–81.
2. Waltz, E.L., and D.M. Buede, "Data Fusion and Decision Support for Command and Control," *Proc. IEEE Trans. on Systems, Man and Cybernetics*, Vol. SMC-16, No. 16, November–December 1986, pp. 865–879.
3. Stone, N.L., "ASW, the Soviet View," *Military Electronics/Countermeasures*, June 1979, pp. 34–73.
4. Starkey, R.J., Jr., "Anti-Submarine Warfare Oceanography," *Military Electronics/Countermeasures*, July 1981, pp. 76–72.
5. Wohl, R., "Ocean Transparency: Impossible or Inevitable?" *Defense Science*, February 1984, pp. 20–28.
6. Anderegg, Maj. Dick, and Lt. Col. Bill Payne, "Tactical Relevance—A Fighter Pilot's View," *Defense Systems Rev.*, September 1983. pp. 36–40.
7. Mahaffey, MGen F.K., "Intelligence Surveillance and Target Acquisition in the 80's—The Army Concept," *Signal*, October 1979, pp. 39–42.
8. Waltz, Edward L., "Data Fusion for C3I Systems," *International C3I Handbook*, EW Communications, Palo Alto, CA, 1986.
9. Fortman, T., Y. Bar-Shalom, and M. Scheffe, "Multitarget Tracking Using Probabilistic Data Association," *Proc. 19th IEEE Conf. Decision and Control*, Vol. 2, December 1980, pp. 807–812.
10. Kosaka, Michitaka, *et al.*, "A Track Correlation Algorithm for Multisensor Integration," *Proc. Digital Avionics Systems Conf.* (DASC), IEEE, 1983, pp. 10.3.1–10.3.8.
11. Goodman, I.R., "An Approach to the Data Association Problem through Possibility Theory," *Proc. 5th MIT/ONR Conference*, August 1982.
12. Nahin, P.J., and J.L. Pokoski, "NCTR Plus Sensor Fusion Equals IFFN," *Proc. IEEE Trans. Aerospace Electron. Systems*, Vol. AES-16 No. 3, May 1980, pp. 320–337.
13. Ruspini, E.H., "Possibility Theory Approaches for Advanced Information Systems," *Computer*, September 1982, pp. 83–91.
14. Shafer, G., "A Mathematical Theory of Evidence," Princeton University Press, Princeton, N.J. 1976.
15. Bonasso, R.P., *et al.*, "ANALYST II: A Knowledge-Based Intelligence Support System," MTR-84W00220, The Mitre Corp., April 1985.
16. Ruach, H.E., "Probability Concepts for an Expert System Used for Data Fusion," *AI Magazine*, Fall 1984, pp. 55–60.
17. Cohen, P., *et al.*, "Representativeness and Uncertainty in Classification Systems," *AI Magazine*, Fall 1985, pp. 136–149.
18. Taylor, John W.R., ed., *Jane's All the World's Aircraft 1983–84*, Janes Pub. Co. Ltd., London, 1984. (*See also* latest edition.)
19. Luo, Ren E., and Michael G. Kay, "Multisensor Integration and Fusion in Intelligent Systems," *IEEE Trans. Systems, Man, and Cybernetics*, Vol. 19, No. 5, September–October 1989, pp. 901–931.

Chapter 4
SENSORS, SOURCES, AND COMMUNICATION LINKS

The purpose of this chapter is to introduce the characteristics of sensors and other sources that provide raw and preprocessed data to fusion systems for composite processing. Signal detection, estimation, and classification principles are first introduced to develop a general sensor model. This model is then related to the situation where multiple sensors contribute observations or decisions to a central fusion node to perform composite detection or classification. Alternative choices for implementing sensors are then considered, along with the influence these choices have on overall system performance. In particular, a distinction is made between sensors that make and report only hard decisions (declarations) and those that report multiple hypotheses with associated measures of uncertainty for each hypothesis. This basic understanding of sensors is a prerequisite to the descriptions of the state estimation and attribute combination processes described in Chapters 6 and 7, respectively. Following these discussions of general sensor principles, representative military sensors are briefly described to illustrate practical implementations of the sensor model, with different processing characteristics unique to each application. Finally, sources of data and links for conveying data to fusion systems are described with the considerations necessary to integrate networked data into a fusion system.

The chapter is organized to distinguish among three means by which data are supplied to the data fusion process:

1. *Sensors* are those devices that detect or measure physical phenomena. *Remote sensing* describes those techniques in which the sensor measures a discriminable phenomenon transmitted through some medium (the atmosphere, earth, ocean, *et cetera*). Other sensors, such as switches, thermistors, and transducers require near-physical contact with the detection phenomena and are not considered remote sensors. Military C^3I systems generally employ relatively long-range remote sensing techniques or remotely deployed sensors that may require near-physical contact with targets (e.g., battlefield trip-wire switch sensors with RF reporting links). In contrast, robotic data fusion appli-

cations may use a suite of sensors that allow near-physical contact and remote sensing sensors (e.g., TV cameras) with a relatively short range from the object of observation.
2. *Sources* describe a variety of originators of data, such as human observations, interception of communicated information, *a priori* information known about future plans or intentions (e.g., battle plans, flight paths) or the state of conditions at a point in time (e.g., map data, force sizes). In the military context, processed intelligence data regarding hostile actions, plans, or intentions are usually considered source data.
3. *Links* are those communication connections from sensors or sources to data fusion processing nodes that pass data from a remote location for fusion. The sensor of source data received from a second party by such means is referred to as *indirect data* [1].

4.1 DETECTION AND ESTIMATION FUNCTIONS OF SENSORS

The primary functions of sensors are the *detection* of the presence of known signals and the *estimation* of parameters of those signals. In some cases, the estimation process classifies the detected signal into one of many possible class hypotheses, each being a known signal class. The detection-estimation process in classical single sensor systems is contrasted with multisensor systems in this section to illustrate the unique characteristics of detection and estimation processes in such systems. The single-sensor model is first introduced to describe classical detection and estimation approaches, followed by multisensor approaches to detection and estimation.

4.1.1 A General Sensor Model

Figure 4.1 depicts the primary elements of a general sensor model, which illustrates the signal detection and estimation functions.

Target and Phenomena

The target or event is the source of remotely sensed energy caused by a physical phenomena such as radiation, reflection, emission, or absorption. Energy unique to the target or event is a time-domain parameter (variable), $E(t)$, and permits discrimination from background or other targets or events.

Transmission Medium

The medium of transmission can be the atmosphere, ocean, or earth for acoustic, electromagnetic, or nuclear particles that transmit energy from the source to the

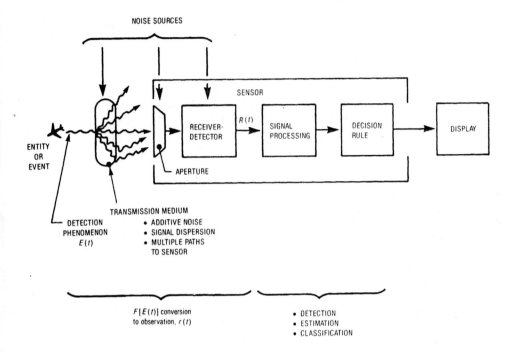

Figure 4.1 Elements of a typical sensor system.

receiver. The medium generally introduces random noise that contaminates the source signal, complicating detection and estimation. This noise is usually considered to be additive and its spectral characteristics must be determined to minimize its influence through signal processing. Dispersion, absorption, and reflection (providing multiple paths from source to sensor) also distort the characteristics of the transmitted signal.

Sensor Aperture

The aperture captures the signal energy using a physical device sized and constructed to efficiently transfer sensed energy to the receiver or detector with a minimum of loss. Apertures for RF energy (antennas), IR-UV-visible light energy (optics), and acoustical energy (microphones, hydrophones) are well-known examples in which physical characteristics directly influence the aperture's spatial coverage, gain, and bandwidth.

Receiver-Detector

The captured signal energy is usually translated to electrical signals by a detector prior to amplification and spectral or temporal filtering to eliminate noise energy outside the bandwidth of the desired signal energy. In this process, the signal may also be conditioned to normalize (or emphasize) amplitudes and translate the signal to different spectral regions for more convenient processing.

Signal Processing

Sensor preprocessing can accomplish a number of necessary signal conditioning functions prior to detection or estimation. Such processing, as well as reception and transmission, transforms the originally transmitted energy variable, $\mathbf{E}(t)$, into an observation space in which the variable is represented by an observation vector $\mathbf{R}(t)$. Signal features such as amplitude, frequency components, or time-series characteristics often represent the components of this observation (or measurement) vector.

Detection-Estimation

Detection and parameter estimation decisions are made at the end of the processing chain and results are reported to the data fusion process. These processes make decisions about the presence (detection) or observed characteristics (parameter estimation) of $\mathbf{E}(t)$ on the basis of $\mathbf{R}(t)$, reporting results to a display in the form of a decision vector, $\mathbf{D}(t)$.

4.1.2 Classical Detection and Estimation

The classical statistical approach to detection is now summarized to introduce the essential characteristics of detection and estimation processing in sensors. A general mathematical model of the sensor process previously described is presented in Figure 4.2. The source energy $\mathbf{E}(t)$ is transformed from a k-dimensional space to the N-dimensional observation space by the mapping function, $F[\mathbf{E}(t)]$. Additive noise, $\mathbf{n}(t)$ is introduced to the observed parameters, resulting in the observation vector, $\mathbf{R}(t)$. Definitions of detection, estimation, and classification may then be given as follows.

Detection is the process of testing the observed $\mathbf{R}$ vector against established hypotheses to determine which most closely matches the observation. The observation space is effectively partitioned into distinct regions (one for each hypothesis) with the partitions forming a decision rule. For simple detection for target presence, the decision is binary with two hypotheses (H_0 = target detected, H_1 = no target

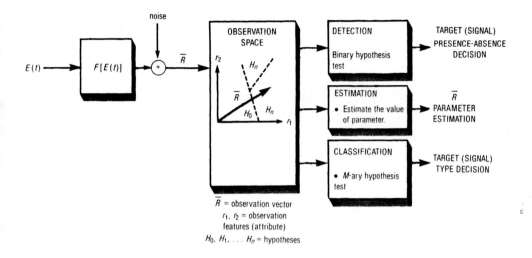

Figure 4.2 Mathematical model of sensor functions.

present). Multiple hypothesis (*m*-ary hypothesis) testing may be applied to detect presence with varying degrees of belief, each hypothesis representing a relative degree of belief in the presence of the target.

Estimation is the process of determining the value (in some optimal sense) of the parameter $\mathbf{E}(t)$, based upon the measured variable $\mathbf{R}(t)$. The determination process that transforms $\mathbf{R}(t)$ into an estimate of $\mathbf{E}(t)$ is referred to as the *estimator*. (This process is described in some detail in Chapter 6, where the estimation of target state as a function of time is described.)

Classification is a special case of *m*-ary detection, in which there are *m* hypotheses (classes) against which $\mathbf{R}(t)$ is compared. This may be used to detect characteristics of the target that are described by the observation vector. In this case, the observation space is partitioned into M regions, corresponding to M hypotheses ($H_0, H_1, \ldots, H_M$). (This process is described in some detail in Chapter 7.)

These processes are closely related to, and can be viewed as, multiple-hypothesis testing problems that seek to minimize some objective function in the observation space: decision error for detection or classification problems and estimation error for estimation problems. To illustrate the formulation of solutions to these problems, we now summarize the development of the fundamental statistical decision criteria that forms the basis for decision and classification theory. Using the binary hypothesis testing problem (summarizing the rigorous development by Van Trees [2]) for signal presence or absence, we assume that two distributions (Figure 4.3) represent the two possible states of an observed signal vector, $\mathbf{R}$. The decision process must consider two possible hypotheses, H_0 (no signal present) and H_1 (sig-

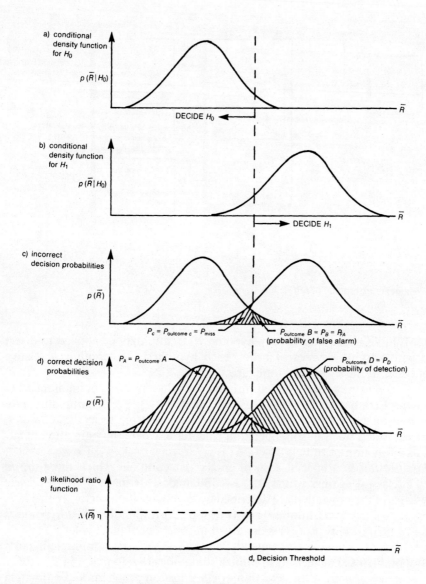

Figure 4.3 Decision functions for two gaussian random variables: a = conditional density function for H_0; b = conditional density function for H_1; c = probability of incorrect decision; d = probability of correct decision; e = likelihood ratio.

nal present; i.e., detection). The *a priori* probabilities for these states are given by P_0 and P_1, respectively.

A decision criterion for the signal receiver (sensor) must then be developed that provides for four possible outcomes:

Outcome	True State	Decision Made	Description of Decision
a	H_0	H_0	Correct, no signal
b	H_0	H_1	False alarm error
c	H_1	H_0	Miss error
d	H_1	H_1	Correct detection

The signal-dimensional example illustrated in Figure 4.3a shows the conditional density function for H_0; and Figure 4.3b shows the conditional density function for H_1. The boundary that defines the decision rule is given by

$r > a$, decide H_1

$r < a$, decide H_0

The decision rule effectively partitions the two distributions into four regions whose areas give the *conditional* probabilities of the four possible outcomes. Figure 4.3c gives the two regions (probabilities) that define the incorrect decisions, and Figure 4.3d gives the two regions (probabilities) for correct decisions. The probability for outcome *b* is referred to as the false alarm probability (P_{fa}) and the probability for outcome *d* is the detection probability (P_d). Two fundamental methods for establishing the decision criteria, *a*, are now developed using the model.

Bayes Criteria

A cost is established for each decision outcome and the average cost over all four decisions is minimized to derive the classical Bayesian criteria. These costs may be based upon the receiver (sensor) application and presume a greater cost for the incorrect decisions than correct decisions. In this binary case the four costs, C_{00}, C_{01}, C_{10}, and C_{11} are assigned to outcomes *a*, *b*, *c*, and *d*, respectively. The expected value of cost, *C* (risk), is defined as the sum of the product of costs and the conditional probabilities for each outcome:

$$C = C_{00}P_0P_a + C_{01}P_0P_b + C_{10}P_1P_c + C_{11}P_1P_d \tag{4.1}$$

This expected cost function is partitioned in fixed and variable terms, and the components of the variable term are separated to establish the test at which cost is minimized. This occurs when the ratio of the values of the two conditional density functions (the *likelihood ratio*) is related to a constant (the threshold test, η), where

$$\Lambda(\mathbf{R}) = \frac{P(\mathbf{R}|H_1)}{P(\mathbf{R}|H_0)} \qquad (4.2)$$

$$\eta = \frac{P_0(C_{10} - C_{00})}{P_1(C_{01} - C_{11})} \qquad (4.3)$$

The value of η can be computed directly from cost and *a priori* probability values. The likelihood ratio function (shown in Figure 4.3e) can be computed directly from the two conditional density functions, so that the detection decision threshold in the observation space, a, can be determined. Note that the likelihood ratio is a scalar function, independent of the dimensionality of $\mathbf{R}$. In practice, it is often more convenient to use the logarithm of the likelihood ratio and threshold test for calculation of the decision threshold. In this case, the likelihood ratio is computed and directly compared with a threshold to choose the detection decision.

Neyman-Pierson Criteria

If cost functions and *a priori* probabilities are not available, a test can still be developed using the likelihood function. In this criterion, the detection probability, P_D is maximized for a specified false alarm probability, P_{fa}. This criterion leads to a test of the likelihood function against a scalar threshold, which is a function only of the conditional probability distribution.

The results can be extended from the binary hypothesis case to m hypotheses [3] by partitioning the decision space into m decision regions, in which case there are m^2 possible outcomes of each decision. As in the binary ($m = 2$) case, costs are assigned to each outcome and the average cost is minimized. This results in likelihood ratio tests that partition the $m - 1$ dimension decision space into decision regions for each hypothesis. The most common m-ary decision rule, which minimizes the total error probability, is the *maximum a posteriori* (MAP) process. In this case, the *a posteriori* probability is computed for each hypothesis, and the largest value is chosen as the correct decision. As we shall see later, the vector of *a posteriori* probabilities provides a convenient report format for sensors to m-ary hypothesis data for fusion with other sensors.

4.1.3 Sensor and System Performance Measures

Sensors have traditionally been specified by detection, state estimation (tracking), and object identification (classification) performance. The most general performance measures are often expressed statistically:

- Detection probability (as a function of false alarm probability for a specified ratio of signal to noise energy).
- State estimation error distribution.
- Classification accuracy (expressed as probability of correct classifications for specified signal to noise conditions).

The fundamental description of sensor detection performance is expressed graphically as the *receiver operating characteristic* (ROC), which plots P_D versus P_{fa}. Figure 4.4 illustrates the characteristic graph for a binary decision between two Gaussian random variables. Four curves are plotted for different statistical distances between the two variables, representing hypotheses H_0 and H_1. As the distance, d, between the means of the two densities increases, the detection performance improvement can be seen as an increased P_D for a given P_{fa}. The variable along each curve is the threshold test value, η, which establishes the operating point for P_D and P_{fa}. The example operating point illustrates the case where a given value of $\eta = \eta'$ provides a $P_D = 0.8$ and $P_{fa} = 0.4$.

Multiple sensor data fusion systems must be specified for total performance, and the selection of the set of sensors (the sensor suite) and their individual levels of performance levels must be properly defined to achieve the specified total, integrated performance using the detection, estimation and classification measures just defined. We now consider the relation between individual sensor ROCs and the composite or system ROC.

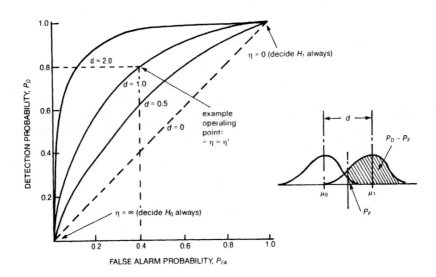

Figure 4.4 Receiver operating characteristic for binary decision between two gaussian variables (after [2]).

4.1.4 Detection and Estimation in Data Fusion Systems

The advent of multisensor systems has caused a reevaluation of the methods by which sensors should be implemented and their role in the data measurement and decision processes altogether. In contrast to single sensor systems, where the detection-estimation process is an integral part of the sensor, multisensor systems can allocate the detection-estimation decisions between the sensors and the fusion node. Each sensor is only a *contributor* to a composite decision process, which may include a central decision node to which all sensor observations are passed for a decision based on the observation space formed by all observations (Figure 4.5a). Alternatively, each sensor may first make a sensor-level decision on the basis of its own observations, passing these decisions to the fusion node where a composite (global) decision is made (Figure 4.5b). As decisions are distributed to the sensors and less information is transmitted to the fusion node, the sensor-to-fusion bandwidth is reduced with a loss of decision performance relative to a completely centralized approach. The capability to distribute the detection-estimation function raises several issues:

1. What alternative architectures are available to allocate these functions between sensors and the central fusion node?
2. To what degree should the processes be centralized or distributed? How does the suboptimal distribution of detection-estimation to the sensors degrade performance relative to the optimal centralized approach?
3. If detection-estimation is distributed, how can performance be related to design factors such as sensor-to-fusion communication link bandwidth (amount of decision data transferred from sensor to fusion), intercommunication among sensors, and system reliability, survivability, and security?

The study of distributed (or decentralized) decision-making systems is the subject of team decision theory [4] and numerous studies have developed applications of the theory to the practical data fusion issues raised earlier. Tenney and Sandall [5] Chair and Varshney [6] have compared the performance of distributed detection architectures with fixed sensor decision rules, while Reibman and Nolte [7] have developed optimal rules that couple sensor decisions and the global decision. Aalo and Viswanathan [8] report the influence of noise correlation on distributed detection performance, while Tenekitzs [9, 10] analyzes the performance of *sequential* distributed detection, applying the methods of Wald [11]. Improved detection can be achieved by the use of soft decision approaches, as shown by Thomopoulis, Viswanatan, and Bouboulias [12], and Lee and Chao [13]. Further extension of these concepts to distributed, m-ary decisions is described by Sadjadi [14], Demirbas [15], and Allen [16].

The results of these application studies emphasize that each sensor is only a contributor to detection-estimation, and this will influence the way in which sen-

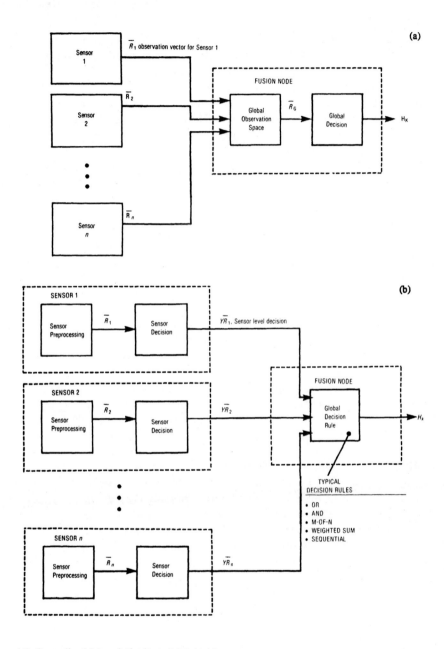

Figure 4.5 Centralized (a) and distributed (b) decision structures.

sors are selected for use in operational systems. In multisensor systems, combinations of sensors may be able to achieve performance requirements individual sensors cannot: only when the observations of all sensors are combined can the requirement be met. Some [17] have warned that designers cannot afford to add sensors without considering penalties, but in many cases where stealth, *low probability of intercept* (LPI), high confidence of target classification, or high detection probability in low signal-to-noise ratio is required, multiple sensor suites may be the appropriate (or only) answer.

Many first generation fusion systems have been designed to utilize a collection of existing sensors, each designed to single-sensor requirements. These systems were developed to integrate the *existing* sensor suite for the best possible performance using sensors that were not designed for complementary operation. Fully integrated systems, in which both sensors and fusion processing are designed together must apply the principles described earlier to most effectively meet *collective* performance requirements for the data fusion system. In Chapter 11, we further describe how collective performance must be partitioned to derive performance measures and requirements for individual sensors or sources.

4.2 ALTERNATIVE SENSOR IMPLEMENTATIONS

The two principal methods for implementing sensors for distributed decision making are now introduced. As discussed in the previous section, these implementation alternatives influence the means by which sensor data is fused in subsequent processing for detection, state estimation, and classification. Models for these two classes of sensors are distinguished by the means by which the detection-estimation process reports data to the data fusion functions, which perform composite detection and estimation. The terminology for classifying sensors is borrowed from coding theory in which signal decoding methods are referred to as *hard* or *soft* on the basis of the methods of decision making and reporting [18]. *Hard-decision sensors* choose a single-hypothesis decision at the sensor and that decision, alone, is reported to the fusion process. *Soft-decision sensors* partition the observation space into multiple detection regions, each representing a different decision threshold. Detection decisions are therefore quantized and the detection report contains a measure of confidence in the decision. In m-ary classification, the multiple hypotheses are all reported (or a selected subset of most likely hypotheses) with a quantitative measure of confidence (or, alternatively, uncertainty) for each hypothesis.

Figure 4.6 compares the functions of these two sensor implementations, contrasting their characteristics. Because the hard sensors do not provide any information to the fusion algorithm for signals below the decision threshold, information at lower signal level is lost and there is no combination of data. Soft sensors, on the other hand, provide a measure of confidence for decision hypotheses at any

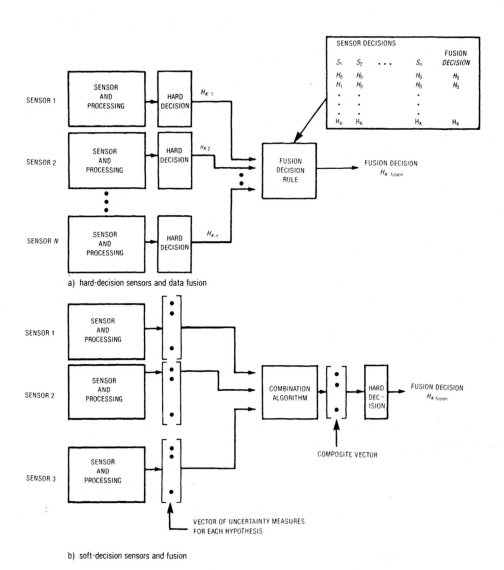

Figure 4.6 Comparison of hard-decision (a) and soft-decision (b) sensors and data fusion.

sensor signal level and permit these values to be combined over a wider range of signal levels. Under certain circumstances, this soft data combination permits global decisions to be made at lower signal levels than those at which hard sensor fusion systems can achieve sensor-level decisions. As shown in the last section, soft-decision sensors must be designed to provide hypotheses with related uncertainty measures that are compatible with the global estimation-decision algorithms used to combine the multisensor data. Figure 4.7 depicts three common methods of implementing multiple-hypothesis soft-decision sensors that provide measures of hypothesis uncertainty.

The simplest sensor (Figure 4.7a) includes multiple decision thresholds (e.g., high, medium, and low-confidence levels) for each hypothesis (H_0 means no signal present, H_1 means signal present). In this example, the resulting output is a vector in which each hypothesis is represented by a measure that takes on one of three values. This is typical of many binary decision sensors (e.g., detection of presence or absence) in which the detection decision is quantified by a low- or high-confidence measure.

A second level of complexity (Figure 4.7b) adds multiple hypotheses (H_i), each with a hard-decision rule to determine which hypothesis is selected for any given measurement. Parametric data is provided in the form of conditional probabilities [19] that model the behavior of the sensor, relating the probability of the presence of each hypothesis (e.g., a target class), conditioned on decisions made by the sensor processor. This category of sensor is called *parametric* [19] because the parameters contained in the *conditional probability matrix* (CPM) must be known before sensing to accurately describe the performance of the sensor for any given decision. To accurately develop the CPM, the *probability density functions* (PDFs) for each possible target class (hypothesis) must be known. The sensor's operating characteristics, hence the PDFs, may vary as a function of signal characteristics (e.g., signal/noise ratio, waveform type, noise characteristics), making multiple CPMs necessary to represent sensor performance over the wide range of conditions under which it performs.

The addition of a *nonparametric* [20] statistical pattern recognition capability (Figure 4.7c) provides a high level of "softness" in the sensor by quantifying, for each individual measurement, the uncertainty as a measure of the statistical distance between the sensed measurement and a representation of the signal expected for each hypothesis. Nonparametric sensor processors do not require prior statistical representations of the sensing process (PDFs or CPMs) to quantify an uncertainty measure. Statistical distances in a multidimensional feature space (e.g., the Bhattacharyya distance measure [21]) between the measurement and all candidate hypotheses can be used to represent relative measures of uncertainty for each.

In addition to these single-look (sensor decision is based upon a single measurement of the target signal) methods, the measure of certainty can be based upon

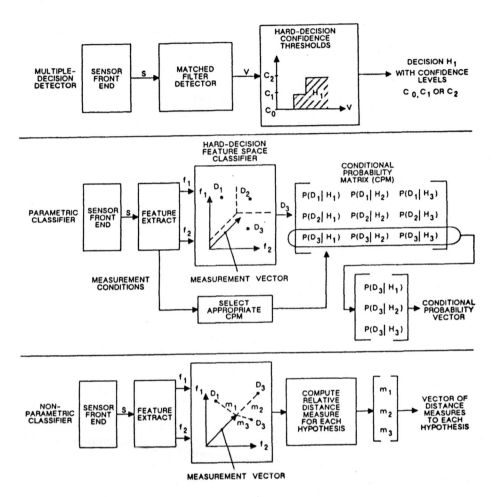

Figure 4.7 Soft-decision sensor implementation approaches: (a) multiple-decision detector; (b) parametric classifier; (c) nonparametric classifier (from [21]).

accumulated evidence from multiple, independent looks at the signals. Common multilook or sequential algorithms include

1. *M-of-N criteria*—The measure is a function of the number of hard-decision detections made, M, over some number of independent measurements, N. If the observation period, N, is fixed and M is a threshold, the *M-of-N* criteria can be used as a hard-decision rule.
2. *Wald sequential observer*—In this approach [11], the number of measurements, N, is variable. After each measurement, test variables are computed

to make one of three decisions: (1) accept hypothesis H_0 (no signal present) and terminate measurements; (2) accept H_1 (signal present) and terminate; or (3) continue testing (indeterminate).
3. *Sequential Bayesian inference*—If sequential measurements are statistically independent, each can be treated as a separate measurement to iteratively compute *a posteriori* probabilities following Bayes rule.

The next section discusses the potential benefits of soft-decision sensing and the conditions under which it can offer improvements over simpler hard-decision approaches. Then, in Section 5.4 we discuss specific methods of implementing soft-decision signal processing structures for both imaging and nonimaging sensors.

4.3 COMPARISON OF HARD- AND SOFT-DECISION SENSORS

Sensors with automatic target recognition capabilities may report their data to the fusion process using a variety of means that influence the design of the data fusion algorithm. One important consideration is the selection among alternative means of representing the degree of belief or uncertainty in processed data and the methods by which this data will be combined with other sensors' data to make final decisions (e.g., detection, track, or classification decisions on targets). Logical algebra's, single-valued probabilistic (Bayesian) inference methods and probabilistic interval methods (Dempster-Shafer) are compared in Chapter 7 as candidate methods for representing uncertainty. Most comparisons of methods of combining uncertain data focus on the characteristics of each approach to accomplish three critical combination functions:

1. Representation of sensed measurements and their inherent uncertainties due to source and measurement (sensor) errors that cannot be removed.
2. Combination of individual measurements and their respective uncertainties into a single measure that minimizes the composite measurement error according to some fidelity criterion (e.g., mean square error). This, in effect, reduces the composite uncertainty in the measurement.
3. Application of a decision rule based on detection or identification probabilities and a false alarm criterion.

Another factor to be considered in comparing approaches includes implementation issues, including the processing burdens, memory requirements, and reporting bandwidths of various approaches. Methods that report and combine multiple hypotheses with uncertainty data require significantly more processing capabilities and data base requirements than simpler, logical methods that combine single-hypothesis declarations from hard-decision sensors.

We now address the operational performance trade-offs between hard- and soft-decision sensors by comparing the *potential* quantitative benefits that can be derived from the data fusion approaches employing soft-decision sensors and probabilistic-based combination algorithms. Criteria must first be established for evaluating the relative benefits of these two sensor-fusion techniques. In general, the benefits of combining soft-decision data should result in composite decisions at lower *individual* sensor signal levels than that achieved by hard-decision methods. Because sensed signal levels can usually be related to range (from sensors to target) two operational *measures of performance* (MOPs) are convenient to compare approaches:

- *Range at decision*—The range at which the data fusion process achieves a decision, against some detection–false alarm criteria.
- *Decision confidence* (measure of uncertainty, degree of belief) *at a given range*—The confidence that the data fusion reports *for the correct hypothesis* at any given range. In the hard-decision case, confidence is not reported, of course, until a data fusion decision is made.

4.3.1 Numerical Results for a Bayesian Example

Quantitative evaluation of a typical sensor suite with both hard- and soft-decision sensors has been reported by Buede and Waltz [21], who have illustrated the benefits that can be achieved by soft-decision sensors. These results also illustrate the conditions that must prevail for a suite of soft sensors to be preferred. Figure 4.8 illustrates two simple cases to compare the two approaches. Consider a target approaching two sensors, each of which is measuring target features and integrating signal energy for target classification in sensor processors. Each sensor processor is computing an *a posteriori* probability for each of several class hypotheses. In the hard-decision case (Figure 4.8a) sensor declarations (*A* and *B* in the figure) are reported only when the sensor classifiers exceed preset decision thresholds, and a combined decision is made only when a voting criterion is met (first sensor in or both must agree). In the simplest first-sensor-in case, the decision is made at range, R_1.

If the sensors' decisions are ambiguous and both decisions are required to remove ambiguity, then the combined decision cannot be made until both sensor decisions are received, for example,

Decision	Decision Range	Target Class Ambiguity Set
Sensor 1	A	4, 5, 12, 18
Sensor 2	B	12, 21, 32, 33
Combination	B	12 (unambiguous)

In contrast (Figure 4.8b) the soft-decision sensors report their partial evidence as soon as the signal processors begin to integrate signal energy. The combination process then combines these values (probabilities, fuzzy variable, probability inter-

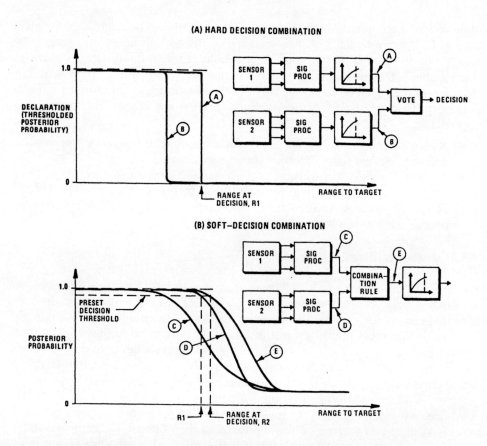

Figure 4.8 Comparison of hard (a) and soft (b) decision combination decision points as a function of range (after [21]).

vals, *et cetera*) to derive a combined measure that may reduce the uncertainty over that of each sensor. As shown in the figure, the decision threshold is applied to the combined measure of uncertainty, achieving an earlier, longer range decision (R_2) than can be provided by either independent sensor or the hard-decision sensor case.

Figure 4.9 presents the numerical results of a three-sensor simulation [21]. In this example, representative overlapping performance of classifiers for the three generic soft-decision sensors (IRST with noncooperative target recognition, IFF, and ESM) were modeled and a Bayesian probabilistic algorithm was chosen for combination of sensor data. Sensor performance levels were modeled to overlap the regions during which detected signal levels provide decisions from the lowest (lowest confidence) to the highest decision thresholds. The figure plots *a posteriori* probability (for the correct hypothesis) *versus* target range for classifiers on each

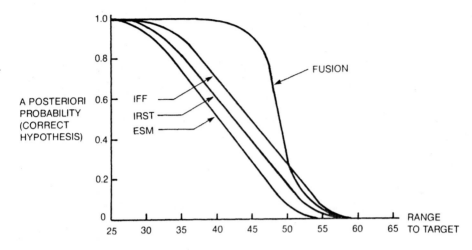

Figure 4.9 Simulated identification range improvements for soft-decision sensors and Bayesian fusion (after [21]).

sensor and for the fused classification decision. For these simulated conditions, the effects of combination are evident: the fused decision reaches a decision threshold of 0.95 at 47 miles range *versus* the first independent sensor, which reaches the same threshold at 37 miles.

4.3.2 Criteria for Beneficial Combination

The range and confidence improvements demonstrated in the example occurred because the "soft regions" of sensor measurements overlapped in range and allowed lower-than-threshold measurements from multiple sensors to be combined for a composite value above the decision threshold. This is the criteria for soft-decision sensors to provide any benefit over hard-decision approaches: *the regions (in time or space) where the sensor measurement signal levels transition from near zero to maximum value must make the transition over the same interval for all sensors.*

Figure 4.8 illustrates this requirement for overlap of the soft regions of both (or all) sensors. We must recognize that both the sensing process (the sensed phenomena, transmission medium, *et cetera*) and the sensor processing influence the slope of the sensor operating curve and the resulting width of overlapping regions. Hard limiters in RF receivers, for example, will drastically reduce the extent of the overlapping regions by introducing steep input-output signal transfer functions.

4.3.3 Operational Benefits of Soft-Decision Sensors and Fusion

The soft sensor suite was shown to provide a range improvement over hard-decision system in the previous example. This range improvement may be of sufficient benefit to some tactical systems to advocate the selection of the soft-decision approach. Each system must be analyzed individually to determine the total system performance improvement achieved for its unique set of sensors and mission conditions. Figure 4.10 compares this potential range improvement (category I) with two additional operational benefits (categories II and III) that may be derived in some systems.

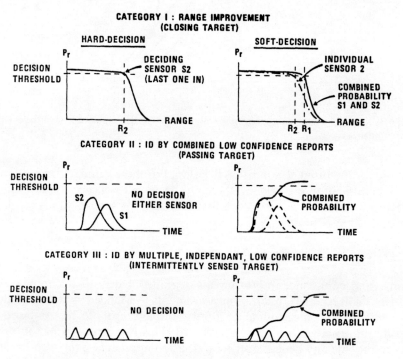

Figure 4.10 Three categories of potential operational benefits provided by soft-decision sensors and fusion.

For simplicity, the previous analysis has been based upon closing targets in which the signal strength is continuously increasing for all sensor measurements. In more complex target scenarios, however, this closing-target situation does not always exist: other target-sensor timelines provide two further opportunities for soft-decision systems to provide operational benefits over hard-decision systems.

Category II improvements occur when multiple sensors view a common target with their soft-measurement regions occurring at *different* periods of time. This condition can occur on turning targets, for example, when sensors differ in response as a function of target aspect. If, in each sensor look, the sensors do not achieve a sufficient signal for hard decisions, the hard-decision fusion system will make no decision at all. In contrast, if the soft-decision data from the first measurement is stored and later combined with the second sensor measurements, the combined data may be able to reach a decision where independent sensors could not.

Category III improvements are similar, but apply only where a series of *independent* looks are made at a target, and these can be combined (e.g., sequential Bayesian inference based upon statistically independent measurements) to reach a decision. In this case, no single look crosses a decision threshold, but multiple looks are combined to achieve a decision.

4.4 SOFT-DECISION SENSOR PROCESSING

A number of alternative signal processing approaches are available to implement soft-decision sensors. This section describes the basic sensor processing functions that have been introduced to a number of military sensors to perform automatic target recognition (ATR) in the presence of cluttered backgrounds. Such processing functions provide the means to implement the multiple hypothesis and uncertainty quantifying characteristics desired in soft-decision sensors. The motivation for these systems has been the desire to detect and classify targets by type, to achieve detection rates higher than those possible by manual (human recognition) means and to permit fully autonomous detection and classification for guidance and weapon targeting (tracking and aimpoint selection) functions [22]. These systems are noncooperative by nature and are also referred to as *noncooperative target recognition* (NCTR) sensor-processing techniques.

Although the ATR terminology has been used primarily for imaging sensors, it is equally appropriate for nonimaging sensors that perform the same functions. Many single-sensor ATR systems have demonstrated the ability to rapidly and accurately detect and classify targets of military interest [23, 24]. This section introduces the primary methods of ATR sensor processing that are capable of providing soft-decision data for composite detection-estimation fusion systems.

4.4.1 Target Signatures

ATR methods are based on analysis of measured signal characteristics that permit an individual target's signal to be distinguished from noise (detection) and, further, estimated for discrimination from targets of different types or classes (classification). The set of unique signal characteristics that discriminate among target classes is often referred to as the *signature*. Three of the most desirable characteristics of signatures are

1. *Repeatability*—Temporal repeatability is required for consistant classification. Factors such as temperature, weather, time of day, and solar angle can strongly influence some candidate signatures. In such cases, the effect must be modeled and compensations applied to achieve the necessary repeatability. Invariance to many of these factors is sought in selecting signature characteristics.
2. *Discrimination from noise*—The signature must be distinguishable from background clutter to ensure that the rate of false alarms remains below some criterion.
3. *Target aspect invariance*—Signature characteristics may vary significantly as a function of sensor-target aspect angle because of target physical characteristics. Aircraft, ground vehicles, and ships are symmetric, but 0°, 90°, and 180° aspect views can be expected to pose very different radar, IR, and visible characteristics to sensors at those angles. Unless the target velocity vector and target signatures are known as a function of all aspect angles, the classification may not be possible.

The signature is described by specific *features,* which are quantifiable signal characteristics that can be extracted (individually measured) from the sensed signal and that permit the discrimination among target classes. Figure 4.11 illustrates how the features may be viewed as dimensions of an N-dimensional space in which tar-

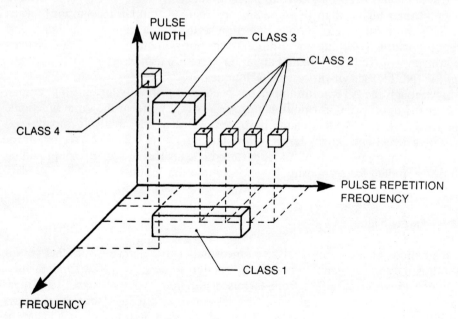

Figure 4.11 Feature space for radar waveforms.

get classes are described as vectors. The figure illustrates an electronic support measure sensor classifier in which radar signature classes are represented by three features: frequency, *pulse repetition frequency* (PRF), and pulsewidth. When the feature measurements are characterized by Gaussian source and measurement noise, the expected class measurements become blurred and form hyperspheres in the feature space.

4.4.2 Features for Imaging and Nonimaging Applications

Nonimaging sensors generally measure a single-dimensional time domain signature (e.g., radar returns, acoustic echoes) that may contain characteristic features of the target. Both time-domain and frequency-domain representations of the signal may be used to extract discriminating target features, such as spectral lines or time series events. The spatial extent of targets in imagery, however, requires the definition of two-dimensional features that describe the size, shape, color, and texture of objects. Figure 4.12 summarizes several basic two-dimensional features [25] based on intensity and shape that are invariant to rotation within the image. One issue in selecting features is the invariance to target aspect as viewed by the sensor or other independent variables. Battlefield tanks, for example, present different shapes to the sensor, depending on the view angle and turret position. Features that are invariant to these conditions are sought to improve classification accuracy and target detection.

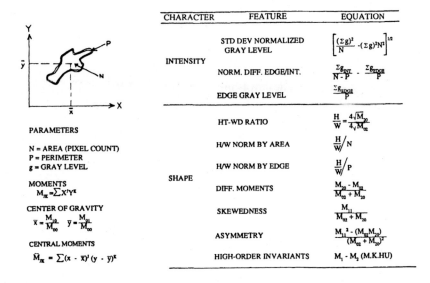

Figure 4.12 Two-dimensional position-invariant features.

4.4.3 Classification Database Requirements

The ATR sensor must contain a predetermined data base of target signature data (models) that describe the types into which detected targets must be classified. The most common classifiers are referred to as *supervised* because prototype target classes are viewed by the sensor prior to classification in a process called *training*. Class separating features are measured and class decision rules are then predetermined by the training process and form the classification database. The database information required for the major categories of classifiers are

- *Syntactic*—Pattern grammar structures that relate target features in time, space, or spectrum.
- *Statistical (parametric)*—Probability distribution functions for each class.
- *Statistical (nonparametric)*—Vector coordinates of each class in feature space for computation of distance to each target feature vector (nearest-neighbor classification procedures).
- *Nonstatistical (distribution-free)*—Discriminant functions that partition the feature space into class regions.

Classifiers may be *unsupervised* if prior training data is not provided and the classifier autonomously identifies feature clusters in the sensed data and separates them into classes that it recognizes as distinguishable by predefined features. Ultimately, such classification methods are self-learning, requiring a teacher only to label the identified classes (e.g., target classes, background, artifact).

4.4.4 ATR Processes

The general sequence of processing stages [26] for imaging sensors follows the flow depicted in Figure 4.13, and descriptions of each stage follow. Nonimaging ATR processes follow similar stages, but the two-dimensional operators described here are replaced with appropriate time, one-dimensional space, or spectral domain operations.

Preprocessing

The sensed imagery must be processed to remove (or reduce) artifacts of the sensing process, including noise, striping (in line scanned imagery using detector arrays), gain or bias errors, and geometric distortion. Two-dimensional operators (digital filters) are often applied to achieve different preprocessing objectives:

- *Low-pass spatial filters*—Removal of high-frequency noise or sharp edges, leaving smoothed background.

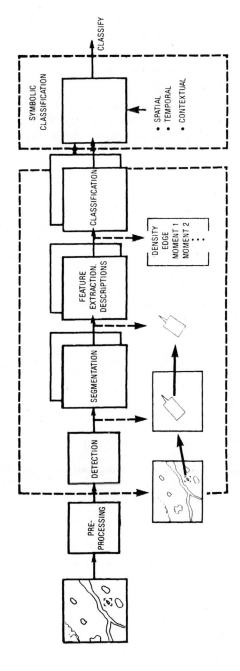

Figure 4.13 General ATR processing flow for imaging sensor data.

- *Median filters*—Noise suppression by a nonlinear sliding window algorithm in which each center pixel is replaced by the median value of all pixels within the window. This is most effective on discrete impulse noise.
- *High-pass spatial filters*—Enhancement of contrast to sharpen edges by removal of low-frequency content.

Successful preprocessing will result in an output image in which characteristics of objects of interest (targets and contextual cues) have been enhanced and image content of no value to the ATR process has been removed or suppressed.

Detection

The location of candidate targets within the image are then identified by a coarse detection algorithm. This is performed as a computational reduction mechanism, eliminating background regions to focus the most computationally intensive processing (segmentation, classification) on regions in which targets are highly probable. Detection criteria usually include contrast relative to background, intensity, multispectral intensities, closed boundary, size, shape, and contextual characteristics. These criteria usually allow the detection of regions ("blobs") that are suitable candidates for final evaluation as targets. Typical detection algorithms include

- Spatial matched filters of the size and shape of expected targets.
- Edge operators that detect closed or nearly closed regions by magnitudes and directions of edges.
- Spatial frequency filters that are based on global estimates of background and small object spectra.

Segmentation

Once candidate target regions have been detected, the specific target data must be defined by identifying the irregular boundary that distinguishes between target and background. The shape, size, and interior of this boundary is the target (battlefield tank), whereas the region outside of the boundary contains potentially helpful contextual information (linear features like a road, track marks, *et cetera* to support the target hypothesis). This boundary identification and target extraction process must precede the computation of target features for classification. General approaches to this segmentation process include

- *Edge linking methods* seek to detect local discontinuities (e.g., edges) and then attempt to construct or link an object boundary of discontinuities.
- *Relaxation methods* iteratively evaluate each pixel for possible discontinuities (creating a decision image) and then evaluate this image of discontinuity deci-

sions to search for clusters of probable edges. The approach is iterative in that surrounding decisions are used to adjust each decision until the decision image can be scanned for objects.
- *Region growing methods* attempt to find the area over which one or more properties remain generally constant.

Feature Extraction and Description

Once the picture elements that specifically define the target have been segmented, specific features that characterize the target must be determined. For statistical classifiers, values for each feature are computed to create an n-tuple (vector in n-dimensional feature space) that quantitatively represents the target. For syntactic classifiers, pattern components are detected individually and relations are determined prior to classification. The extraction process may include features within the segmented target as well as those related local contextual features segmented from the image related to the target.

Classification

The classification process effectively partitions the feature space into m decision regions and determines the region (target class) into which the segmented target falls. In statistical classification, the classification is performed by parametric or nonparametric means. Syntactic classifiers attempt to parse the components of the target using grammatical techniques to determine if a class of targets can be constructed.

Contextual and High-Level Classification

Global context (relationships among all detected entities within the scene), temporal context (e.g., changes, behaviors recognized by analysis of successive sensor frames), and external information may be applied following local target classification to achieve a higher level of scene classification. Syntactic or symbolic (knowledge-based) classification approaches are appropriate at this point to determine relationships among targets, distinguish temporal behaviors, and reinforce or devalue the results of the target classifications based on local scene data.

The results of the classification process can be reported as hard or soft decisions that are passed to the data fusion process. The classifier measures for each target hypothesis (or a subset of the leading candidates) must be transformed to normalized measures that represent the degree of belief (or uncertainty) in the form accepted by the combination process.

4.5 MILITARY SENSORS

In this section, we describe the major operational characteristics of military sensors and survey representative sensors used in military C^3I systems over the wide spectrum of space, airborne, surface, and subsurface military missions. Whereas the applications vary widely, the essential characteristics of these sensor classes have much in common as used in data fusion systems. Figure 4.14 summarizes the functions and classes of sensors employed on seven tactical military platforms. For a more in-depth introduction to sensors, the reader is referred to Hovanessian [27].

Although many of these sensors have traditionally been considered either defensive (threat warning), offensive (fire control), or a combination of both (surveillance), their measurements can be combined in a data fusion system to provide complementary information for all functions. Passive detections by threat warning sensors may be used to cue active fire control sensors, or active fire control sensors may cue threat warning systems to examine unknown tracks for identification. The introduction of data fusion and fully integrated sensor suites will eliminate the application of these mission functions to distinct sensors: they will be applied to operational functions of the total system.

When describing each sensor, three characteristics of sensors are of operational importance to military users. These include the means by which sensors collect information and the effect that the collection process has on other military systems.

Sensor Emissions

One important distinction is between sensors that emit energy (active sensors) and those that do not (passive sensors). Purely passive sensors, such as electro-optical sensors, are desired for stealthy operation where the sensor operation must not be detected. Some active sensors may be operated in low probability of intercept modes, which use a variety of means to reduce detectability of the sensor operation. Some of these LPI techniques include (1) blanking of emissions over selected spatial regions (sectors); (2) using spread-spectrum techniques to widen transmission bandwidths and lower radiated power levels [28]; (3) randomizing transmission sequences to reduce the periodic structure of transmissions; and (4) separating transmission and reception processes to permit active operation without emissions at the sensor platform (e.g., multistatic radar operations [29]).

Cooperation of the Target

Cooperative sensors require that the target cooperate with the sensor in a predetermined manner. An example of this is the air traffic control beacon required on civil

MILITARY PLATFORMS	ROLE	SENSOR FUNCTIONS	TYPICAL SENSORS
	AIR-AIR COMBAT	•DETECT, TRACK AND ID AIRCRAFT (FRIEND-FOE AND TYPE ID) •ENGAGE HOSTILE AIRCRAFT AND VERIFY KILL	•MULTI-MODE RADAR •IRST, TV •IFF •ESM
	GROUND ATTACK	•SEARCH, ACQUIRE, ID HOSTILE GROUND TARGETS (MOBILE, FIXED) •ENGAGE TARGETS: HANDOFF AIRCRAFT-TO-WEAPON SENSORS	•TERRAIN-FOLLOWING RADAR •IMAGING/MAPPING RADAR •FORWARD LOOKING IR •ESM
	ANTI-AIR WARFARE	•CONDUCT SURVEILLANCE FOR MILITARY ATC, HOSTILE TARGET DETECTION, ID •ID, TRACK AND ENGAGE HOSTILE AIRCRAFT (CAP/SAM)	•AIR SEARCH RADARS •FIRE CONTROL RADARS •ESM •IRST •IFF
	SURFACE, SUB-SURFACE WARFARE	•CONDUCT SURFACE /SUB-SURFACE SURVEILLANCE FOR HOSTILE SHIP/SUB DETECTION, ID •COORDINATE AIR, SURFACE, SUBSURFACE ENGAGEMENTS	•SURFACE SEARCH RADAR •HULL-MOUNTED SONAR •TOWED-ARRAY SONAR •ESM

	BATTLEFIELD INTELLIGENCE, SURVEILLANCE, TARGET ACQUISITION	•OBSERVE ALL BATTLEFIELD ACTIVITIES BEYOND AND WITHIN FLOT, TO DETECT AND ID : MOVERS SHOOTERS EMITTERS
		•DETECT, LOCATE HOSTILE TARGETS AND COORDINATE ENGAGEMENT/VERIFY KILL
			•AIRBORNE:
			  ESM
			  MAPPING RADAR
			•GROUND-BASED:
			  ESM (ELINT)
			  COUNTER-WEAPON RADAR FLIR
			  ACOUSTIC, SEISMIC

	AIRBORNE WARNING AND CONTROL	•CONDUCT WIDE-AREA AIR SURVEILLANCE
		  -MILITARY AIR TRAFFIC CONTROL
		  -FIGHTER CONTROL
		  -TARGET DETECTION, TRACK AND ID
		•COORDINATE SENSOR NET DATA
			•SURVEILLANCE RADAR
			•IFF
			•ESM
			•DATA LINK TO GROUND CONTROL NETWORKS

	SUB-SURFACE ATTACK SUBMARINE	•DETECT, TRACK AND ID SURFACE AND SUB-SURFACE TRAFFIC •LOCATE HOSTILE SHIPS/ SUBS AND ENGAGE/VERIFY KILL	•SURFACE ESM •HULL-MOUNTED SONAR •TOWED-ARRAY SONAR

Figure 4.14 Principal military sensor applications.

airliners, also called *secondary surveillance radar*. Ground radars transmit an encoded "challenge" waveform to the aircraft that carries a beacon transponder. The transponder receives the challenge and cooperates by transmitting a reply message. Cooperative methods of identification also include the use of predetermined behaviors of targets (e.g., following prescribed routes) that make them recognizable. Cooperative sensors and identification methods are useful in the military context, of course, only with friendly participants and cannot be used to identify uncooperative hostile targets. In these cases, the cooperative means are secured by electronic coding methods to prevent hostile countermeasures from "spoofing" the sensor into accepting hostiles as friends. The great value of cooperative sensors are that there are highly reliable means of rapidly detecting and sorting friendly participants from unknowns (presumed hostile or neutral), which must be subsequently sensed and refined by other means. *Noncooperative* target recognition sensors generally sense attributes unique to the target for identification, requiring no target cooperation.

Spatial Measurement Characteristics

Another characteristic that distinguishes many tactical sensors is the spatial measuring capability provided, particularly the spatial dimensionality of the target data. Figure 4.15 shows the two most common categories of data provided and the situation assessment functions of each. The first category (I) supports the surveillance and fire control functions by providing two or three dimensions of information on the target (usually azimuth-range or azimuth-elevation-range). The data are generally provided on a regular basis by a scanning sensor, and they provide a high assurance of detection of all targets within the surveillance volume. Active sensors (radar, IFF beacons, active sonar, *et cetera*) usually provide category I data, permitting the location and temporal tracking of targets in time and space. The second category of data (II) include those generally provided by warning and signal intelligence sensors that are passive and restricted to the use of opportunistic data (e.g., intercepted signals under the control of the target of interest). This includes the intercept of IR, RF, or visible light emissions, which are limited in spatial resolution and accuracy. As a result, the data can be either aperiodic (e.g., communication transmissions, artillery shots) or periodic in nature (e.g., a scanning radar or a target emitting continuous operating energy).

4.5.1 Radar Surveillance and Fire Control Sensors

The primary sensor for air target surveillance and missile fire control since the Second World War has been monostatic (single site transmitter and receiver) radar. In

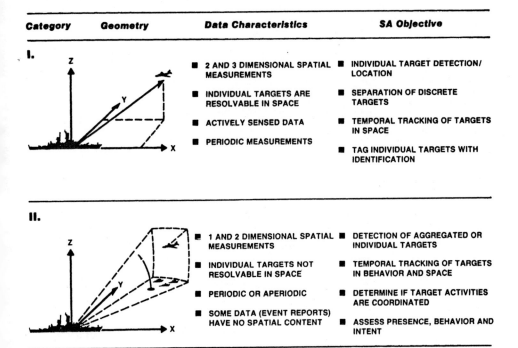

Figure 4.15 Two categories of sensor spatial data measurements.

addition, radar has been applied to ground target, submarine periscope, and other detection as well as terrain following and altimetry applications. Based on the "skin return" reflection (retransmission) of RF energy from the target, radar systems are characterized by two-way spherical radiation, first from the radar sensor source outward to the target and then the reflected radiation back to the sensor. This causes the received signal power (S) proportional to range (R) as a function of R^{-4}, as expressed in the classical radar range equation:

$$S = D_t \times P_{r/i} \times A_r \tag{4.4}$$

where

D_t = power density at the target range, R;
$P_{r/i}$ = ratio of power density of reflected radar energy at the radar receiver to the power density of the illuminating energy at the target range, R;
A_r = effective receiving aperture of the radar;

and, then, inserting usual physical and electrical parameters:

$$S = \frac{P_t G^2 \lambda^2 \sigma}{(4\pi)^3 R^4} \tag{4.5}$$

where

P_t = transmitted power of radar,
G = antenna gain,
λ = wavelength of transmitted wave,
σ = radar cross section of target.

A taxonomy of major radar sensor implementation characteristics that influence the application of radar as a data fusion sensor are summarized in Figure 4.16. The inherently active nature of monostatic radar can be circumvented by three methods shown in the figure:

- Multistatic systems [29], in which the transmitter is not collocated with the receiver, permit the receiver to become a passive sensor. A primary limitation is the requirement that the receiver maintain accurate timing and appropriate geometry relative to the transmitter for proper operation. This requirement often limits the operational usefulness of such systems.
- The use of LPI waveforms (e.g., spread spectrum signals) in place of conventional *continuous wave* (CW), pulsed, or pulsed-doppler waveforms.
- The modification of a pulsed radar's characteristic PRF by randomized, staggered, or intermittent operations.

The introduction of *electronically scanned array* (ESA) antennas has freed radars from mechanically scanning systems in which beam pointing and rotation rates are governed by the inertia of mechanical servo systems. In addition, the regular scan patterns of mechanical systems can be replaced by rapid, concurrent multimode operations between scan, track, and ATR functions. Figure 4.17 compares the typical raster scan pattern of a mechanically scanned airborne radar with that achieved by an ESA. The figure illustrates how target revisit rates, scan areas, and dwell-on-target periods can be controlled by the data fusion sensor manager to optimize the use of such radar.

The general processing flow of radar data is depicted in Figure 4.18, showing the transmitter and receiver processing chain, which produces target detections (nonimaging radar) or return signal power maps (imaging radar). Both categories of radar may be used for ATR when the signal is analyzed for discriminating target signature data. Nonimaging signatures may include temporal (e.g., high range resolution), amplitude (radar cross section), or spectral features [30]. Radar imagery may be collected by real-beam apertures (side-looking airborne radar) or syntheti-

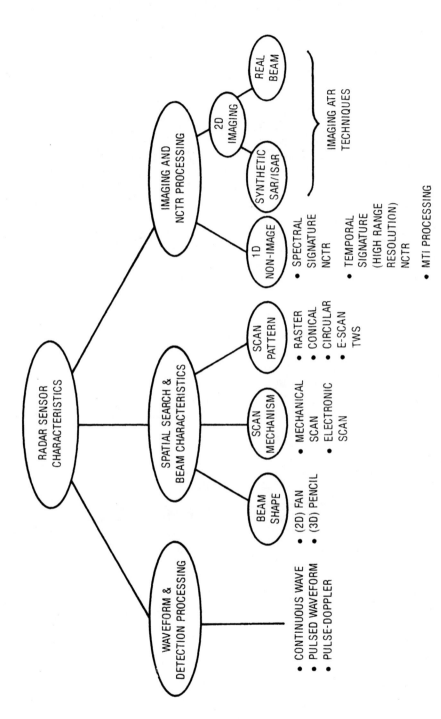

Figure 4.16 Taxonomy of major radar sensor implementation characteristics.

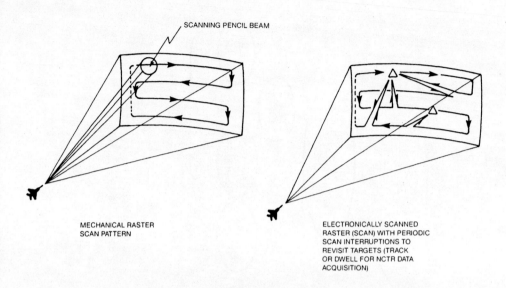

Figure 4.17 Mechanical and electronic scan patterns compared.

cally formed apertures to create two-dimensional radar return maps. Synthetic images may be formed by platform motion as in conventional *synthetic aperture radar* (SAR) [31] or by motion of the target relative to the sensor as in *inverse SAR* (ISAR) [32]. In both cases, two-dimensional ATR techniques have been applied to the imagery to recognize the structure of scatterers or targets for target classification.

4.5.2 Infrared Search-Track and Imaging Sensors

Infrared detectors [33] operating in the 3–5 and 8–12 μm atmospheric windows can detect many targets of military interest by their IR radiation against the background. Multiple bands may be used simultaneously to measure multispectral characteristics of the target to provide classification where target phenomena provide discrimination by spectral means. IR sensors are generally passive, relying on reflected sunlight or emitted energy to provide a contrast between the target and a colder background. Unless supported by an active ranging sensor (e.g., laser, millimeter-wave radar), the IR sensor does not provide independent range measurements.

Nonimaging IR search and track sensors scan a single or array of detectors over a surveillance volume to detect IR point sources by their distinctive contrast against the background [34]. Detections can be measured repetitively to provide

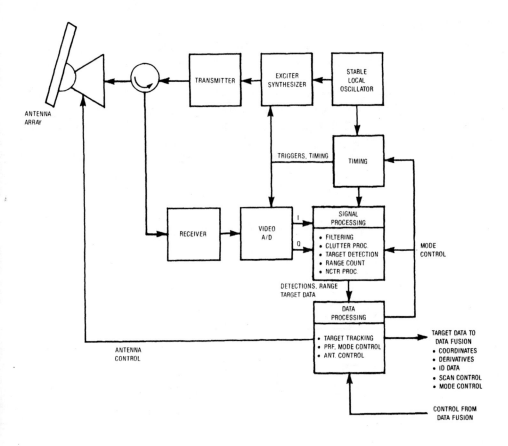

Figure 4.18 Functional elements of a typical airborne radar system.

highly accurate angle tracks, due to the narrow beamwidths achievable (on the order of millirads) compared to radars (on the order of degrees). If the source spectra contains discriminable characteristics that separate target classes (e.g., spectra of rocket, jet plumes), ATR techniques may be applied for target identification. Imaging IR sensors use scanning one-dimensional arrays or staring two-dimensional arrays at the optical focal point to measure two-dimensional radiation data. Resolution relative to the target can be high (e.g., forward-looking IR sensors to image battlefield targets) or low (e.g., spaceborne staring arrays that search for ballistic reentry vehicles against a starry background) depending on the application.

The principal elements of an IR sensor are illustrated in Figure 4.19, which also enumerates the major design alternatives for each element. The optics and

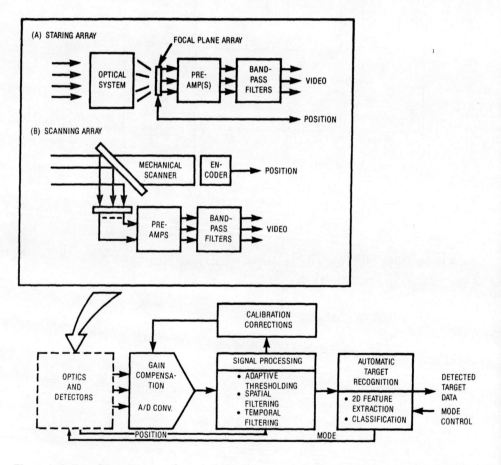

Figure 4.19 Functional elements of a typical IR sensor.

scan-stare mechanism establishes the system resolution and *modulation transfer function* (MTF), which characterizes the spatial filtering introduced between the source and the detected signal. The signal processing flow for an IRST is similar to radar processing, requiring successive detections by the scanning sensor to acquire and track targets by angle. Imaging IR sensors, on the other hand, generally employ different processing functions:

1. Two-dimensional ATR is used to locate and classify targets with spatial extent (not point sources) within the imagery.
2. In time-sequential (framed) imaging sensors, the motion of detected targets must be "tracked" from frame to frame by successively reacquiring the target in each new frame based on last position, motion, and target attributes.
3. In staring arrays, temporal integration and moving target detection may be performed to discriminate impulse noise, stationary point sources (e.g., stars), and targets.

The very wide range of IR sensors [33] have the potential to provide the data fusion system with accurate angle tracking and target identification but without ranging information unless supplemented by another sensor.

4.5.3 Electro-Optical Sensors

Electro-optical sensors are similar to IR sensors in that electronically scanned (television vidicon), mechanically scanned (line arrays), or staring (e.g., *charge-coupled devices,* CCDs) devices may be used to image scenes to permit discrimination of targets from background by their contrast, color, or shape as measured in the visible spectrum. Similar to IR sensors, resolution may be low relative to the target for detection and tracking or may be sufficiently high to permit the target classification [35, 36].

4.5.4 Identification Friend or Foe Sensors

The fundamental cooperative sensor is the IFF or secondary surveillance radar sensor, which requires each cooperating target to carry a beacon transponder to reply to challenge messages transmitted to it by the sensor, referred to as the *interrogator.* Transponder-based air target detection, tracking, and identification systems (secondary surveillance radar) are used in international air traffic control systems in which aircraft transponders are required aboard aircraft operating under certain conditions. These transponders report aircraft identification codes that allow air controllers to uniquely distinguish individual aircraft and aircraft pressure altitude.

Military systems, referred to as IFF, also perform detection and tracking functions for air and ship surveillance, but add a necessary secure identification capability [37–39] that permits high-confidence positive identification of friendly aircraft. This is a primary means of preventing fratricide among friendly forces. High confidence friendly ID serves two critical purposes:

1. Surveillance and control of friendly forces.
2. Rapid means of sorting friends from unknowns that are candidate foes, against which the more limited resources of other sensors can be directed to seek positive hostile identification.

Most military rules of engagement do not permit engagement of targets that do not reply on IFF because several circumstances can prevent a friend from replying (e.g., failed transponder, shadowed transponder antenna, jamming). In spite of this restriction, IFF is generally recognized as the highest-confidence method of military identification. The strength of IFF is its rapid ability to detect and identify friends. Security is achieved by a secure communication uplink message to transmit the challenge and, likewise, a secure downlink to transmit the reply. These cryptographically encrypted and time-varying messages are designed to prevent several forms of countermeasures [40] from influencing the sensor's performance:

- *Spoofing*—The ability of an opponent to present a valid transponder reply message, appearing to be a friend.
- *Exploitation*—The ability to capture a valid interrogation through signal intelligence and then retransmit it to other targets to achieve a positive hostile identification capability.
- *Jamming*—Denial of the transponder to receive challenges or to reply, thereby denying a friend the ability to properly identify itself.

Because the IFF system includes an active transmitter at the interrogator sensor as well as at the transponder, it enjoys an R^2 advantage over radar in range, for a constant effective radiated power. Figure 4.20 shows the functional elements of a typical interrogator and transponder, including the cryptographic elements in each unit that must accept and retain identical keying information.

Although IFF interrogators are generally considered to be hard-decision sensors, the sequential decision criteria may be used to develop a measure of uncertainty for detected replies below the typical probability criteria for detection of a friend. The general design criteria for IFF systems define probabilities for both friend detection and enemy admissability. Interrogator reports provide target range, azimuth, friend detection decision, and any data extracted from the reply message.

4.5.5 Electronic Support Measures Sensors

The functions of passive search, interception, identification, direction-finding, and localization of emitters are included under the general description of ESM sensors,

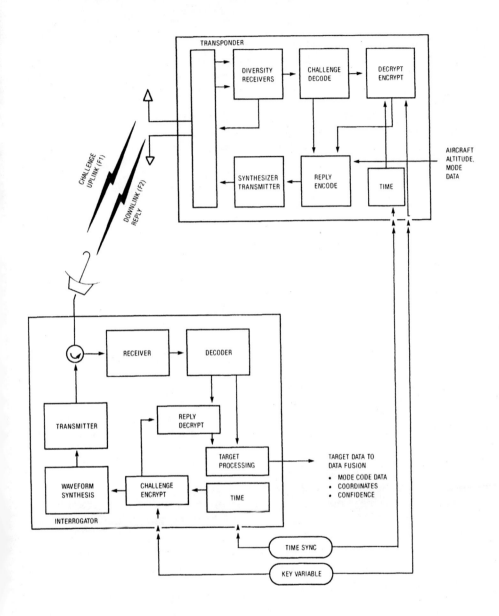

Figure 4.20 Functional elements of an IFF system.

although all functions are not always performed by any one sensor. This should be contrasted with the specific intelligence-gathering functions designated to the two categories of *signal intelligence* (SIGINT) sensors:

- *Communications intelligence (COMINT)*—Sensors that gain technical and intelligence information by the intercept of foreign communications.
- *Electronic intelligence (ELINT)*—Sensors that derive technical and intelligence by the intercept of unknown or hostile noncommunication electromagnetic radiation sources.

Only the most general tactical ESM sensor will be described here, although the principles apply to the wide variety of ESM sensors. Detectable electronic emitters include all active sensors such as fire control or surveillance radars, as well as communications and other systems that rely on active transmission of information to operate. The detection and processing of any radiation from targets can be used for threat warning, sensor cueing, countermeasure control, target location, and target identification (individually or fused with other sensors).

Because ESM sensors depend upon the uncontrollable emissions of targets, performance is characterized by the probability that an emitter is intercepted, P_{int} [41], which is a function of

- *Frequency of emission*—The rate of occurrence of the emitted waveform.
- *Beam intercept*—If either the ESM or emitter antenna beams, or both, are not stationary, P_{int} is determined by the probability that they are coincident when the emission occurs.
- *Frequency intercept*—If either the ESM receiver (frequency sweeping) or the emitter frequency (frequency hopping) is not stationary in time, P_{int} is determined by the probability that the receiver is tuned to the frequency of emission at the time of emission.

Because of these factors, ESM sensors have been extended to broadband operation with wide coverage and increasingly sensitive receivers to counteract the countermeasures of LPI waveforms, frequency hopping modes, and directional transmissions.

The typical ESM sensor (Figure 4.21) generally can be characterized by the methods by which it implements three functions: (1) intercept receivers; (2) direction finding technique; and (3) signal classification processing. The received and detected signals are tagged with time of arrival (TOA) and angle of arrival (AOA), and features are measured to characterize the waveform (e.g., frequency, pulsewidth). These discrete detections of "pulses" are then sorted to deinterleave overlapping pulse trains by using features, as shown in the figure. Once time series of waveforms are clustered, the series are classified by using individual pulse and time series characteristics for comparison with a library of threat waveforms. Identified signals are placed in an active emitter file and an emitter track is maintained. Tem-

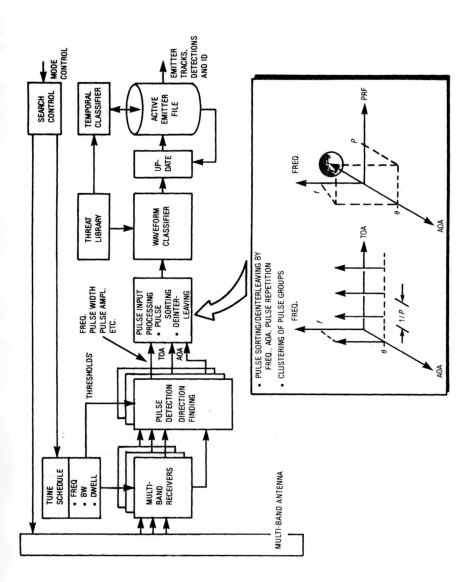

Figure 4.21 Functional elements of a typical ESM sensor.

poral behavior (e.g., scan rates and PRF modes) can also be used to classify the emitter and the degree of threat that it represents.

4.5.6 Acoustic Sensors

Acoustic (pressure) waveforms transmitted through water, the earth, or the atmosphere may be used to detect, track, and classify targets. The most common acoustic sensors include sonar for sensing submarines and ships [42], microphonic sensors for sensing aircraft or ground vehicles [43], and seismometers for detecting and locating events such as nuclear detonations or troop movements by locating the source of seismic stress waves propagated through the earth's surface [44]. Because the general processing is similar for all three sensor applications, only the sonar application will be described in this section.

Military sonar systems include both passive and active systems and are employed in a variety of methods: air-dropped sonobuoys placed individually or in array patterns, helicopter-dipped sensors, ship and submarine hull-mounted arrays, towed linear arrays, or permanent underwater arrays.

In all applications, the basic detection process is dependent on spherical spreading (where propagation loss is a function of R^2) of sound pressure waves that are influenced by several factors:

- Absorption of sound energy is a function of the water transmission medium, sound frequency, and depth.
- Transmission paths include the direct path, bottom bounce, and ducted paths, in which thermal layers influence the direction of pressure waves, as a function of frequency. Multiple paths of transmission is a characteristic of all sonar systems, causing received signals to be a superposition of several source signals, each individually weighted and delayed in time.
- Noise sources, including natural hydrodynamic, biological (e.g., fish), and commercial vessel activity.

Relative to other electromagnetic-based sensors, sonar is characterized by a very low propagation velocities and long wavelengths. Because of the long wavelengths, it is often difficult to achieve sufficiently large apertures to obtain high resolution for applications such as imaging [45]. In addition, the character of the open ocean sonar mission (long detection distances, large surveillance volumes, and dynamic ocean environment) introduces severe noise to the sensor.

A typical passive sonar array sensor system is depicted in Figure 4.22 to show the general processing stages required to form a synthetic beam and detect the presence of a target. Because the accoustic frequency (spectral) characteristics of signals and noise are similar, space-time processing is used to separate the various sources of accoustic energy by arrival angle as well as frequency content. These stages are described in the following paragraphs.

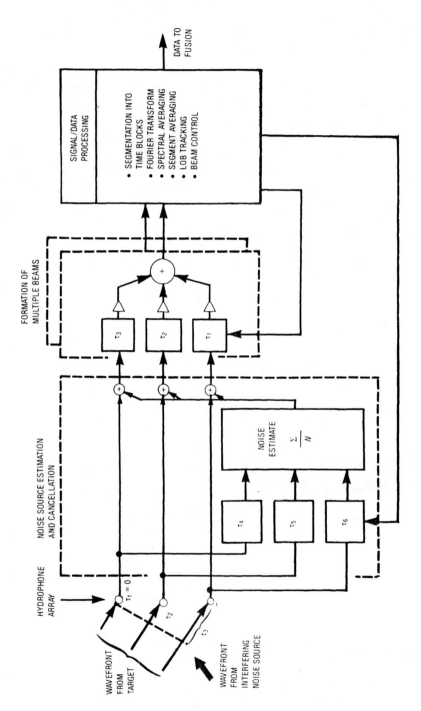

Figure 4.22 Functional elements of a typical sonar sensor.

Hydrophone Array

An array of hydrophones is used to detect the planar wavefront received from an assumed distant point source. As illustrated, the time-of-arrival of signals to each element of the array depends upon the source's direction of arrival.

Beam Forming

A conventional time-based beam former is illustrated in which time delays for each hydrophone can be adjusted individually to effectively "steer" the beam to bring all energy from a constant wavefront angle into time-coincidence. The output of the beam former is then summed, filtered, and squared to provide an integrated signal for further processing. Steering of the array to optimize signal strength provides a bearing estimate to the target source.

Null Steering

Multiple, simultaneous beams can, of course, be formed to observe target and noise sources at the same time by the implementation of parallel beam formers. By observing noise sources and finding their directions, the target beam can be formed to subtract the noise wavefronts, therefore minimizing the effects of noise. In addition, multiple, simultaneous, target-observing beams can be formed to search or track targets.

Noise Cancellation

If directly measured or accurately estimated, additive noise sources can be subtracted from target beam signals to cancel their effective contribution to target frequencies.

Spectral Analysis

Received accoustical signals can be analyzed for spectral content to remove noise components and discriminate targets by spectral means. Statistical or syntactic classification of spectral features can provide ATR capabilities to the acoustic sensor, for fusion with other sensors or sources.

4.6 MILITARY SOURCE AND DATA LINK CHARACTERISTICS

Sources of data other than sensors and information passed from second parties are important suppliers of knowledge to many data fusion systems and require special consideration for properly utilizing their contribution to the fusion process. In this section we introduce source data and the links that permit the passage of data from sensors and sources to data fusion nodes.

4.6.1 Source Data

Human observations and the results of intelligence analyses form the basis of most source information, usually considered to be intelligence information that can be formatted for input to automated fusion systems. In addition, information about own force plans is a valuable source of data for situation assessment. The primary forms of source information include the following.

Human Intelligence (HUMINT)

Data derived from human sources, including personal observations, interrogations of prisoners, inferences and deductions from conversations, *et cetera* [46]. An example of HUMINT in the tactical air mission area includes visual observations, by pilots, of tactical targets (aircraft, ground targets, hostile tactics, *et cetera*) in the course of combat. The use of this information in an automated fusion system requires two considerations. First, a means must be provided for immediate manual input of target designation data. This may include manual identification of target tracks or manual cueing of other sensors to search for a target at given coordinates. In this case, the human senses are directly coupled with system sensors. The second use of such information is in preparing the intelligence data base for subsequent missions. This, in effect, translates the traditional preflight intelligence briefing into the data base of the fusion system.

Communications Intelligence (COMINT)

Data derived by the intercept and analysis of communications [47]. Both the *internal* content of messages and the *external* characteristics of the transmissions (e.g., rate and number of transmissions, interactions between communication nodes) provide useful source data from which other intelligence data may be logically inferred. COMINT data must be used with caution because of the possibility of deceptive communication activities that have the potential of rapid contamination of the fusion data base.

Friendly Force Plans

The military objectives and planned actions of friendly forces (including predefined routes, orders of battle, contingency plans, *et cetera*) are invaluable sources of data for situation assessment. These data form the *a priori* conditions for assessing the behavior of and threat to own forces, as well as for inferring identity of unknown (or presumed friendly) targets on the basis of behavior. The categories of planning data maintained in situation data bases is summarized in Chapter 10.

The elements of source data are generally fused at the situation and threat assessment levels of fusion. They are therefore not directly combined with sensor data and are usually partitioned from incoming sensor data. In addition, the security level of source data may be problematic.

4.6.2 Communication Links for Sensor and Source Data

The ability of military platforms to transfer data from platform to platform depends on reliable, jam-resistant, and secure communication links. These links permit the exchange of information about friendly force activities (locations, plans, status, *et cetera*) as well as observations by friendly forces' sensors. Although voice communication provides a means of exchange of human intelligence (HUMINT), the digital data link is the primary means of rapidly exchanging sensor data between data fusion nodes. A typical three minute fighter-to-fighter voice interchange to engage two maneuvering targets, for example, may require over 300 words [48] between pilots (expressed in abbreviated pilot dialect). The categories of information that must be transferred in such an exchange include sensor detections and tracks (airborne radar), own-ship locations and headings, and target assignments as well as changes. In contrast, the use of tactical information distribution networks permit much more rapid and accurate transfer of higher volumes of such information between platforms. In such links, digital data is transferred and displayed to operators after all data are associated for common presentation. The common display combines associated data from all sources, and provides means to exchange target designation as well as command and control data between platforms without verbal communication.

The use of data links to exchange information introduces specific terminology to distinguish data types and network parties:

- *Participant* refers to any node of the network of data links that sends information, including data on self (location, status, *et cetera*) as well as sensor data.
- *Local data* is that data collected and processed by a participant and does not include information received from other participants.
- *Global data* includes data from more than one participant, which may be fused together to provide a composite set of data.

Information Exchanged on Data Links

The general categories of information desired to be transferred across data links between tactical platforms include

- Participant location and status
- Participant sensor (locally fused) data
- Target assignments
- Command and control messages
- Relative navigation data

Tactical Data Links

The US and NATO have developed a variety of digital data links for the transfer of information between C^2 nodes and weapons platforms, each with its own unique characteristics to meet specific applications. The most complex of these links is the *Joint Tactical Information Distribution System* (JTIDS), which uses the NATO message protocol known as TADIL-J. JTIDS [49, 50] is a *time division multiple access* (TDMA) system that requires time synchronism among all participants in the network. Each participant is preassigned time slots during which it may transmit data in one of two forms:

- *Formatted messages* are structured messages that contain specific data formats to convey standard information (e.g., location, target data, navigation data, sensor data).
- *Unformatted messages* permit nonstandard data such as digital voice or TTY messages to transmitted in fixed-length segments. This also permits non-TADIL-J messages (TADIL A, B, C, Link 1, *et cetera*) to be transmitted to retain commonality and gain link security, antijam performance.

Any participant may receive all messages from other participants and may apply selective filters to extract only that information of interest (e.g., targets within some volume). Participant identification within messages as well as message-type identifiers permit this filtering and extraction process as well as the association and updating of data from multiple participants. This permits the system to be effectively implemented as a nodeless network in which there can be an all-way exchange of data.

Data Link and Fusion Design Considerations

The use of data links to transfer information between data fusion nodes requires the consideration of several system characteristics that directly relate to the capacity required of the data link. These considerations include

1. The self report rate of each network participant must be defined to maintain track of all participants (a function of each participant's mobility and speed) or status update rates.
2. The rate at which sensor data are reported is likewise dictated by revisit rate required to maintain track of sensor targets.

3. The amount of data contained in each sensor report transmitted has a significant impact on network capacity requirements. Hard-decision sensors demand the least capacity, providing short messages (target declarations) only when target track or ID updates are made. Soft-decision sensors, on the other hand, demand large messages to report multiple hypotheses and associated quantitative confidence data.
4. The interconnection architecture of the network also influences the capacity of individual links between participants. The use of subnets or network hierarchies can limit the capacity of information flowing between any two fusion nodes while ensuring that each node receives all the information necessary to perform its fusion function.
5. The interchange of sensor and fused information between fusion nodes must follow specific procedures to avoid two categories of errors unique to multisensor systems that attempt to fuse data linked information

    - *Gridlock errors*—When the relative spatial measurement errors between sensors cannot be reduced to an acceptable value, association of targets common to overlapped sensors cannot be performed and multiple targets detections or tracks are spawned for real individual targets. This results in target multiplication and loss of any of the potential benefits of fusion.
    - *Fusion data feedback*—If sensor and fused data are not properly identified and segregated, local sensor information can be transferred to other sites, only to be returned and fused with the original data thereby falsely increasing confidence and accuracy estimates. This results in inaccurate estimates of uncertainty in data and can lead to incorrect detection decisions or parametric estimates with exaggerated statistics.

4.7 A SIMPLE FUSION NETWORK EXAMPLE

Figure 4.23 illustrates a two-platform data link system in which two aircraft exchange own-ship locations and status data as well as sensor data fused prior to transmission. The sensors on the two aircraft view common targets (in the sensor overlap regions) as well as targets exclusive to each. In this system, sensors on each aircraft are fused (sensor-to-sensor association and combination) and detections, tracks, and identifications are maintained in a local data base.

The fused sensor data from the lead aircraft are filtered (selected target data is extracted) and transmitted to the wingman along with own-ship status and position information. The own-ship position is based upon absolute or relative navigation measurements and a grid alignment process that attempts to correct spatial measurements from all sensors into a common coordinate system (grid). This is performed using continuous measurements, from multiple sources, of common coordinates (e.g., net participants, stationary ground stations) to maintain a recur-

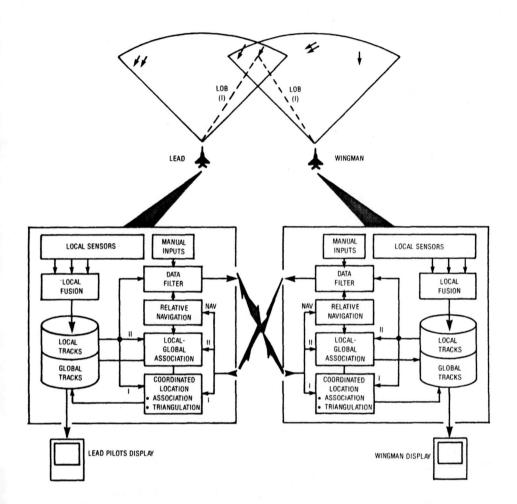

Figure 4.23 Functional elements of a two-aircraft data link of sensor data and data fusion.

sive navigation filter that estimates corrections to absolute and relative grid coordinates. These corrections are applied to all location data received over the link to refine spatial accuracies.

The wingman similarly fuses, filters, and transmits sensor data to the lead aircraft, along with navigation and status information. The lead aircraft attempts to associate own category II data (detections and tracks) with wingman's data, using grid correction data to transform that data to the common grid. *Lines of bearing* (LOBs) to passively sensed targets (category I data) from both aircraft are similarly associated (by relative angles of arrival, *times of arrival* (TOA), and measurement attributes) and, if possible, triangulation or TOA techniques are used to derive target locations. The data base maintains separate local and global files, and the pilots may view either data base on a display in grid coordinates. In this simple example, assuming equal sensor and navigation performance of both platforms, both aircraft will derive common global data bases and exchange of global data is unnecessary.

In complex networks with many participants and multiple fusion nodes, it may become necessary for the exchange of global information to avoid the necessity of all-way exchange of sensor data and complete fusion of data at all nodes. In these cases, some participants (e.g., C^3 sites or major surveillance platforms) may be designated as fusion nodes to collect participant sensor data, perform fusion, and disseminate a common global data base (or selected portions) to participants.

REFERENCES

1. Klass, Philip J., "Defense Department Urges Acceleration of Mk. 15 IFF System for NATO." *AW&ST*, August 12, 1985, pp. 67 f.
2. Van Trees, H.L., "Detection, Estimation and Modulation Theory," Vol. 1, John Wiley and Sons, New York, 1968, pp. 19 ff.
3. Ibid., pp. 46 ff.
4. Ho, Yu-Chi, "Team Decision Theory and Information Structures," *Proc. IEEE*, Vol. 68, No. 6, June 1980, pp. 644–654.
5. Tenney, R.R., and N.R. Sandell, Jr., "Detection with Distributed Sensors," *IEEE Trans. on AES*, Vol. AES-17, No. 4, July 1981, pp. 501–510.
6. Chair, Z., and P.K. Varshney, "Optimal Data Fusion in Multiple Sensor Detection Systems," *IEEE Trans. on AES*, Vol. AES-22, No. 1, January 1986, pp. 98–101.
7. Reibman, A.R., and L.W. Nolte, "Optimal Detection and Performance of Distributed Sensor Systems," *IEEE Trans. on AES*, Vol. AES-23, No. 1, January 1987, pp. 24–30.
8. Aalo, V., and R. Viswanathan, "On Distributed Detection with Correlated Sensors: Two Examples," *IEEE Trans. on AES*, Vol. AES-25, No. 3, May 1989, pp. 414–421.
9. Teneketzis, D., "The Decentralized Wald Problem," *Proc. IEEE 1982 Int. Large-Scale Systems Symp.*, October 1982, pp. 423–430.
10. Teneketzis, D., and P. Varaiya, "The Decentralized Quickest Detection Problem," *IEEE Trans. Auto. Control*, Vol. AC-29, No. 7, July 1984, pp. 641–644.
11. Wald, A., *Sequential Analysis*, John Wiley and Sons, New York, 1947.
12. Thomopoulis, S.C.A., R. Viswanathan, and D.P. Bouboulias, "Optimal Decision Fusion in Multiple Sensor Systems," *IEEE Trans. on AES*, Vol. AES-23, No. 5, September 1987, pp. 644–653.
13. Lee, C.C., and J.J. Chao, "Optimum Local Decision Space Partitioning for Distributed Detection," *IEEE Trans. on AES*, Vol. AES-25, No. 4, July 1989, pp. 536–543.
14. Sadjadi, F.A., "Hypothesis Testing in a Distributed Environment," *IEEE Trans. on AES*, Vol. AES-22, No. 2, March 1986, pp. 134–137.
15. Demirbas, K., "Maximum A Posteriori Approach to Object Recognition with Distributed Sensors," *IEEE Trans. on AES*, Vol. AES-24, No. 3, May 1988, pp. 309–313.
16. Allen, John B., "Statistically Optimum Methods of Combining Multiple Sensor Data for Automatic Target Acquisition," *IEEE Proc. 1984 NAECON*, pp. 237–244.

17. Zembower, Andrew, "The Challenge of Automatic Target Recognition," *Tactical Weapon Guidance and Control Information and Analysis Center, GACIAC Bulletin*, Vol. 11, No. 4, July 1988, pp. 1-2.
18. Taub, H., and D. Schilling, *Principles of Communications Systems*, 2 ed., McGraw-Hill, New York, 1970, p. 536.
19. Nahin, P.J., "NCTR Plus Sensor Fusion Equals IFFN or Can Two Plus Two Equal Five?" *IEEE Trans. on AES*, Vol. AES-16, No. 3, May 1980, pp. 320-337.
20. Andrews, H.C., *Introduction to Mathematical Techniques in Pattern Recognition*, Wiley-Interscience, New York, 1972.
21. Buede, D.M., and E.L. Waltz, "Benefits of Soft Sensors and Probabilistic Fusion," *Proc. SPIE*, Vol. 1096, *Signal and Data Processing of Small Targets*, 1989, pp. 309-320.
22. Bhanu, Bir, "Automatic Target Recognition," *IEEE, Trans. on AES*, Vol. AES-22, No. 4, July 1986, pp. 364-379.
23. Helland, A.R., "Application of Image Understanding to Automatic Tactical Target Acquisition," *Proc. SPIE 281*, 1981, pp. 26-31.
24. Soland, D.E., "Prototype Automatic Target Screener: Smart Sensors," *Proc. SPIE 178*, April 1979, pp. 175-184.
25. Milgram, D.L., "Results in FLIR Target Detection and Classification," *Proc. DARPA Image Understanding Workshop*, May 3 1978, p. 118.
26. Bhanu, "Automatic Target Recognition," p. 365.
27. Hovanessian, Shahan, A., *Introduction to Sensor Systems*, Artech House, Norwood MA, 1988.
28. Dixon, R.C., *Spread Spectrum Systems*, John Wiley and Sons, New York, 1976.
29. Elson, Benjamin J., "Bistatic Airborne Radar Use Studied," *AW&ST*, August 13, 1979, pp. 57-58.
30. Dalle Mese M., "Target Identification by Means of Radars," *Microwave J.*, December 1984, pp. 85-103.
31. Ausherman, Dale A., "Digital Versus Optical Techniques in Synthetic Aperture Radar (SAR) Data Processing," *Optical Engineering*, Vol. 19 No. 2, March-April 1980.
32. Prickett, Michael W., "Principles of ISAR Imaging," *Proc. EASCON*, 1980, pp. 340-344.
33. Hudson, R.D., Jr., and J.W. Hudson, "The Military Applications of Remote Sensing by Infrared," *Proc. of IEEE*, Vol. 63, No. 1, January 1975, pp. 104-128.
34. Klass, Philip J., "Introduction of Stealth Will Change Avionic Needs," *AW&ST*, March 18, 1985, p. 235.
35. "Tomcat Sees through Long-Range Eyes," *Defense Systems Rev.*, November 1983, p. 36.
36. "Target Identification Upgrade Sought," *AW&ST*, August 6, 1979, p. 59.
37. Grammuler, Harold, "NATO Identification System (NIS)," *Signal*, December 1981, pp. 87-91.
38. "The New NATO Identification System," *Special Electronics*, No. 2, 1984, pp. 41-43.
39. "Identification of Friend or Foe in Air Warfare—A Capability Long Neglected and Urgently Needed," House Rep. 99-354, Committee on Government Operations, November 1, 1985.
40. Bridge, W.M., "The Role of Precise Time in IFF," *Military Electronics/Countermeasures*, April 1982, pp. 53-60.
41. Hatcher, B.R., "Intercept Probability and Intercept Time," *EW*, March-April 1986, pp. 95-103.
42. Urick, R.J., *Principles of Underwater Sound for Engineers*, McGraw-Hill, New York, 1967.
43. "Distributed Sensor Networks," MIT Lincoln Laboratory, ADA-131245, 24 June 1983, pp. 17 ff.
44. Wood, Lawrence C., "Seismic Signal Processing," *Proc. IEEE*, Vol. 63, No. 4, April 1975, pp. 649-661.
45. Keating, Patrick N., "Signal Processing in Acoustic Imaging," *Proc. IEEE*, Vol. 67, No. 4, April 1979, pp. 496-510.

46. *Department of Defense Dictionary of Military and Associated Terms,* JCS Pub. 1, 1 June 1987, p. 175.
47. *Ibid.,* p. 80.
48. Lubkin, Yale Jay, "The Need for INEWS," *Defense Science and Electronics,* November 1986, p. 11.
49. Leondes, C.T., ed., "Principles and Operational Aspects of Precision Position Determination Systems," AGARD-AG-245, NATO Advisory Group for Aerospace Research and Development, ADA 075208, July 1979.
50. Toone, Joseph, "Introduction to JTIDS," *Signal,* August 1978, pp. 55–59.

Chapter 5
SENSOR MANAGEMENT

Not only important is it in multiple sensor systems to include a complementary suite of sensors that individually achieve robust detection and discrimination, but also to properly manage these sensors to optimize their *collective* performance relative to several operational criteria. These collective criteria include such items as detection and track performance, search coverage and rate, emissions control, continuity of coverage over time, and target identification. The use of multiple sensors in a multiple target environment requires that the targets also be ranked so the sensors can be allocated to targets on the basis of factors such as threat lethality, engagement opportunity, and sensor needs. Sensor management requirements vary from sensor to sensor because of the unique characteristics of sensors and their mission applications, but the general reasons for management include

- *Spatial management*—Pointing is required on sensors that are not omnidirectional. For surveillance applications, the sensor's *field of view* (FOV) must be systemmatically moved (scanned) to search and acquire new targets or repositioned periodically to maintain a track of moving targets.
- *Mode management*—Mode selection is required on sensors having variable parameters that affect operation. This can include selection of apertures (FOVs), search patterns, signal waveforms, power levels, or processing techniques.
- *Temporal management*—Timing of sensor operations is required where the sensor must be synchronized with other sensors or with events in the target environment (e.g., target detections, track losses, countermeasures activities).

The principles of sensor management are applicable to three primary categories of application. The first category includes complex, multimode sensors such as electronically scanned radars that permit rapid pointing and mode (e.g., pulse repetition rate) changes. This capability requires a high degree of flexibility to adjust the time dedicated to scanning and revisiting targets to maintain track and dwell to acquire signature measurements. The management of such a sensor requires the control of pointing between search, track update, and signature collection priorities, specified by detection probability, track accuracy, and classification accuracy requirements, respectively. The most efficient use of *time* to perform all tasks is the objective of this sensor management function. Electronically scanned strategic, ground-based radars, shipboard surveillance radars, and airborne radars are some users of this category of sensors. The second application category includes multiple sensor suites located on a common platform. The most effective use of the *different characteristics* of the sensors, as well as time, is the function of the sensor manager in this case. The third application includes networks in which the sensors are geographically distributed. Coordination among all sensor sites is required to cause (1) multiple sensors to observe a common target for localization data (see Chapter 6 for a description of multiple-site localization methods), and (2) one sensor to acquire and take over track control from another when the target passes from the field of view of one sensor to the next.

In all three applications, management of sensor resources requires optimization of specific performance measures (detection probability, intercept probability, emissions from own sensors, track accuracy or loss probability, *et cetera*) in the face of demands from multiple targets and scan volumes that contend for limited resources. This chapter examines the interfaces required to control sensors, the factors that impose restrictions on sensor operations, the specific criteria for allocating sensors to targets, and methods of sensor management for military applications. The principles presented here are applicable to all three categories of application just described, with emphasis on collocated multiple sensors.

5.1 SENSOR MANAGEMENT FUNCTIONS

The most general categories of sensor management functions are described in this section, with the following sections detailing the methods of implementing the major functions. Although this general model may not be applicable to all data fusion systems, it does encompass the principal functions required to manage multiple sensors in a multiple target environment. Each specific application will dictate the sensor management architecture necessary to efficiently allocate sensors to service targets and search volumes. The criteria for managing sensors must be defined and quantified to specify the meaning of optimal allocation of sensors for any given application. Criteria generally include quantifiable parameters such as new target detection probability, track continuation probability or state estimate accuracy, target revisit (update) rate, identification accuracy, *et cetera*.

Sensor control management is based on manual inputs, data fusion information (track file data, situation assessments), and external cues or hand-off requests. The goal of such control, or allocation of sensors to targets or search volumes, is the optimization of an objective function that quantifies the various just criteria stated. In general, the sensor suite may not be able to service all of the targets or achieve all of the sensing performance objectives (e.g., detection, resolution, revisitation) desired by the system, but the optimal compromise among conflicting demands is sought.

Figure 5.1 illustrates the relationship between the functions that constitute a general sensor manager in a data fusion system. Inputs, management functions, and sensor controls are shown. Section 5.2 describes typical sensor interfaces, and Sections 5.3 and 5.4 describe the methods of performing the two functions required in virtually all applications: target ranking and sensor assignment. The following paragraphs describe each of the management functions in the figure.

Target Ranking

The relative priority of detected targets (and possibly volumes to be searched) must be determined to establish a quantitative basis upon which to determine the trade-offs for using sensor resources. The inputs can include manually assigned priorities, current target states (from the track file) as well as future target states (from the event prediction function), which may indicate possible future threats or events. Section 5.2 describes the basis for different ranks and methods of ranking targets.

Event Prediction

Knowledge of current events, target states, and tactical doctrine can be used to predict future events and manage sensors to detect or verify anticipated events. In target tracking, the sensors are directed to revisit a target under track, based on the prediction of its future location at the time planned for revisit. In Chapter 6 we describe target tracking algorithms that estimate the kinematic state of targets, based on sequential sensor observations. The tracking algorithm's state estimate (target position and higher order derivatives) permits extrapolation forward in time to the next scheduled revisit time to project the sensor pointing geometry to reacquire the target. This process of look angle generation minimizes the time to reacquire targets and the loss of tracks. The estimator also predicts the state estimate error covariance: a large variance in the state estimate could mean that, upon revisit, a small search may be required. In a similar manner, sensors can be managed on the basis of predicted threats or opportunities, determined by the situation assessment function.

Look angle generation involves basic observation prediction; that is, given a sensor that observes parameters α and β (e.g., azimuth/elevation, range/range-rate), look angle prediction involves a transformation of the form:

$$\begin{bmatrix} \alpha(t_{\text{obs}}) \\ \beta(t_{\text{obs}}) \end{bmatrix} = \bar{f}[\mathbf{x}(t_{\text{obs}}), \mathbf{Z}] \quad (5.1)$$

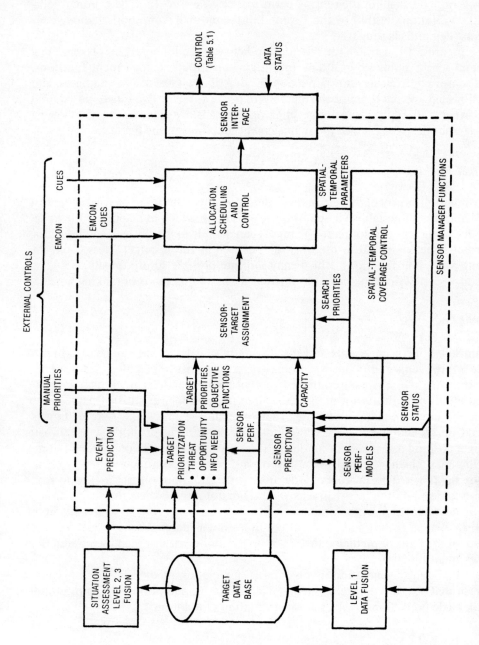

Figure 5.1 General sensor manager functions.

where $x(t_{obs})$ is the predicted position of a dynamically moving object at the observation time, t_{obs}, and constant parameters, Z (such as the location of the sensor, boresite direction). To perform this observation modeling requires

1. A dynamic model (differential equation of motion) that allows the state vector of the tracked object to be updated from an *a priori* time, t_0, to the observation time, t_{obs}.
2. An observation model that transforms the state vector at time t_{obs} to a predicted observation at the same time, t_{obs}.
3. Possibly, a noise and bias model to account for the effect of observational noise or bias.

This look angle generation is one part of the tracking problem described in Chapter 6. Obviously, the specific form of equation (5.1) is dependent upon the sensors being modeled.

Sensor Prediction

The *ability* of sensors to service targets must be determined prior to the generation of assignment alternatives. The ability for any sensor to be assigned to a target is a function of availability (i.e., failure or dedication to other targets may render a sensor unavailable) or capability (predicted detection, identification performance against a target). Bier and Rothman [1] use a Boolean matrix to define permissible sensor-target pairings based on capability and availability. Sensor performance models may also be used to predict the performance of sensors against targets to quantify the marginal benefits of various candidate sensor assignments. Models of detection, track, and ID ranges can use available data from the track files to predict performance of future measurements. In this case, the value of each possible pairing can be predicted for subsequent use in the calculation of an assignment objective function.

Target-Sensor Assignment

When multiple targets and multiple sensors are present, the management of sensors requires an assignment of sensors to targets (or surveillance volumes or cued positions to accept a target being handed off by another sensor). Assignment is usually based on optimizing some composite objective function that is a measure of the performance of the collection of sensors against the entire target complex. The objective function is a function of both target priorities and sensor-target performance values. Section 5.3 describes the application of linear programming methods to perform assignment computations.

Spatial and Temporal Coverage Control

The spatial volume covered by the suite must be managed by positioning the fields of view of pointing sensors to cover detected targets (to update tracks) and to search for new, as yet undetected targets that may enter into the suite's surveillance volume. The amount of time spent searching in each spatial volume or dwelling on targets under track must also be managed to balance several criteria: detection probabilities (new targets), tracking and identification performance (targets in track), and own-ship detection probability due to emissions in the direction of possible hostile sensors. This control is established as a function of mission mode, operator manual inputs, and current situation.

Allocation and Control Strategy

The allocation of sensors requires converting assignment solutions into commands to sensors, while incorporating a number of additional control factors:

1. *Counter-countermeasure*—The sensor suite may be managed to optimize its immunity to countermeasures (e.g., jamming, deception, exploitation) by controlling the spectral, spatial, or temporal use of sensors to avoid the effects of measures against the system.
2. *Emission control (EMCON)*—Because some active sensors emit energy that can be detected by adversaries and are subject to exploitation, it is often necessary to manage the use of active sensors, controlling the power, duration, spatial coverage, or modes of such sensors. This control is often imposed as a function of operational considerations where the objective function to be minimized is detectability or identification of the sensor suite.
3. *Cueing and hand-off*—When any sensor (B) must be directed (cued) to search for and acquire a target designated by another sensor (A), special search commands may be required for sensor B. These commands must define the search pattern, duration, and designated target identification such that sensor B can validate that an acquired target is, indeed, the designated target. Section 5.5 describes these functions and the methods of analyzing cueing performance.
4. *Scheduling*—The time-sequencing of specific sensor controls is required for the efficient operation of sensor modes, pointing, and active operations. The constraints considered in scheduling include
   - Concurrent target measurements by multiple sensors to minimize time-space alignment errors.
   - Overall reduction in sensor repositioning time *versus* time on targets.
   - Synchronization of cueing and hand-off times between sensors.
   - Time of active emissions relative to hostile sensors' scan patterns.
   - Rates of target revisitation by sensors to maintain track continuity.

Numerous sensor manager architectures that use expert systems to coordinate these functions have been reported in the literature. Rothman and Bier [1, 2] have described a sensor manager for the tactical aircraft application described at the introduction to this chapter. Leon and Heller [3] (also see the description in [4]) have described a sensor management and control expert system for strategic sensor network applications. Slagel and Hamburger [5] have described a similar weapons-to-target resource allocation system that uses an expected total destruction objective function to perform the allocation process. A similar expert management system for aircraft navigation sensors is described by Hui [6], which controls the use of sensor measurements to optimize navigation performance. Cowan [7] has simulated an expert system for management of typical tactical aircraft sensors and reported on the tracking performance with and without expert management.

5.2 SENSOR INTERFACES

As indicated in the last chapter, sensors are an integral part of the entire fusion system architecture. As such, they must be designed to provide interfaces that permit management by the fusion system to coordinate all sensors. In this section, we summarize the most general functional requirements that must be considered while preparing *interface control documents* (ICDs) for sensors to be used in data fusion systems. Although sensor controls vary greatly among sensor types (radar, sonar, IRST, *et cetera*) and sensor missions (spaceborne, airborne, battlefield, *et cetera*), there are common functional parameters that must be defined in each case to permit real-time sensor management. Table 5.1 summarizes these general parameters.

The operation of each sensor requires the sensor manager to command the sensor to select modes, positions, and timing. These operations include four types of control processes.

Global Control

This command control establishes the overall parameters for the sensor operation. In sensors that do not perform concurrent or time-multiplexed operations, this is the single control state for the sensor. In sensors that do permit concurrent operations (sectoring, track-while-scan, *et cetera*) this may be the primary mode for all but special sectors or designated target activities.

Sector Control

Some sensors permit the surveillance volume to be partitioned into *sectors* over which sensor parameters can be selected individually. For a mechanically scanning surveillance radar, for example, the 360° surveillance coverage may be partitioned into four sectors as follows:

No.	Sector (degrees)	Mode	PRF (Hz)	Power	Comments
1	0–65	OFF	OFF	OFF	Reduce emissions toward LOB 30°
2	65–175	Search (doppler)	600	High	Long-range search
3	175–225	Pulsed doppler	250	Med	Track targets
4	225–0	Search (doppler)	600	High	Long-range search

The spatial coordinates of each sector must be provided (e.g., start azimuth, end azimuth) as well as the operating modes and parameters for the sector. The storage of sector control data and control of mode or parameter switching may reside locally in the sensor, or this control may be performed in real-time by the data fusion sensor manager if the interface allows.

Designated Target Control

In the same sense that the sensor may be directed to perform special functions over sectors, individual targets may be designated for special sensor activities, such as periodic revisits for tracking or long-duration dwell to collect signature data. The sensor is issued requests in which the sensor manager must provide an estimated location or bearing for the target and search limits about the estimate as well as mode parameters to control sensor activities on the target. Each designated target is generally assigned an index (unique number) to permit the sensor to identify measurements by index number.

Search Control

The sensor manager may request the sensor to search a specific volume in which the limits may be defined (e.g., range, azimuth, elevation limits) as well as the sensor modes. Filter parameters may also be defined to limit the kinds of detected targets that should be reported, based upon target attributes (e.g., target type, altitude, velocity).

5.3 ESTABLISHING TARGET PRIORITY

Setting the priority of targets (and search volumes) must preceed the allocation of sensors to tasks (assignment to designated targets or search volumes). When sensor resources are limited and multiple targets are present, this process ranks the importance of each target for sensor viewing. In first-generation radar fire control systems, the radar sensor was locked on a single target at a time and tracked that target

Table 5.1
General Sensor Controls

Category	Control Functions Input to Sensor

Mode Control Functions	• On/off control
	• Sensor mode selection
	   Power level (active sensors)
	   Waveform or processing mode (long-range search, high resolution, NCTR, *et cetera*)
	   Scan, track, or track-while-scan
	• Sensor processing parameters:
	   Decision thresholds
	   Detection, track, ID criteria

Spatial Control Functions	• Pointing coordinates (center of FOV)
	• Field of view selection
	• Scan/search rate
	• Scan/search pattern select
	• Parameters to control individual sectors:
	   Sector coordinates
	   Modes within sector
	• Parameters for designated targets:
	   Target or track index (no.)
	   Coordinates or search volume
	   Mode to be used
	   Predicted time of appearance
	   Dwell time on target

Temporal Control Functions (timing)	• Start/stop times for modes, sector control • Specified sensor look time • Specified dwell time on target, search • Maximum permissible emission duty cycle

Reporting Control	• Report filters based on target attributes:
	   Friend, foe, or both
	   Filter by class or type
	   Filter by lethality
	• Report filters based on spatial attributes:
	   Min/max range limits
	   Altitude layer filters
	   Spatial region filters (sectors, volumes defined by geometry)
	• Priority of designated targets (by index)

throughout the period of acquisition, identification, and engagement. As track-while-scan processing and electronically scanned capabilities have been introduced, ranking of track and search tasks was required, as more than one target could be serviced concurrently. In multiple sensor systems, often using track-while-scan sensors, the relative values of viewing each target and searching for new targets must be quantified to rank and select sensor-to-target assignments.

5.3.1 Establishing Priority Autonomously

We consider first the problem of setting target priority as viewed by a central data fusion system or by a single, autonomous multisensor platform. Several factors form the basis for setting target priority prior to the assignment of sensor resources to them. These factors include

- *Identity*—The identity of military targets in terms of allegiance (friend, foe, neutral) and target type (e.g., fighter, bomber, missile-carrier, surveillance, transport) is a critical factor in determining priority, as is the lack of identification (unknown targets). Unknown and identified or presumed hostile targets (foes) may pose a threat to be defended against or an opportunity for engagement by offensive weapon systems.
- *Information need*—The need for additional sensor information to establish, refine, or update location and identification of detected targets is another factor in establishing priorities. The state of knowledge about any target can be used to determine what additional sensor data may be required to resolve ambiguities, refine measurement accuracies, or update data on dynamic targets. A number of quantitative measures of information state may be used:
  1. State covariance data or other score functions that characterize the accuracy of state estimates in position location and tracking systems.
  2. *A posteriori* probabilities or likelihood ratio measures that quantify the confidence in soft-decision identification systems.
  3. Entropy functions [8] that measure the gain in identification information between the entropy with only *a priori* data and the entropy with current measurements.
- *Threat*—The degree of threat posed to friendly forces (or the host weapon's platform on which the fusion system resides) by each target is a *defensive* factor in setting target priority. Threat is usually measured in terms of opportunity (a measure of the time-space window over which one's own ship becomes vulnerable to detection and engagement), lethality (a measure of the ability of the target to engage or destroy), and imminency (time remaining until the target, proceeding on its nominal course, will achieve some lethal value). These values may be determined by inference of the target's possible or probable intent from its current or past behavior.
- *Opportunity*—This *offensive* factor is analogous to the threat factor, with the own-ship target roles reversed. The factor is a measure of ability to engage and destroy a target with own-ship weapons. For missile systems, this is determined by target geometry relative to the missile launch envelope that permits prediction of kill probability as the measure of opportunity.
- *Fire control need*—If sensors used by the data fusion system also perform fire control functions, the priority of these operations must be considered relative

to surveillance needs. If the fire control sensor (e.g., radar) is needed to illuminate targets or provide a data link for command-guided missiles in-flight, for example, this activity may override all surveillance needs until the missile has reached autonomous end-game or until a kill-assessment has been completed.

These factors may be computed numerically by the sensor manager for each target, and a composite priority determined to rank targets for allocation. In addition, manual inputs from human operators may also be incorporated to weight targets or overide computed priorities.

Table 5.2 summarizes many of the specific parameters that may be used to quantify the factors in each of the five categories. The factors in the table can be

Table 5.2
Factors in Setting Target Priority

Priority Factor Categories	Specific Factors & Criteria
Identity	• Target allegiance (friend, foe, or neutral) • Target type or class • Target lethality

Information Need	• Spatial location accuracy
	• Target identification status
	• Track state estimate accuracy
	• Detected targets:
	  Track filter covariances
	  Influences revisit rates required to achieve a specified tracking accuracy
	• Search volumes, sectors:
	  Scan or dwell period requirements to achieve a specified detection/intercept probability

Threat (defensive factor relative to hostile, unknown presumed hostile targets)	• Target type identity • Range to target (R) • Range rate ($\dot{R}$) • Range/range rate (time-to-go or imminence) • Target lethality (a function of target type) • Geometry of own ship relative to target's weapon envelope(s)
Opportunity (offensive factor relative to positive identified hostile targets)	• Geometry of target relative to own (or other friendly) weapon envelopes • Range relative to own weapon envelopes • Time-to-go until target reaches detection/engagement point • Probability of own ship detection by target
Fire Control Needs	• State of sensor/weapon commitments against target (lock, track, in-flight, *et cetera*) • Time-to-go for command-guided weapons in flight

used individually or combined numerically to provide quantitative values by which targets may be ranked.

Figure 5.2 illustrates one such ranking of fourteen targets into four major categories:

- Category 1 targets are hostile targets that have been engaged with missiles in flight. These targets require sensor support for command-guidance and kill assessment. Such targets remain a top priority until (1) the weapon achieves autonomous guidance, (2) the predicted impact time plus a margin, or (3) a kill assessment is completed.
- Category 2 targets are hostile targets against which missiles are scheduled to be launched.
- Category 3 includes targets that have been identified as hostile, as well as unknowns (which can be considered to be potential or presumed hostiles). These targets are ranked according to a function of potential threat (T), opportunity (P), and information need (I).

Priority	Category	Target Rank	Prioritization Factors				Ranking Critera
			ID	T	P	I	
1	Hostiles with missiles in flight	1	H	—	—	.7	Targets are ranked by command guidance and kill assessment information need (I_m)
		2	H	—	—	.5	
2	Hostiles with missiles scheduled for engagement	3	H	.7	.8		Targets are ranked by opportunity (P), then by launch information need (I_l)
		4	H	.5	.7		
3	Confirmed hostiles and unknowns	5	H	.6	.9	.6	Targets are ranked by ID, then threat (T), then opportunity (P), then by sensor information need (I_s)
		6	H	.5	.5	.6	
		7	H	.3	.2	.7	
		8	U	.7	.9	.6	
		9	U	.7	.2	.6	
		10	U	.5	.4	.5	
		11	U	.2	.3	.5	
4	Friends	12	F	—	—	.6	Targets are ranked by information need (I)
		13	F	—	—	.5	
		14	F	—	—	.3	

NOTE: H-hostile, F-friendly, U-unknown

Figure 5.2 Example: setting priorities of aircraft targets by multiple factors.

- Category 4 is the lowest priority because it includes targets that have been positively identified as friends and that pose no threat. These targets are ranked by information need (I), to maintain sufficient sensor coverage to maintain tracks.

Note that in this simple example, the T, P, and I quantities are maintained, as appropriate, for each target. Within each category, the computation of priority may be a different function of these values. The combination of criteria can be performed by algebraic expressions, Bayesian inferencing, or other combination methods described in Chapter 7. Baldwin [9], for example, has proposed an approach using fuzzy variables to account for uncertainties in the criteria for ranking aircraft targets.

5.3.2 Establishing Priority Cooperatively

When two or more multisensor platforms view the same target complex, cooperation between platforms is required to *jointly* rank the targets and allocate sensors. Taber [10] has described one approach to the tactical counterair cooperative ranking problem. In this case, multiple blue (friendly) fighter aircraft (i) with overlapping sensor and weapons coverages mutually rank red (hostile) target tracks (j) to optimize combined target coverage and capability to engage targets. The cooperative ranking and assignment must consider the autonomous priorities of each fighter-track pairing as well as the relative geometry of each pairing and the availability of missiles on each fighter.

The algorithm flow illustrated in Figure 5.3 requires the following steps be performed, on the common track file data base available aboard each fighter:

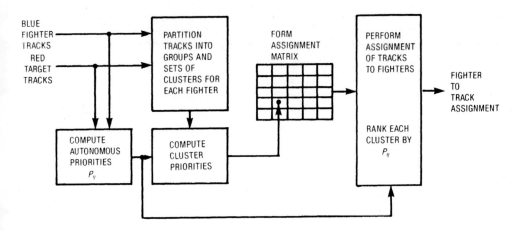

Figure 5.3 Cooperative ranking and assignment algorithm flow (after [10]).

1. An autonomous priority, P_{ij}, is computed for each fighter-track pair, based upon an initial priority, as modified by four factors according to the following steps:
   a. *Target class*—A predefined priority is stored for all target classes. In the defensive counterair mission, for example, bombers may have higher mission priority than fighters.
   b. *Attack status*—Targets against which missiles are committed are raised in priority by a factor that accounts for the status of the missile mode (e.g., command guidance required, autonomous but requiring kill assessment).
   c. *Potential threat*—Hostile targets whose missile launch envelopes (MLEs) contain blue fighters are also raised in priority as function of the threat to own forces.
   d. *Fighter-target geometry*—Each priority is finally multiplied by a geometric coefficient that can be a function of range, range over range rate, or altitude.
   e. *Manual designation*—The pilot may designate any track as highest priority or may lower the priority of any track.
2. The tracks are grouped into n distinct geometric sets (spatial sectors) on the basis of their location relative to the location of the blue fighter computing the cooperative priorities. This simplifies the clustering process that follows.
3. Within each of the n geometric sets, targets are formed into clusters. For each blue fighter, a set of clusters is formed that includes all targets and is mutually exclusive. The clustering process is constrained in two characteristics: (1) number of targets in a cluster is limited to the number of remaining missiles available by a blue fighter (i) that may be assigned to engage the cluster, and (2) angular spacing between any two targets in a cluster is limited.
4. The set of clusters for each blue fighter is then individually ranked by computing a combined priority that is the sum of all priorities for those pairings in each cluster.
5. An $i \times i$ matrix of fighter/cluster priorities is formed and an assignment algorithm chooses the assignment that achieves the greatest benefit (measured as the sum of all cluster priorities for a given assignment of fighters to target clusters).
6. Finally, the targets assigned to each fighter are ranked in order of their autonomous priorities.

The primary function of this algorithm is to set target priority for weapon-to-target assignment, rather than sensor-to-target assignment. Therefore, the priority factors were related to weapons rather than sensors. Both autonomous and cooperative priority algorithms provide a means of quantifying the relative value of targets. This priority value, although based upon many factors, provides a single convenient parameter for assigning sensors or weapons to targets.

5.4 SENSOR-TO-TARGET ASSIGNMENT METHODS

The process of assigning sensors to targets requires a means of solving the general transportation or assignment problems of operations research [11], which may be solved using a variety of linear optimization methods. The assignment methods may be used in two similar sensor management applications. First, m similar sensors (e.g., radars in a network) can be assigned to n targets in some optimal sense. In this case, the optimization criteria is based on coverage of all targets to minimize the loss of detection (or "leakage") of any target. Second, m dissimilar sensors can be assigned to n targets to optimize some objective function based on one or a combination of the factors, such as,

1. Total coverage of all targets or surveillance volume (with targets and search regions each having unique revisit rates to achieve track maintenance or new target detection performance levels, respectively).
2. Continuity or state accuracy of track estimates.
3. Control of sensor emissions.
4. Classification of targets or reduction of classification ambiguity.

The general sensor assignment problem can be solved by the methods of linear programming when the assignment criteria are established to meet the following criteria: (1) a well-defined *linear* objective function must describe the value to be minimized or cost to be maximized for each possible set of mutually exclusive assignments; (2) linear constraints to sensor uses must be defined mathematically; and (3) sensor resources must be finite and economically quantifiable.

The methods of formulating and solving the assignment problem are developed here by progressing from an intuitive example where $m = n$ (one-on-one assignment) and the optimal assignment is made by exhaustive consideration of all alternatives, to the general formulation that considers $m \neq n$ and includes practical sensor constraints, while permitting single *or* sensor combination assignments. This will show how the problem becomes comutationally intensive, even for small suites of sensors, requiring the use of efficient algorithms developed for this class of problems. In all cases the basic problem is formulated, for n sensors and m targets, with the construction of an $n \times m$ assignment matrix, in which each element of the matrix is a value that quantifies the value (or cost) of that single sensor-to-target assignment. A *joint* objective function of those elements (e.g., the sum of elements for each possible set of assignments) can then be computed to choose an assignment that optimizes the objective.

We begin with a simple example in which the problem is the assignment of three antiballistic missile radars ($m = 3$) to three ($n = 3$) multiple, independently targeted reentry vehicles [12]. In this application, the predicted detection and tracking performance for each radar-warhead pair is the basis of optimization. In its simplest form, the problem is formulated as a square matrix problem with m sensors and n targets, in which each sensor must be paired with but one target based upon some optimum pairing criteria.

Typical performance measures for pairing include a composite, weighted function of sensor-target slant range, predicted signal strength, or predicted tracking coverage if the target continues on present trajectory. The measures are quantitatively computed in this example such that lower values are associated with better performance. The $m \times n$ measures, a_{ij}, are organized in a square matrix of real numbers, as in this 3×3 example:

$$a_{ij} = \begin{bmatrix} 15 & 12 & 13 \\ 16 & 18 & 9 \\ 15 & 11 & 8 \end{bmatrix} \quad (5.1)$$

The set of permutations, p (p_i, $i = 1, 2, \ldots, n$) of the integers $1, 2, \ldots, n$ defines all possible means by which the matrix elements may be chosen to select sets of n pairs at a time. The overall objective function or measure of performance, M, for any set of n pairs is then given as

$$M = \sum_{i=1}^{n} d_{ip_i} \quad (5.2)$$

The set of permutations for the 3×3 example may be exhaustively computed as follows:

P	Permutation Sequence	Measure of Performance, M
1	123	$d_{11} + d_{22} + d_{33} = 41$
2	132	$d_{11} + d_{23} + d_{32} = 35$
3	213	$d_{12} + d_{21} + d_{33} = 36$
4	231	$d_{12} + d_{23} + d_{31} = 36$
5	312	$d_{13} + d_{21} + d_{32} = 40$
6	321	$d_{13} + d_{22} + d_{31} = 46$

In this example, the permutation sequence $p = 2$ provides the minimum value of M, and it is the solution that results in the following assignment:

Coefficient	Sensor-to-Target	
d_{11}	S_1	T_1
d_{23}	S_2	T_3
d_{32}	S_3	T_2

We now extend this simple example to the more general application where targets exceed the number of sensors, (i.e., one-on-one assignment is not possible), sensors can be assigned to more than one target, and combinations of sensors can be assigned to targets. The basic example just illustrated provided for assignments of sensor-target pairs to ensure that each target would be covered by at least one sensor. In practical applications, however, it is reasonable to expect that combinations of sensors should be considered for assignment to certain high priority targets for tracking or identification. Nash [13] has formulated the structure of this more general assignment approach that considers the use of multiple sensors against individual targets as well as the constraints of sensor capacity and required target coverage (i.e., each target must be viewed by at least one sensor). The approach uses an augmented assignment matrix that includes all single sensor-target pairings as well as all possible sensor combinations that can be assigned to each target. The augmented matrix, shown in Figure 5.4, includes cost (or utility) values c_{ij} for each assignment and identifies the maximum track capacity (a_i) for each sensor, i. Compared to the single-sensor matrix, in the augmented matrix the sensor dimension is expanded from n to n' to account for the sensor combinations:

$$n' = \sum_{i=1}^{n} \binom{k}{i} = 2_n - 1 \tag{5.3}$$

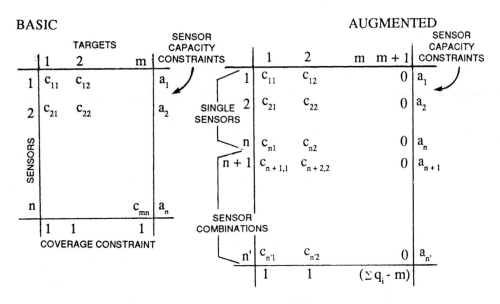

Figure 5.4 Augmented sensor-target assignment matrix (after [13]).

In addition, a zero-cost target row is appended (slack variables) to aid the assignment process to develop initial feasible solutions and combined sensor capacities are defined as the minimum capacity of the sensors used in the combination. In this case, candidate solutions can be viewed as a matrix of decision variables, x_{ij} (0 = no assignment, 1 = assignment) that meet the target coverage and track capacity constraints, while maximizing the total overall objective function:

$$M = \sum_{i=1}^{n'} \sum_{j=1}^{m+1} c_{ij} x_{ij} \tag{5.4}$$

The constraints are stated as a function of the decision variable matrix:

1. Maximum track capacity constraint (each row sum must equal the track capacity of the sensor):

$$\sum_{j=1}^{m+1} x_{ij} = a_i \quad i = 1, \cdots, n' \tag{5.5}$$

2. Target coverage constraint (each column must ensure that a single sensor or combination of sensors is assigned to each target):

$$\sum_{i=1}^{n'} x_{ij} = 1 \quad j = 1, \cdots, m, m+1 \tag{5.6}$$

The objective function for assignment can be a composite function of the factors previously enumerated in Table 5.2. Parametric values for each of those factors provide measures of utility or cost that can be combined (in a weighted product) to form a single objective function to be optimized for assignment. Blackman [14] has described the use of marginal utility as the assignment objective function. *Utility* is defined as the ratio of current performance (in the refence application it is tracking performance, measured as estimation error standard deviation) to desired performance. The desired performance is a standard criteria established by the data fusion system. *Marginal utility* is defined as the difference between current utility and the *expected* utility that would be achieved if the sensor is allocated to the target. Computation of expected utility requires a model of sensor performance to predict performance if the sensor is assigned to each candidate target.

In the first example, the small size of the matrix made it practical to compute all n^2 permuatations to choose the optimum assignment. In most real applications of the assignment problem, where n is large, the exhaustive computation of permutations (proportional to $n!$) becomes prohibitive, and direct solutions have been developed to reduce the computation required to solve for optimal pairing. A number of general linear programming algorithms have been developed to more efficiently perform the matrix operations involved in the soultion of the assignment (or transportation) problem. The Munkres Algorithm [15] is one method that reduces the solution to a process proportional to n^3 by sequentially reducing the assignment matrix to "equivalent" matrices that have the same optimal assignment. The process results in successive reductions until an independent set of n zero values is reached in the equivalent matrix: this matrix then provides the optimal assignment directly.

The Munkres algorithm has been extended to rectangular matrices [16] to accommodate realistic sensor-to-target assignments in which m and n are not equal. To further speed computation of the algorithm, Marshall [17] has shown that parallel implementations of the algorithm can take advantage of parallelism to achieve execution times proportional to n^2. The earlier algorithms of Kuhn (the "Hungarian" method [18]) and Ford-Fulkerson [19] are also applicable to the assignment problem, although most published applications have focused on the Munkres approach. Bier and Rothman have applied the greedy service algorithm [20] and proposed the application of more general search algorithms to sensor assignment applications.

5.5 SENSOR CUEING AND HAND-OFF

The management of sensors may also require that different sensors cooperate to acquire measurements on a common target. The two primary cooperative functions are cueing and hand-off. *Cueing* is the process of using the detection (or track) from one sensor (A) to point another sensor (B) toward the same target or event. *Hand-off* occurs when sensor A has cued sensor B for transferring surveillance or fire control responsibility from A to B.

Cueing and hand-off are required for a number of operational reasons including the following conditions:

1. A target may be passing out of the coverage (spatial or temporal) of one sensor (A) of a network and it may be possible to cue other sensors ($B, C, \ldots$), into the coverage of which the target will appear, to pick up and resume the track.
2. A sensor (A) may detect a target and provide only limited spatial measurements (e.g., passive sensors with bearing-only data) or identification information (e.g., detection-only data). In this case cueing another sensor (B) to acquire and collect data on the target may be necessary to refine spatial or identification data.
3. A passive sensor (A) may intercept bearing-only data from an emitter and may cue another, remotely located sensor (B) to collect additional bearing data for cooperatively locating (e.g., Stansfield analysis, triangulation) the tar-

get. In this case, the cueing sensor may provide specific data to the cued sensor to aid the search (emitter attributes, timing, predicted search bearings for sensor *B, et cetera*).

4. A primary weapon system sensor (*A*) may detect and identify a target to be engaged by a weapon with an autonomous seeker, which must acquire and lock-on to the target. In this case the target attributes (image, emitter characteristics, *et cetera*) as well as location or sensor pointing angles must be transferred from *A* to the sensor (*B*) in the weapon's seeker. In such cases, the seeker's sensor usually has lower performance (resolution, sensitivity, *et cetera*) than the primary sensor.

Two processes must occur for cueing or hand-off: the cueing sensor must provide the cued sensor data that contains sufficient information to point to the target and identify it as the specific target being cued, and the cued sensor must search for the target of interest and verify that it has been acquired.

Haskins [21] developed a cueing performance analysis methodology for a typical ESM-to-radar cueing operation where both sensors have rectangular fields of view. In this case, the larger FOV ESM sensor 1 detects a target and attempts to cue the smaller FOV radar (sensor 2) to the target (Figure 5.5). The analysis of such cueing operations must consider the speed with which the cued sensor can acquire and lock onto the detected target, and the probability that the target can, in fact, be detected. (Factors that can prevent cued detection include a target out of the cued sensor's detection range or target motion that moves the target out of the search FOV prior to detection by the cued sensor.) Haskins's analysis illustrates the

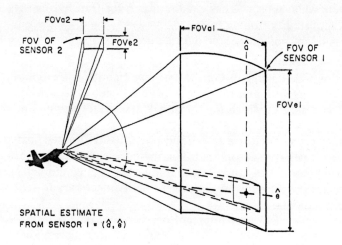

Figure 5.5 Typical airborne radar-to-ESM sensor cueing geometry (from [21]).

necessary procedures to evaluate feasability and performance of cueing operations. He assumes that the target is within range, target motion relative to the search is negligible, and the target estimate error due to motion during the period of slewing the cued sensor to the search FOV can be predicted from target and own-ship velocity vectors. Haskins's analysis was performed to determine the performance for two approaches summarized in the following paragraphs.

Static Cueing Performance

In the simplest case, the small FOV sensor is simply cued to the center of the cueing sensor's larger FOV. In this situation, the probability of correct cueing, $P(Q)$, at that single location is computed as a function of FOVs given the error statistics of sensor 1. Figure 5.6 shows the cueing probability as a function of sensor 1 azimuth and elevation error for the condition where both azimuth and elevation estimates are provided by sensor 1, to cue sensor 2. Three cases were considered.

Sensor	FOVs	Case 1	Case 2	Case 3
1	Azimuth			30° for all three cases
	Elevation			60° for all three cases
2	Azimuth	2°	5°	0.5°
	Elevation	2°	5°	0.5°

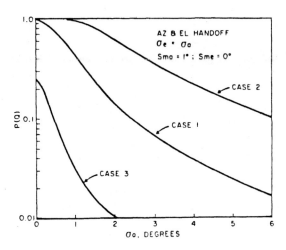

Figure 5.6 Static cueing performance (from [21]).

The figure illustrates that simple static cueing provides generally unacceptable performance (i.e., $P(Q)$ below 0.9) except where the cueing sensor's estimation error is comparable to the FOV of the cued sensor. As the estimation error increases, the probability of detecting the target at the center of the sensor FOV rapidly diminishes. For this reason, a dynamic search of the entire FOV is required to achieve a satisfactory $P(Q)$.

Dynamic Cueing Performance

To achieve an acceptable $P(Q)$, then, a spatial search of the sensor 1 FOV is performed by sensor 2 using a predetermined search pattern chosen on the basis of error statistics. The analysis is performed by computing the static $P(Q)$ at each step of the search and computing the cumulative $P(Q)$ as the search continues. Figure 5.7 shows the percent of sensor 1 FOV that must be searched to attain a $P(Q) \geq 0.99$ as a function of sensor 1 azimuth and elevation error for the three cases.

More general search problems described in the operations research literature (see, for example, Washburn [22] for the analysis of typical military search problems) can also be applied to the cueing and hand-off process. In addition to search, dense target environments may require the cued sensor to verify that a detected target is, indeed, the target that has been cued or handed off. This verification may be performed using track state characteristics or target attribute data to compare the detected target with the target previously designated for cueing.

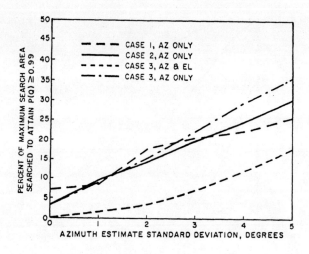

Figure 5.7 Dynamic cueing performance (from [21]).

5.6 SENSOR MANAGEMENT APPLICATIONS

To illustrate the application of the concepts introduced in this chapter, two military examples will be described to show the practical situations that develop requirements for sensor management.

5.6.1 An Air-Combat Sensor Management Example

The air-combat example shown in Figure 5.8 illustrates the typical sensor management problems presented by autonomous multiple sensor systems. The fighter (own-ship) is assigned to a defensive counterair mission whose objective is to detect, track, identify, and engage multiple hostile aircraft penetrating friendly airspace. Primary targets are bombers, with escort fighters secondary, unless threatening own-ship or other friendly aircraft. The tactical aircraft in this example [1] is equipped with a suite of five sensors:

- *Radar with NCTR*—The radar is an electronically scanned array on a mechanically positioned mount. Azimuth coverage (FOV) of the array is ± 45°, and the mount can be positioned to permit coverage over a full *field of regard* (FOR) of +60° (azimuth) and ±25° (elevation). Several operating modes are provided: (1) the high-PRF mode permits long-range detection of targets by doppler sorting methods, providing velocity measurements without

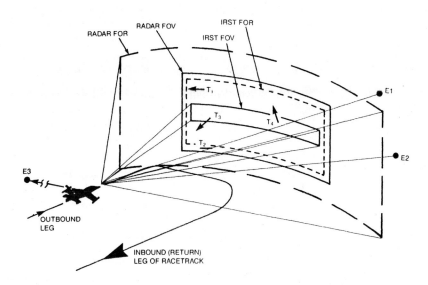

Figure 5.8 Air combat sensor management scenario.

range; (2) the medium-PRF pulsed-doppler mode provides accurate range, azimuth, elevation detection, and track-while-scan operation in three selectable frame patterns; (3) the noncooperative target recognition mode requires a long-duration dwell on target to collect signature data to classify airframe type; (4) the high resolution mode also requires a dwell on the target to resolve multiple, closely spaced aircraft.

- *Omnidirectional ESM*—Full 360° azimuth coverage ($\pm 80°$ vertical aperture) is provided with a low-accuracy ($\pm 20°$) angle of arrival measurement to emitters.
- *Directional ESM*—A high-accuracy forward-looking ESM covers the forward 180° ($\pm 45°$ vertical aperture), providing a $\pm 5°$ angle of arrival accuracy. This sensor must be cued to steer an electronically scanned array to form a beam at any particular azimuth position for collecting emitter data.
- *IFF interrogator*—An electronically scanned IFF antenna may be scanned or positioned at any azimuth for interrogating a target to determine if the target can respond to secure challenge messages, authenticating its identity as a high-confidence friend. The IFF FOR is $\pm 60°$ (azimuth) and the vertical aperture is $\pm 45°$.
- *IR search and track*—The IRST mechanically scans an 8° vertical swath over an azimuth of $\pm 45°$. The vertical swath can be positioned to permit coverage over a full FOR of $\pm 20°$ (elevation) and $\pm 45°$ (azimuth).

In the example shown in Figure 5.8, the scenario presents a number of sensor inputs to the data fusion process, upon which the sensor manager must act. The omnidirectional ESM detects three emitters: E1 at 18° (an emitter that is ambiguous between neutral and hostile aircraft), E2 at +43° (a known hostile fire control radar), and E3 at 210° (a friend). Only the first emitter is within the directional ESM sensor FOV. The IRST and radar are concurrently tracking three target complexes:

- T_2 is a complex of two unidentified targets at a range of 32 nmi and azimuth 22° that are at the right edges of the FOVs of both the radar and the IRST. The targets are on course to leave the FOVs within 15 sec, unless they are repositioned to the right.
- T_3 is a target complex inbound at Mach 2, range 39 nmi, and azimuth $-27°$. The target is unidentified, and IRST is unable to determine if the complex is a single target or multiple targets.
- T_4 is an inbound complex of multiple targets (number not resolved by IRST) at 65 nmi, azimuth $\pm 17.5°$. Identity is unknown and targets are climbing and increasing in velocity.

The radar is tracking a single target, T_1, at $-22°$ azimuth, range 73 nmi. The target is at the left edge of the radar FOV and on a course that will cause it to depart from radar coverage within 35 sec. Its course will take it into restricted airspace within 180 sec, making it a potential threat if it is a hostile bomber with stand-off weapons.

Target complexes T_2 and T_3 have been interrogated by IFF and have not replied to challenges. This may indicate that they are hostile. Target T_1 has been challenged, but a strong jamming strobe at that azimuth prevented validation that replies are being returned.

This example illustrates several conflicting demands that are placed on the sensor manager to optimize information to the pilot for effective mission performance. The sensor manager is faced with the following demands and issues:

1. Emitter E1 may be from targets T_2 or T_4, or from a different target beyond the range of the IRST or radar. The emitter may be a hostile threat, and therefore it is important to determine if is associated with T_2 (which would place one's own ship close to T_2's launch envelope) or with T_4, which is much farther in range. This situation calls for the use of the directional ESM sensor to acquire and track E1 emissions to an accuracy that would permit association with T_2 or T_4, if possible.
2. Because E1's identity is ambiguous (hostile or neutral), it is desirable to use the radar NCTR mode to resolve the ambiguity. If T_2 can be positively identified as hostile, it can be engaged with medium-range weapons.
3. Emitter E2 poses a potential threat to one's own ship because it is a fire control radar of a hostile fighter at an azimuth outside of the radar or IRST's FOVs. Range is not known. This threat demands a repositioning of the IRST or radar FOVs to detect the target elevation and range and determine if it poses an imminent threat.
4. Target complex T_4 is climbing and will require a repositioning of the IRST FOR or a change of the radar scan pattern to maintain the targets in track. Identities are unknown, and we want the targets to be interrogated with IFF to determine if they are a returning friendly strike package.
5. In addition, because one's own ship is flying at Mach 1 along an air combat patrol racetrack course on the outbound leg (toward the threat axis or forward edge of the battle area), flight control indicates that 175 sec remain in the outbound leg of the racetrack. At that point the aircraft reaches the turning waypoint and must turn 180° to begin the inbound leg. All tracks and identifications must be handed off to the trailing aircraft via data link at that point so it can acquire, associate, and continue tracking.
6. The missile launch envelope prediction for target complex T_2 and target T_3 indicate that, if they remain on course, either may be engaged with medium-range missiles within the remaining outbound leg time if launched within 120 sec of the turning waypoint. This will permit sufficient time for midcourse guidance update to provide a 0.95 kill probability (P_k) for the final autonomous flight phase. This means that identification of positively hostile aircraft

must be achieved within 55 sec if the beyond visual range engagement is to be carried out.

In summary, conflicting demands exist in several areas, and operational considerations must rank by priority the demands posed by targets:

- E2 demands immediate IRST, radar repositioning, or course change.
- T_2 requires radar repositioning and dwell time on complex to identify.
- Directional ESM is needed to search for E2 as well as E1 to refine emitter azimuth-elevation data to support or cue IRST or radar sensors.
- T_1 is a potential high-value threat to friendly ground forces and continued tracking for hand-off requires repositioning of the radar (or IRST). The required repositioning is in direct conflict with the demands of E2 and T_2.
- T_3 is a potential multiple target complex and demands radar dwell time for a high-resolution assessment as well as for NCTR to determine identity.

These issues require a management function that considers operational restrictions (doctrine), the value or threat of each target immediacy, and the need for sensor data on each target, as well as physical limitations of sensors to select among limited sensor resources. The function must also apply manual (pilot) inputs and overrides, including sensor controls and flight controls that govern sensor coverage as the flight path is changed.

5.6.2 A Distributed Sensor Network Example

Next, consider the three-site network of ground sensors depicted in Figure 5.9, whose mission is air traffic control of friendly aircraft and detection and tracking of hostile aircraft entering their joint surveillance volume. The sensor sites are located within the national border shown as the dashed line and coastline on the terrain grid. Characteristics of the sensors at the three sites are

1. Site 1 (S1) installation includes a continuously scanning (mechanically scanned) surveillance radar with IFF to interrogate friendly aircraft for identification. The circular coverage of the scanning fan-beam antennas are illustrated in the figure. Electronic support measures passive sensors include an omnidirectional emitter sensor (low angular resolution) and a directional, high angular resolution sensor that requires an antenna mechanically pointed in the direction of signal arrival.
2. Site 2 (S2) includes a phased-array radar with IFF that has 160° coverage as depicted in the figure. Coverage overlaps with that of S1 and S2 radars. A high resolution ESM sensor is also located at this site.
3. Site 3 (S3) includes the same sensor suite as S1.

The surveillance tasks assigned to the three sites include maintenance of tracks for all aircraft within the combined surveillance volume, control of military

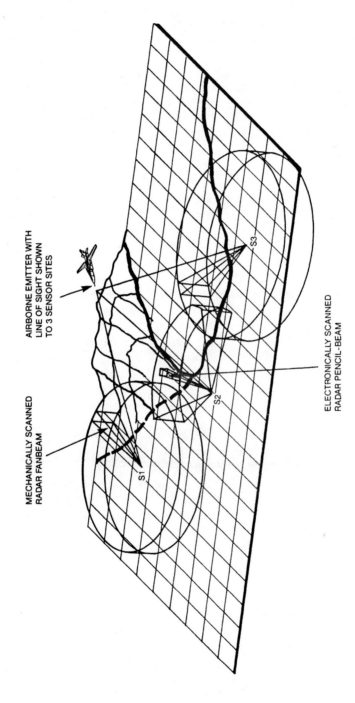

Figure 5.9 Distributed sensor network management example.

air traffic, detection and identification of hostile targets that do not respond to IFF, and location of hostile emitters beyond the surveillance volume using multiple-site geometry, with high-resolution ESM sensors. These tasks require a coordination of several functions among the three sites to utilize the sensor resources.

Site surveillance is by continuous radar scanning of each surveillance volume and identification of aircraft, commercial and military, using IFF. Commercial aircraft codes are compared with flight plan data supplied by the aviation authority. This autonomous function is performed independently by each site.

Site-to-site target hand-off is required when any site determines that a target is passing out of its own surveillance volume into that of another. In this case, the originating site predicts the time at which the target will reach the edge of the receiving site's volume. The receiving site is cued to the look angle at which it will acquire the target and, once detection or track is confirmed, the receiving site takes control of the track. For friendly targets, IFF aircraft codes are used to confirm that the proper track has been handed-off. For unidentified targets, hand-off confirmation is based on comparison of kinematic state estimates of both sites during the time that both sites are tracking the target in the overlap region.

Emitter cueing is performed whenever any site detects a potential hostile emitter with its omidirectional ESM sensor. The *angle of arrival* (AOA), *emitter identifier* (EI), and time of arrival are forwarded to the other sites for association on the basis of EI and approximate TOA (which will be different for each site). The course AOAs for two or more sensors associating the same emission can be used to derive a coarse location, and the directional ESM sensors can be cued along the predicted AOAs for each site to listen for subsequent emissions. Two or three sites can track a continuous or frequent emitter using passive location techniques based on coordinated AOA and TOA measurements.

The electronically scanned radar between the two mechanically scanned sites provides flexibility to handle targets between S1 and S3, which cannot adjust their target revisit times and therefore require little management. (In this case, the only management of these radars includes sector control of emission or PRF to minimize interference or maximize detection performance.) The S2 radar, therefore, requires the most management, sequencing between surveillance scan, hand-off, and tracking tasks. Targets in track for hand-off to S1 (or S3) must be projected forward to the predicted S1 (or S3) radar scan-by times. Coordination of ESM sensors requires setting priorities among detected emitters to allocate the three directional sensor antennas to only the higest priority threats. These sensors can track only one target at a time, and this priority must be traded off against the localization of newer threats. The objective of all ESM tracking is to pass on the tracks to radar sensors for sustained tracking.

These functions must be performed concurrently and as the target density increases, the apparent importance of sensor management increases. Of particular note in this example is the necessity to set emitter input priorities to assign the very limited directional ESM sensors. The hand-off from ESM tracks to radar sensors must occur as efficiently as possible to free up the ESM sensors. The example also illustrates the inherent flexibility of electronically scanned sensors over mechanically scanned systems. If an electronically steered directional ESM sensor were used, several emitters could be serviced concurrently by looking at each target only during the predicted pulse repetition intervals and interleaving the tracking of multiple targets with search operations, as with the radar at S2.

REFERENCES

1. Bier, Steven G., and P. Rothman, "Intelligent Sensor Management for Beyond Visual Range Air-to-Air Combat," *Proc. NAECON 1988*, IEEE, pp. 264–270.
2. Rothman, P., and S. Bier, "Intelligent Sensor Management Systems for Tactical Aircraft," paper presented at SPIE Second National Symp. on Sensor and Data Fusion, 1989.
3. Leon, Barbara D., and Paul R. Heller, "An Expert System and Simulation Approach for Sensor Management and Control in a Distributed Surveillance Network," *Proc. SPIE* Vol. 786, *Applications of Artificial Intelligence V,* 1978, pp. 41–50.
4. Addison, Edwin R., and Barbara D. Leon, "A Blackboard Architecture for Cooperating Expert System to Manage a Distributed Sensor Network," *Proc. Data Fusion Symp. 1987,* pp. 669–675.
5. Slagel, J.R., and H. Hamburger, "An Expert System for a Resource Allocation Problem," *Comm. of the ACM,* Vol. 28 No. 9, September 1985, pp. 994–1004.
6. Hui, Patrick J., "EXNAV: An Intelligent Sensor Processor," *Proc. NAECON 1988,* IEEE, pp. 1214–1219.
7. Cowan, Rosa A., "Improved Tracking and Data Fusion through Sensor Management and Control," *Proc. Data Fusion Symp. 1987,* pp. 661–665.
8. Pugh, G.E., D.F. Noble, "An Information Fusion System for Wargaming and Information Warfare Applications," Decision Science Applications Rep. 314, August 1981, AD-A106391.
9. Baldwin, M.J., "The Pilot's Associate: Ranking Targets in a Multi-Target Environment," *Proc. NAECON 1978,* Vol. 4, IEEE, pp. 1360–1366.
10. Taber, Norma J., "Concepts for Beyond Visual Range Engagement of Multiple Targets," *Proc. NAECON 1983,* IEEE, pp. 434–451.
11. Marty, K., *Linear and Combinatorial Programming,* John Wiley and Sons, New York, 1976.
12. Silver, R., "An Algorithm for the Assignment Problem," *Comm. Assoc. Computing Machinery,* Vol. 3, November 1960, pp. 605–607.
13. Nash, Jeffrey, "Optimal Allocation of Tracking Resources," *Proc. IEEE Conf. on Decision and Control,* 1977, pp. 1177–1180.
14. Blackman, Samuel S., *Multiple-Target Tracking with Radar Applications,* Artech House, Norwood MA, 1986, pp. 387–391.
15. Munkres, James, "Algorithms for the Assignment and Transportation Problems," *J. Soc. Indust. Applied Math.,* Vol. 5, No. 1, March 1957, pp. 32–38.
16. Bourgeois, Francois, "An Extension of the Munkres Algorithm for the Assignment Problem to Rectangular Matrices," *Comm. Assoc. Computing Machinery,* Vol. 14, No. 12, December 1971, pp. 802–805.
17. Marshall, Duane D., "The Multisensor Data Correlation and Handover Problem," *Proc. IEEE Parallel Proc. Conf.,* August 21–24, 1979, p. 147.

18. Kuhn, H.W., "The Hungarian Method for the Assignment Problem," *Naval Research Logistics Quarterly,* Vol. 2, 1955, pp. 83–97.
19. Ford, L.R., Jr., and D.R. Fulkerson, "Notes on Linear Programming—Part XXXII: Solving the Transportation Problem," *Management Science,* Vol. 3, 1956, pp. 24–32.
20. Rothman, P., and S. Bier, *op. cit.* [1, 2].
21. Haskins, Thomas G., "Sensor Cueing Performance Analysis," *Proc. NAECON,* IEEE, 1984, pp. 262–265.
22. Washburn, Alan R., "Search and Detection," Operations Research Society of America, Ketron, Arlington, VA, 1981.

Chapter 6

DATA FUSION FOR STATE ESTIMATION

The fusion of multiple measurements to derive target position or kinematic state information is required when single or multiple sensors provide time-sampled information on dynamic targets or when multiple sensors provide independent measurements of a common target. Two essential processes are involved in the derivation of position or kinematic information: data association and state estimation.

Data association is the process of relating individual sensor measurements (data) to other measurements to determine if they have a common source (e.g., target or event). Although the measurements may be referenced to different coordinate systems with different views of the source, at different times, and with different spatial accuracies or resolutions, the association process must relate each measurement to a number of possible sets of data, each representing a hypothesis to explain the source of the measurement:

- The false alarm set, indicating that the measurement is unreal and to be ignored.
- The new target set, indicating that the measurement is real and relates to a target for which there are no previous measurements.
- An existing set of previous measurements related to a single target. A set exists for each previously detected target.

Association is implemented as an m-ary decision (correlation) process that quantifies the spatial (and attribute) relationships between observations and predicted target states to determine which of m hypotheses best describes the observation. Two kinds of decision processes can be performed: a hard decision (an "assignment" to one and only one set) or a soft decision (permitting the data to be assigned to multiple sets, with each candidate assignment having a measure of uncertainty). The soft decision results in multiple hypotheses that may be combined into a single hypothesis or deferred for subsequent hard decisions when uncertainty is reduced by additional data. The association may be based upon spatial measurements (e.g., position, kinematic state), nonspatial target attributes (e.g., emitter characteristics, past behavior, signature features), or both.

State estimation is the process of mathematically determining an estimate of the state (e.g., position, velocity) of a static or dynamic target based on measurements related to the state. Each discrete sensor measurement is often referred to as an *observation,* although the terms *report, detection,* or *hit* are also used interchangeably in the literature. The measurements used in the estimation process are those *assigned* to a common target or track as a result of the association process. The batch mathematical model of the time history of a dynamic target or a discrete-time (recursive) math model is referred to as the *track.* This model of behavior uses previous observations related to target state to estimate the temporal equations of state, permitting extrapolation forward in time to predict the state at which future observations may be expected to be observed.

A large and diverse set of association, estimation, and tracking approaches have been developed and reported in the literature, resulting in surveys [1–3] and texts [4–6] that categorize and describe the various algorithms. In this chapter, we provide an overview of the data association and tracking problem, the many unique application issues in data fusion systems, and the principal mathematical methods used to associate sensor data for deriving target position and kinematic information.

6.1 ASSOCIATION OF DATA AND TRACKING OF DYNAMIC TARGETS

The requirements for association and estimation for target tracking are as diverse as the characteristics of the sensors and sources used to collect data and the uses of the resulting target data. In this section, we provide an overview of the different categories of applications that require association and tracking. The development of classical aircraft target tracking technology [4, 5] has followed the technological development of radar sensors and increased capacity of digital signal processing. Radar developments have provided ever-increasing performance in terms of revisit rate (sampling rate), spatial resolution, and detection performance. In addition, the development of improved passive sensors, such as sonar [6], electronic support measures, and IR sensors, has stressed developing association methods to improve passive ranging and localization capabilities. The development of high-resolution imaging sensors has also promoted research in two-dimensional tracking algorithms for applications such as battlefield target tracking within FLIR (forward-looking infrared) imagery [7] and reentry vehicle tracking against stellar backgrounds.

6.1.1 Static Data Association for Target Localization

Sensor measurements from stationary targets or events, as well as intermittent measurements (detections of opportunity) from dynamic targets, require the association of individual data measurements (reports) that have no model to represent temporal behavior (no track). This class of problems is usually referred to as *direction finding* (association in one or two angular dimensions or lines of bearing) or *localization* (association of multiple lines of bearing to derive a region of space within which the target exists with a specified probability). Figure 6.1 illustrates the primary categories of static association applications.

Multiple Sensor, Common Dimensionality

In this case, the spatial measurements from similar or dissimilar sensors located at a common site have a common number of dimensions. Measurements may require conversion to a common coordinate and time base, and association is made in a single spatial decision space. The following table shows examples.

Sensors	Coordinates	Association Space Dimensionality
Radar-to-radar	Range-AZ-EL	3
IRST-to-precision ESM	(AZ, EL)	2
ESM-to-ESM	AZ or bearing only	1
Sonar-to-sonar	AZ or bearing only	1

Multiple Sensor, Different Dimensionality

Association may also be required between sensors whose measurements have different numbers of spatial measuring dimensions. This requires the association metric to account for uncertainty, not only in the measurement process in the common dimensions, but also in the dimensions that are not common to the sensors. Common examples are shown in the following table.

Sensors	Coordinates	Association Dimensionality
Radar-to-IFF	(Range, AZ, EL), (range, AZ)	3, 2
Radar-to-IRST	(Range, AZ, EL), (AZ, EL)	3, 2
Radar-to-radar	(Range, AZ, EL), (AZ, EL)	3, 2
IRST-to-ESM	(AZ, EL), (AZ)	2, 1

Multiple Site, Multiple Sensor

When multiple sensors are spatially separated, measurements of angle only, frequency only (for doppler techniques), or time of arrival only can be used to derive

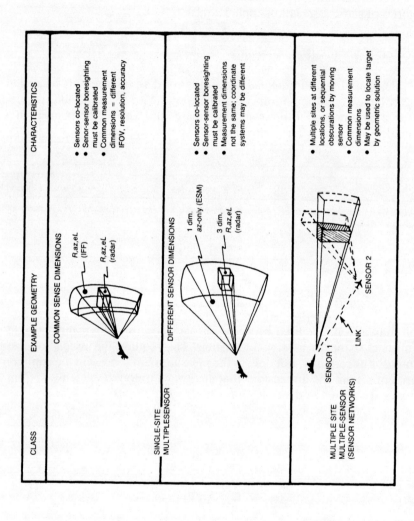

Figure 6.1 Classes of static association and localization applications.

target state data using geometric techniques. The reports from different sites must be associated to determine that they have a common source prior to locating the source with geometric solutions. These techniques, known as *localization, passive ranging,* or *fixing solutions,* can use single-event measurements from multiple sites on a known time or space baseline. Alternatively, this process can be performed by a single sensor that makes sequential measurements as it moves along a known baseline.

Some target classes (e.g., surface-to-air missile batteries) are known to be relatively stationary once deployed, and association without dynamic tracking can be applied directly. In other cases, frequency of sensor data cannot support tracking, but intermittent association is required when measurements can be made from multiple sensors.

6.1.2 Dynamic Data Association and Target Tracking

Dynamic targets require continuous or discrete, time-sampled measurements of the target location and an ability to estimate the kinematic behavior of the target to predict future positions for continued sensor coverage. This process requires the iterative association of each new set of sensor data with predicted locations of known target tracks to determine which sensor detections are current tracks, new targets, or false alarms. Four broad areas of application are usually distinguished, as depicted in Figure 6.2.

Single-Sensor, Single-Target Tracking (STT)

The most basic tracking application is for a single sensor to track the motion of a single target, as in a single-target tracking fire control radar. In this case, the process is focused on a single target, whose state is continuously updated, and the predictions are used to reposition the sensor to track the target motion. The single-target assumption greatly simplifies the process by eliminating the need for complex assignment logic: each detection is assumed to be from the single target or a false alarm.

Single-Sensor Multiple-Target Tracking (MTT)

The addition of multiple targets complicates the association problem by requiring each detection to be labeled as one of the existing tracks, a new track, or a false alarm. As the target complex becomes dense and targets cross, split, or remain close together (relative to the sensor resolution), the assignment process becomes complex, limiting confidence in assignments and introducing potentially severe position

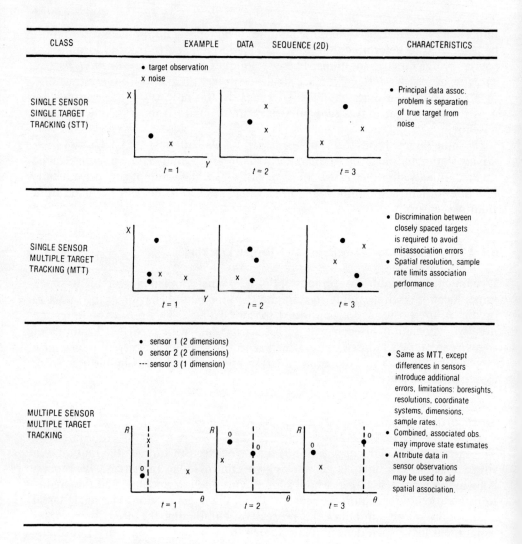

Figure 6.2 Classes of dynamic data association and target tracking applications.

estimation errors due to misassociation. This is the critical difference between STT and MTT systems, which requires more complex association processes for most MTT applications. Sequentially framed imaging sensors form a special case of target tracking applications in which targets must be detected and dynamically tracked in the two-dimensional imagery from frame to frame. Autonomous acquisition sensors, missile seekers, and celestial telescopes fall into this application area, where targets (as well as the background in the case of the star field for celestial searches) must be tracked from frame to frame. The emphasis in these applications is association of the target image segment from frame to frame as the target may be changing scale, perspective, or intensity over time.

Multisensor, Multiple-Target Tracking (MTT)

The most complex mutliple-target tracking problem includes multiple sensors with different target viewing angles, measurement geometries, accuracies, resolutions, and fields of view. Differences in any of these sensor characteristics further complicate the problem of associating measurements, although inherent attribute data in the different sensor observations may aid the association process by considering parameters beyond spatial state. In such cases, closely spaced targets may be accurately associated using nonspatial attribute data (e.g., target discriminants) rather than relying solely on spatial data association.

6.1.3 The Roles of Association and Estimation

Association and state estimation are the two fundamental functions integrated in different ways in each of the tracking algorithms described in this chapter. Classical recursive automatic target tracking systems, for example, place the data association and state estimation processes as separate, distinct functions in a closed-loop feedback system whereas other batch and recursive algorithms more tightly integrate the association and estimation functions. To introduce these functions, the classical recursive tracking loop is described here. Later, in Sections 6.2 and 6.4 we describe association algorithms, and in Section 6.6 we provide a more detailed introduction to the principles of state estimation. Throughout this chapter, the discussion of association and estimation are necessarily intertwined and this brief introduction serves to help the unfamiliar reader to distinguish between their respective functions. Figure 6.3 illustrates a recursive tracking loop that automatically translates time-sampled sensor measurements into sets of false alarms, new detections, and tracks. The target state, **x**, is a vector of components that fully describe the kinematic state of each target. The sensors do not directly measure **x**, but some set of observation variables represented by **y**. This observation vector, **y**, is directly related (usually geometrically) to **x** by the observation matrix, **A**. Each observation vector

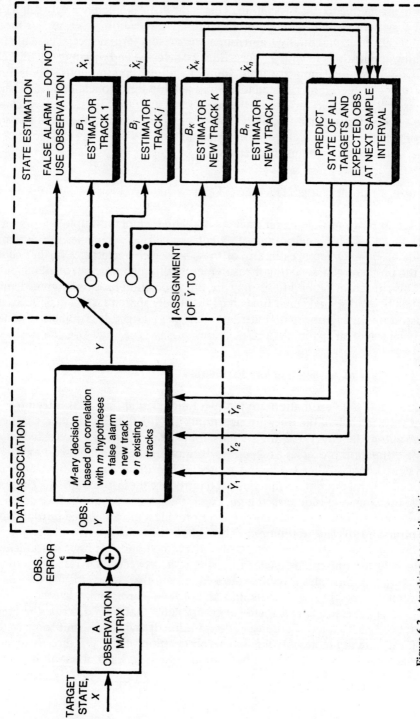

Figure 6.3 A tracking loop with distinct association and state estimation functions.

from a detected target is provided as input data to the tracking loop for association prior to estimation of state.

Data association between each sensor observation and existing tracks precedes an assignment of that observation to one or more of the following data sets: (1) an existing track or tracks for estimation updating (track "maintenance" or "continuation"); (2) a new target set to start a new track (track "initiation"); or (3) the false alarm set (in which case the data is deleted from further processing). Sets of measurements associated with each target are then used to derive a new or updated state estimate. The association function performs an m-ary decision (correlation) process, comparing each incoming observation with criteria for n existing tracks, a "new track detection" criterion, and a false alarm criteria to determine which assignment to make. The correlation criteria can be based on spatial and kinematic similarity as well as the similarity of other measured target attributes. The introductory principles of m-ary detection described earlier in Chapter 4, Section 4.1, apply to this correlation process.

State estimation is performed subsequent to association, when sets of observations have been assigned to distinct target sets. In the simple model of Figure 6.3, the set of assigned observation vectors associated with a single target track (k), $\mathbf{y}_k$, is each entered into an estimation matrix, $\mathbf{B}_k$, which estimates the state of track k. These state estimates are used to predict the future state of each track at the next observation period, and they are then fed back for subsequent association with future observations. The estimator matrix is developed to optimize some criteria, using all of the available knowledge of the target state, dynamic limitations, and measurement statistics, to achieve the "best" estimate of the true target state. The rich history of the development of estimation theory was summarized in Chapter 1, Section 1.2, and the principal estimators used in this chapter are summarized in Section 6.6.

6.1.4 Issues in Association and Tracking

The association of data and estimation of target dynamics is complicated by a number of factors that limit the performance of data fusion in a number of ways. These factors include
1. Number of targets and target densities;
2. Sensor detection performance; (expressed by ROC, described in Section 4.1.3);
3. Target revisit rate and target dynamics;
4. Sensor measurement accuracy and process noise;
5. Sensor or model biases;
6. Background noise sources (e.g., star fields, ground clutter, ocean noise);
7. State estimator performance.

The design of association and tracking algorithms attempts to use the sensor measurement data in the most efficient method possible, within processing constraints, to minimize misassociation and state estimation criteria. In the next section, we provide alternative methods used to implement systems for the wide range of association and tracking applications.

6.2 STATIC DATA ASSOCIATION AND TARGET POSITION LOCATION

The basic association problem considers the targets to be stationary (in actuality or at least at the instant of measurement). Under these static conditions, the reports are associated to each other for the three general categories of applications previously described in Section 6.1. In this section we summarize association-localization approaches for each case.

6.2.1 Multiple Sensor, Common Dimensionality

The most straightforward static association requirement is between measurements from similar or dissimilar sensors that have a common dimensionality. The principal approaches include using "distance" measures to quantify the "closeness" of the sensor measurements for association. These measures have the following uses:

1. Ranking the relative distance of candidate pairs (or n-tuples) of measurements to select the most likely associated sets by a decision rule.
2. Establishing a hypothesis test (or gate) criteria to determine whether the measurements are of two association decisions: H_0 (not associated) or H_1 (associated).

Association tests are performed by using spatial distance measures, statistical distance measures, and geometric or statistical hypotheses [8, 9].

Spatial Distance Measure

The simplest geometric measure is the magnitude of the vector difference between observations, expressed as vectors **x** and **y**:

$$d = |\mathbf{x} - \mathbf{y}| \tag{6.1}$$

This measure considers each observation location to be a point in space without error in measurement and does not take into account the individual measurement characteristics of each sensor.

Statistical Distance Measure

The measurement error statistics of individual observations can be accounted for when a means is provided to normalize the spatial distance between observations by the relative variance of the measurements. The generalized statistical distance measure between **x** and **y** [8] is given by the quadratic form:

$$d^2 = \mathbf{A}^T \mathbf{S}^{-1} \mathbf{A} \tag{6.2}$$

where

$\mathbf{A} = (\mathbf{x} - \mathbf{y})$, vector difference between observations

$$\mathbf{S} = \begin{bmatrix} \sigma_{11}^2 & \sigma_{12}^2 & & \\ \sigma_{21}^2 & \sigma_{22}^2 & & \\ & & \ddots & \\ & & & \sigma_{nn}^2 \end{bmatrix} \quad \text{matrix of variances for the observations } \mathbf{x} \text{ and } \mathbf{y}$$

and when the errors are Gaussian, the probability density function of **A** is given by

$$f(A) = \frac{e^{-d^2/2}}{(2\pi)^{M/2}\sqrt{|S|}} \tag{6.3}$$

where M = the measurement dimensionality.

Hypothesis Test (Gate)

Instead of computing a distance function between observations, as earlier, a hypothesis test can be developed to determine if the observations are associated within a specified decision confidence level. Figure 6.4 illustrates two hypothesis tests where distances are computed and tested against constants to determine H_0 and H_1. If the measurement error distributions are Gaussian, the statistical test can be considered to be a Chi square test [9] in which the statistical distance becomes a Chi square (χ^2) statistic. The test determines the equality of the means of the two distributions, **x** and **y**. The value of χ^2 can be chosen from standard statistical tables (using the measurement dimension, M, to define the number of degrees of freedom) to establish the probability that the distributions are equivalent.

Each of these approaches can be applied to the iterative association tests performed in dynamic data association, described in subsequent sections of this chap-

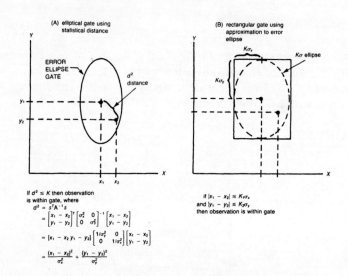

Figure 6.4 Typical association hypothesis (gate) tests in two dimensions.

ter. In fact, dynamic association processes simply apply these kinds of association tests for gating decisions to select "neighboring" observations to predicted target locations and to quantify the "goodness" of each candidate neighboring observation to (predicted) observation pairing. Scores may then be derived for competing hypotheses that partition the sets of pairings into mutually exclusive assignments of all observations from one sensor to the observations from another sensor. These methods are further described in the Section 6.3 for dynamic data association.

6.2.2 Multiple Sensor, Different Dimensionality

The association process is significantly complicated when attempting to associate measurements from sensors whose measurements have different numbers of spatial measuring dimensions. A common example is radar (three dimensions) to ESM (one dimension: line of bearing) association. The radar measures a point in space, whereas the ESM sensor measures a volume, often defined as a wedge-shaped "fan-beam" volume.

In these cases, the spatial association must be made in the common dimensions of measurement and, where possible, extension to a dynamic association problem may be required for satisfactory performance. In the radar-ESM example, static association by azimuth is generally considered unreliable because the ESM volume is so large that misassociation probabilities will be unacceptable in even low target densities. Temporal tracking of the ESM line of bearing and the radar target position has been explored by Coleman [10] and Trunk and Wilson [11] to achieve acceptable association using a time sequence of measurements. Figure 6.5 shows the Bayesian multiple hypothesis testing process developed by Trunk and Wilson, which develops an *a posteriori* association probability for each ESM track to a given radar track:

1. For a given set of ESM data, $\theta_e(t_i)$ sampled at the times t_i ($i = 1, 2, \ldots, n$), each of $j = 1, 2, \ldots, m$ radar tracks are smoothed and the radar angles θ_j are estimated at the same times, t_i.
2. A one-dimensional (angle-only) generalized distance measure d_j^2 is computed for each set of corresponding radar and ESM angle measurements (at common time samples).

$$d_j^2 = \sum_{i=1}^{ni} [\theta_e(t_i) - \theta_j(t_i)]^2/\sigma^2 \qquad (6.4)$$

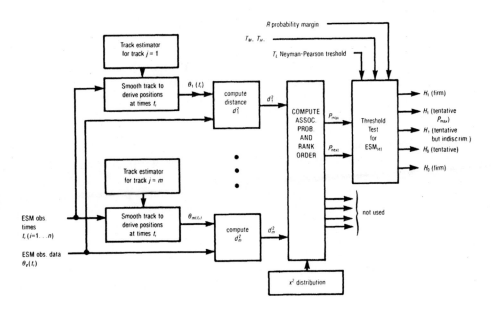

Figure 6.5 Functional flow of ESM-radar association algorithm (from [11]).

3. The distribution of the distance measure is chi-square because the angle measurements are assumed to be Gaussian. This permits a decision rule to be based upon *a posteriori* probabilities. A probability of association is computed for each possible radar track to the ESM track using the chi-square distribution (with n_j degrees of freedom) to apply a hypothesis test. The probability is given by

$$P_j = P_r[z \geq d_j], \text{ where } z \text{ is } \chi^2(n_j) \qquad (6.5)$$

4. The association probabilities are ranked and a decision threshold ratio test is applied to the largest (P_{max}) and next-to-largest (P_{next}) values to make the assignment. A lower threshold based on the Neyman-Pearson criteria establishes the value below which the ESM signal is determined to be unassignable to any radar track. High and medium thresholds and a probability margin are also used to establish three levels for the H_1 decision and two levels of the H_0 decison.

6.2.3 Multiple-Site, Multiple-Sensor

When mutliple sensors are spatially separated, angular or temporal measurements can be used to derive target position data using geometric techniques. The most common applications include ESM (classical radio direction finding) and sonar sensors that passively sense target emissions or reflections to derive target location. The reports from different sites must be associated to determine that they have a common source prior to searching for the source through geometric solutions. These techniques, known as *passive ranging* or *position fixing solutions*, can use single-event measurements from multiple sites on a known time or space baseline. Alternatively, this process can be performed by a single sensor making sequential measurements as it moves along a known baseline.

These techniques are based upon well-known navigation principles developed to derive own-ship position from multiple transmitting sites (e.g., LORAN, Omega) at known locations. We summarize here the principal methods of associating multiple-site data to localize targets, providing references to the literature where detailed derivations may be found. The principal characteristics of this class of problems include the following:

- A geometric solution exists in which the observation measurements (bearing only, range only, time or range difference only) and known sensor locations can jointly solve for a target location. Each observation results in some *surface of position* (SOP) in space in which the target must lie. The intersection of SOPs establishes the location of the target.
- Location errors can be attributed to uncertainty in sensor locations as well as measurement noise, which is generally assumed to be zero mean, white gaus-

sian noise. Systematic or bias measurement errors have been considered [12] for specific sensors in which this is a significant factor. In addition, real target motion (the target is generally assumed to be stationary in these problems) and sensor coordinate conversions may introduce errors to the process.
- As a result of these uncertainties, the multiple noisy observations do not result in a single intersection of SOPs, providing a region of uncertainty within which the target is located.
- A characteristic of these solutions is the dependence of location error upon the measurement geometry. This *geometric dilution of precision* (GDOP) is generally quantified as the ratio of root mean square (rms) position error to rms range or angle error. Although this property limits the locating effectiveness for certain geometries, Kelly [13] has shown that biased estimators (e.g., ridge regression techniques) can be used to reduce the total mean square error beyond that achievable by conventional unbiased estimators.
- A numerical estimator is applied to reduce the effects of noise and minimize the resulting location error. Typical estimators applied to the problem include linear and nonlinear least squares [14] and maximum likelihood [15] for each batch processing of data sets and Kalman filters [16] for sequential data.

Figure 6.6 illustrates the most common position location geometries and the SOPs described by the geometric functions, $f(x)$. *Direction of arrival* (DOA) or bearings-only measurements result in *lines of position* (LOPs) along which the target must lie, without range along any given measurement. Triangulation and circulation [17] geometric solutions are used to solve for target position. The intersections of these LOPs, in the presence of noise, provide a region referred to as the *cocked hat* in which the target may lie. The emphasis of early analyses by Stansfield [18] was to determine the probability of the true target being within this region. Statistical estimators, on the other hand, use all measurements and estimates of error statistics to derive the most likely target location independent of the cocked hat geometry. In Section 6.6 we describe a maximum likelihood estimator that linearizes the inherently nonlinear geometric functions in these categories of problems.

Time difference of arrival (TDOA) or range-difference measurements for a common signal detected at two sites provides hyperbolic SOPs that have the two stations as foci. The figure illustrates the requirement for three stations to derive the intersection of two hyperbolas, assuming sensors and target lie in the same plane. In this case, note that two intersections are possible, but *a priori* information is often used to eliminate one solution (e.g., general knowledge of hostile forces).

Mixed-mode systems may combine DOA, TDOA, or other measurements (e.g., doppler data [19], range-only data) to enhance the estimate by using the best available data. Wax [20] has described a unified maximum likelihood formulation of the problem that accepts AOA, TDOA, range-only, or range-sum data from a single site or multiple sites.

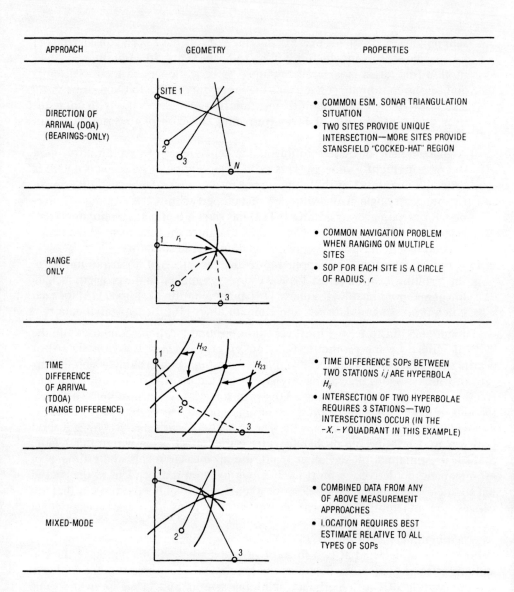

Figure 6.6 Representative two-dimensional, multiple-site position location applications.

6.3 TAXONOMY OF DYNAMIC DATA ASSOCIATION AND TRACKING ALGORITHMS

A large and diverse set of mathematical approaches and algorithms are applicable to the problem of associating data and dynamic tracking targets. In this section, we introduce the primary association-tracking functions and the alternative approaches that face the designer. The following sections will then describe the predominant algorithms used to perform data association and tracking.

6.3.1 A General Association and Tracking Loop

Figure 6.7 illustrates the basic functions in the most general recursive data association and target tracking loop. The inputs to the loop are sensor reports, also called *detections, observations,* or *measurements*. The following functions must be performed to develop the file of target tracks at the output of the loop.

Spatial Alignment

Incoming sensor data is aligned by a transformation into a common spatial reference for association with other measurements in the target data base or track file. Two categories of data are stored in the file:

1. Reports are individual measurements that have been determined valid but have not been associated with previous measurements to initiate a track.
2. Tracks are models of the motion of targets derived from sets of previous sequential measurements that have been associated with each other.

Prediction or Temporal Alignment

Each track must first be aligned in time with the current sensor measurement, to account for the difference in the time of the last estimate to the time of the incoming sensor measurement. For tracks, this requires propagation forward in time using the estimate of state (target kinematic derivatives and position) to predict the state at the time of the new sensor measurement.

Candidate Selection

The location of the spatially aligned sensor measurement and its attribute data are used to select from the target data base all detections or tracks "adjacent" to the measurement as candidates for association. This adjacency criteria eliminates an

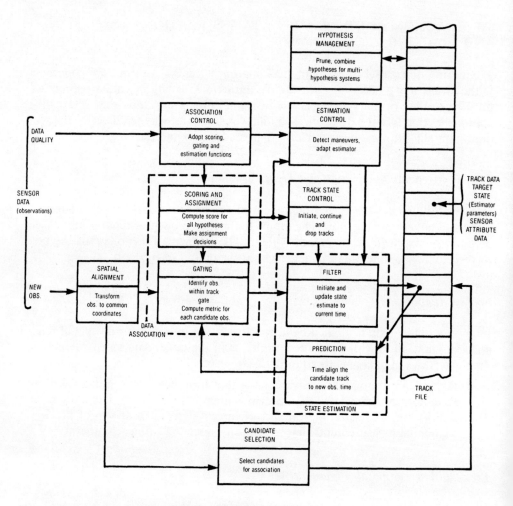

Figure 6.7 General recursive association and tracking functions.

exhaustive association of each new measurement with all detections and tracks and focuses the process on the most reasonable candidate associations.

Gating

The association process then compares each time-aligned candidate track with the new sensor measurement. An association "gate" is placed about each predicted target state to determine if the new measurement falls within a spatial (and possibly attribute) criteria to declare it a "neighbor" of the existing data. The gate for target

T_i is a region of measurement space within which the current sensor measurement is expected to fall if its source is, indeed, T_i. The gates are validation regions, unique to each detection or track in the track file. The data association process must then determine the optimal way to assign observations that fall both within and outside of the gates to the appropriate existing track or to initiate a new track. Several gating conditions must be considered:

1. Some observations may fall outside of all gates ("outliers") and be determined to be noise (false alarms), new track starts, or existing tracks that have changed behavior (e.g., target maneuvering) sufficiently to exceed the predicted gating region developed by the tracking model.
2. Some gates may contain a single observation ("singleton"), which is generally considered to be the expected subsequent observation for the gate's track; it is usually assigned to that track.
3. Some gates may include more than one observation, forcing a selection of the single "best" observation or a means to consider all the observations within the gate.
4. Where gates overlap due to close spacing of targets, observations may fall within the overlap region, causing a conflict in the candidacy of the observation between multiple tracks.

Figure 6.8 illustrates two-dimensional association gates for report-report and report-track associations. The figure illustrates how the gate must consider the variances for both the estimator and measurements and may require compromises to model the true shape of the error envelope of measurements (e.g., rectangular representation of elliptical error distributions).

Scoring and Assignment

For each existing track or detection for which the new observation is a neighbor (i.e., it falls within the association gate), an association metric based upon location, kinematics, measured target attributes, or other parameters is computed. The metrics for individual candidate pairing are then combined into a *score* for each possible way in which mutually exclusive sets of pairings can be taken jointly to explain the observations from sensors. The scores for each of these hypotheses are then used to make one of several possible assignments of the new sensor data:

- *False alarm*—The measurement is assumed to not be a real target, and it is deleted from further processing.
- *New detection (track initiation)*—The measurement cannot be associated with any current target, and it is assumed to be the first detection (or Nth if multiple, sequential detections are required to confirm a track) of a new target not previously seen. It is stored to determine if subsequent measurements can confirm a new track.

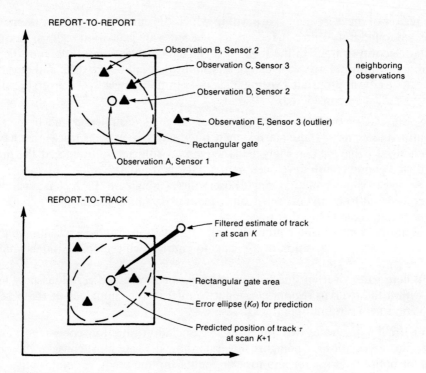

Figure 6.8 Typical two-dimensional association gates.

- *Track confirmation*—The measurement is associated with sufficient previous detections to assume that the set belongs to a target: an estimate of motion is initiated and the resulting estimate is stored as a new track.
- *Track continuation*—The measurement is associated with a current track, and can be used to update the estimate of that track.

Track Filtering or Estimation

For each pairing of a new measurement with an existing detection (causing a track initiation) or an existing track (causing a track continuation), the mathematical estimate of state is recomputed. The estimator may be an iterative or batch model that performs two functions: smoothing of a sequence of data measurements to minimize some error criterion (e.g., mean square prediction error compared to future measurements); and predicting future position of the target for ment with subsequent measurements. Several additional functions m to support the processes of this tracking loop.

Association Control

The measurement variance, time between measurements, or other factors may be used to adjust the scoring process or the dimensions of the association gate. This permits adaptation to changing data quality with time or control of a general association process that must accommodate observations from various sensors.

Estimator Control

Real changes in the track behavior outside of the range of steady state behavior in the track estimator model (e.g., target maneuvers) must be detected to adjust the model parameters to follow the transient behavior.

Track State Control

The results of scoring and assignment are used to control the state of target tracks. These control functions include initiation (detection), continuation, and deletion. Figure 6.9 illustrates a simple state transition diagram to provide control for four track states.

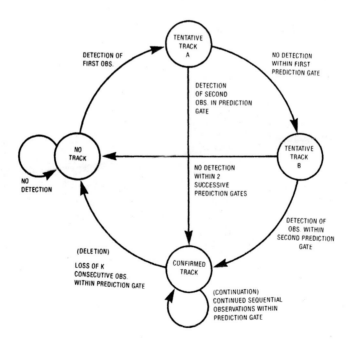

Figure 6.9 Track state transition diagram.

- *No track*—Observation is declared to be false alarm.
- *Tentative track A*—A single observation is detected.
- *Tentative track B*—A tentative track failed to be confirmed because an observation did not occur within the first prediction gate.
- *Confirmed track*—Two successive or two-of-three observations have fallen within prediction gates, confirming the sequence of observations to be a track. Subsequent observations with no more than k successive losses continue the track.

The state transition criteria can be statistically chosen to balance the speed with which association takes place and the association performance to minimize misassociation.

Hypothesis Management

In the case of deferred decision multiple hypothesis systems, the ever-growing hypotheses must be managed by pruning or eliminating hypotheses with scores that fall below some criterion of acceptance; combining hypotheses determined to be a common track that has been maintained as separate hypotheses; splitting hypotheses determined to be separate tracks.

6.3.2 Taxonomy of Design Approaches

Figure 6.10 summarizes the taxonomy of major design approaches that must be considered to construct association and tracking systems. The major alternatives presented in the figure are discussed in the following paragraphs.

Search Direction

Most common tracking systems are sensor driven, in that each incoming sensor report causes a search through the track file for data that can be associated with it. Target-driven systems, on the other hand, use a primary sensor for tracking and use the target track to direct other sensors to acquire data or search data bases for reports that can be associated with particular tracks.

Levels of Association and Tracking

The architectural structure used to associate and merge data from multiple sensors into a single track can take on several forms. Figure 6.11 illustrates three basic approaches to combine data from two sensors:

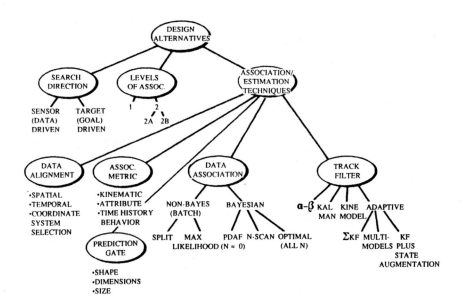

Figure 6.10 Taxonomy of data association and tracking design approaches.

- Single level, or centralized, systems associate all data at a single node, maintaining a single estimator for each track. Association is performed between each new sensor report and the central track file.
- Two-level, or autonomous, systems maintain separate sensor-level and fusion-level trackers. Each sensor-level tracker independently acquires, initiates, continues, and drops tracks. Association is performed between the tracks reported by each sensor tracker.
- Two-level with track-selection, or hybrid, approaches maintain separate sensor-level tracks and perform track-track association as described. Once the association is made between tracks, the sensor reports for those tracks are then used to compute the track estimate or the "best" track is selected as the estimate.

Association-estimation techniques include all of the means to align, associate, and estimate the state of data in the tracking loop. These include the four functions that follow.

Data Alignment

Sensor measurements must be aligned both in time and space. Spatial alignment approaches must consider the measuring coordinate systems of all sensors and the

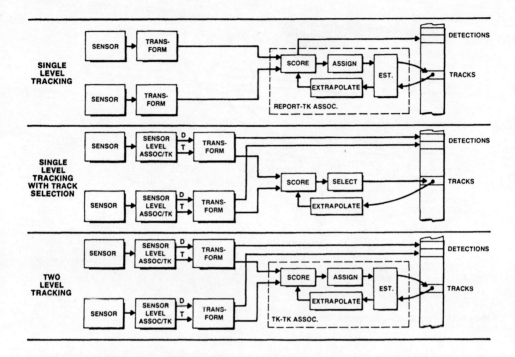

Figure 6.11 Levels of partitioning tracking and data association.

effects of transforming coordinates as well as error volumes (i.e., uncertainty regions about measurements, such as ellipsoids describing gaussian measurement uncertainty in two-dimensional Cartesian space) from system to system. The selection of a common space for association of all measurements influences both the complexity of computation and the accuracy of association.

Prediction Gating

The design of methods for temporal alignment of measurements requires a consideration of prediction performance to size the gates selected for association. Figure 6.12 illustrates various types of static and dynamic gates that may be chosen for various tracking applications. The primary considerations that influence gate selection are based upon the balance between two application-dependent factors: maximum single observation detection probability (tending toward large gates to ensure that subsequent observations are declared neighbors in both steady state and

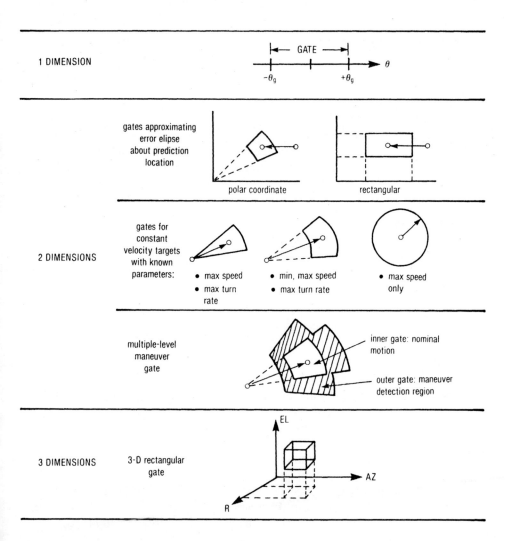

Figure 6.12 Typical association gate characteristics.

maneuvering conditions), and minimum misassociation probability (tending toward small gates to exclude observations from other tracks and noise).

Association Parameters

The selection of spatial parameters (target spatial coordinates and derivatives), attribute parameters (sensed target characteristics such as IFF mode codes or emitter parameters), or both defines a metric space in which observation-track pairing scores are computed. Although the scores are often computed in terms of likelihood or probability, the scores may be based on several quantitative measures:

- Spatial distance or statistical distance measures, which quantify the spatial correlation ("closeness") between observations and predictions, as previously described in Section 7.2.
- *Figure-of-merit* (FOM) metrics based on spatial, statistical, or attribute data or a combination of these.
- Measures that quantify the realism or an observation or track based on *prior* assumptions, such as track lengths, target densities, or target (track) behavior.

Data Association and Assignment

The taxonomy of major methods for performing association of measurements are based on two general alternative means of assigning observations to tracks In the first method, measurement is assumed to be associated with a single "nearest neighbor," based on a metric criterion that quantitatively ranks each of the candidate pairs. This hard decision forces the selection of the single, "best" pairing of the new data and does not account for ambiguity when multiple, closely ranked choices are present. In the second method, measurement is considered to be associated with "all neighbors" that fall within some criterion and one of two methods of handling this multiple-neighbor situation is selected: a deferred, multiple hypothesis or combined, probabilistic data association.

In a deferred multiple hypothesis, each candidate pairing is considered a viable hypothesis and these multiple hypotheses are retained in time until a decision criterion can eliminate or confirm the hypotheses. This permits assignment decisions to be deferred until sufficient information is available to increase confidence in the hypotheses. As successive measurements are made, the set of hypotheses will rapidly grow until tracks (hypotheses) are combined or deleted. The assignment logic that permits multiple hypotheses may permit the following:

1-to-K assignments—Each measurement may be assigned to up to K possible hypotheses.

J-to-K assignments—Multiple measurements (up to *J*) may be assigned to up to *K* hypotheses. In this case, multiple measurements may update the same hypothesis.

In combined, probabilistic data association, each candidate pairing is used to contribute to the update of the track estimator, weighted by a quantitative factor that describes its probability of being the "correct" measurement. In this approach, all neighboring measurements contribute to the track and deferred decision making is not required.

The primary design decisions that influence which of these approaches may be selected include the selection of four processing parameters:

- The number of previous scans of observations used in data association (0; a finite number, n; or all scans). This is the batch or recursive processing decision.
- The selection of maximum likelihood or Bayesian methods for computing and comparing hypothesis scores.
- The use of hard decisions to assign observations to tracks at each recursion or soft-decision methods to combine or maintain multiple hypotheses for deferred hard decisions.
- The method of using multiple neighboring observations that fall within the gate.

Track Estimators

Numerical estimates of true target state are computed by algorithms that use time sequences of associated measurements to develop an estimate of the target state. The estimators include *a priori* models of track dynamics and use observations to refine the estimate of state to minimize some objective function. In addition to this observation filtering function, the estimator can predict the state at the next observation interval for gating. The alternative fixed and adaptive coefficient methods for modeling dynamic tracks are provided in Section 6.5.

6.3.3 Key Design Parameters

The performance of association and tracking systems is dependent upon a large number of factors that influence the two key mathematical processes: association and estimation. Figure 6.13 shows the interrelationship among the sensor and scenario factors and the design parameters that drive the key measures of association and tracking performance:

1. Association decision performance, defined by the correct decision probability (P_{ca}) or misassociation probability ($1 - P_{ca}$).

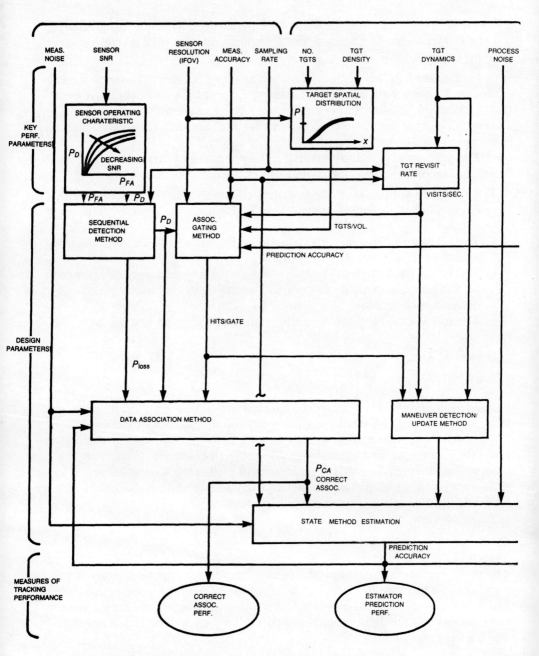

Figure 6.13 Relationships between key application factors and design parameters that influence measures of tracking performance.

2. Estimation accuracy, defined by prediction error measures for each element of the state (x, y, R, θ, etc.).

These parameters are closely interrelated in that erroneous association decisions will contaminate the estimation performance by providing incorrect observation data and, alternatively, poor estimation performance will create inaccurate prediction of new observations, causing association errors. These factors require careful selection of sensor, gating, association, and estimation design parameters to optimize the tracking performance.

As shown in the figure, the sensor's *signal-to-noise ratio* (SNR) and operating characteristic define the detection and false alarm probabilities (P_d and P_{fa}), which determine the sequential track initiation, continuation, and deletion performance. The gating approach (size and shape) as well as numerous sensor and scenario factors determine the number of hits (observations) per prediction gate for any one track:

- Sensor measurement accuracies and resolutions.
- Target revisit rate.
- Single observation detection probabilities.
- Estimator's prediction performance.
- Target spatial distribution (expressed as a density or targets per scan per unit volume for homogeneous distributions of targets).

As multiple hits occur within the gates (primarily due to undersampling or large target densities), the association process is forced to choose among multiple hypotheses (immediately or deferred) and the potential for miscorrelation increases. The association performance directly influences estimator performance. Misassociations degrade the ability to perform proper future associations because erroneous observations are inserted into the estimation process, causing prediction errors to be fed back to the gating process for subsequent observation association. In the case of Kalman filters estimators, misassociation leads to divergence in estimation solutions unless the misassociation errors can be modeled by the covariance matrix [21, 22]. Maneuver detection and adaption methods also influence estimation performance as a function of their ability to detect and adapt to transitions from steady state behavior.

In the next sections we discuss the principal algorithms that implement these functions for report-report, report-track, and track-track association and tracking.

6.4 REPORT-TRACK DATA ASSOCIATION FOR TARGET TRACKING

In this section, we provide an overview of the variety of methods that may be used to associate data for single- and multiple-sensor MTT systems. These algorithms associate sensor observations with other observations (report-report) to initiate tracks as well as with established tracks (report-track) to continue tracking of targets. For detailed derivations of these algorithms, with results of performance analyses in many cases, the reader is referred to the texts by Blackman [4] or Bar-Shalom and Fortmann [5].

Figure 6.14 illustrates the typical data association problem in which a time sequence of sensor observations (measurements) can be interpreted in alternate ways. The figure shows four sensor scans, in which each scan produces a number of observations in the square sensor field of view (the "scene"). The overlay of the four scenes illustrates the sequence of observations (represented by different shapes for each time interval) produced after four scans. Three possible hypotheses are shown in the figure, illustrating the different ways in which observations can be associated. Each hypothesis describes a possible solution to the association of observation data into possible sets that represent feasible time series of sensor observations. The association process must determine which set of track solutions (i.e., sets of association hypotheses) best explains the observed data, based upon some fidelity criterion.

Figure 6.15 illustrates the two primary approaches to implement association processes for sensor observations:

- *Recursive processes* operate on each new set of observations from a single observation interval (scan) of the sensor as they are received (Figure 6.15a).
- *Batch processes* (Figure 6.15b) operate on a time sequence of measurements, all at one time, taking advantage of the time history of all measurements to seek long-term spatial or temporal relationships. These processes require storage of data for delayed, subsequent processing and may not be appropriate for all real-time applications.

The figure illustrates the major alternatives that must be considered in the design of any association algorithm, which are detailed below.

Number of Scans of Data Used (N)

The number of (previous) scans of data used in association can be in one of three possible categories:

$N = 0$, only the current scan is used in recursion;
$N = n$, a finite number of previous scans of data are used to defer decisions by as many as n scans;
$N = \infty$ the complete time series of data is used in association (batch process).

Association Metric

The quantitative metric computed for gating tests and ranking possible association pairings can be chosen from a variety of functions, as shown in Table 6.1. The

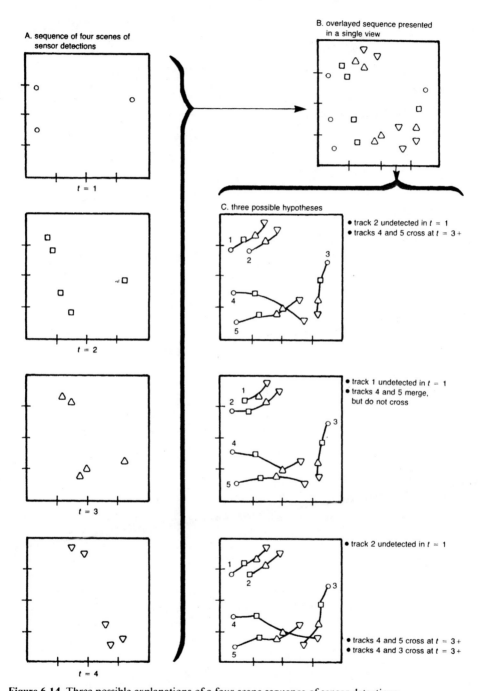

Figure 6.14 Three possible explanations of a four-scene sequence of sensor detections.

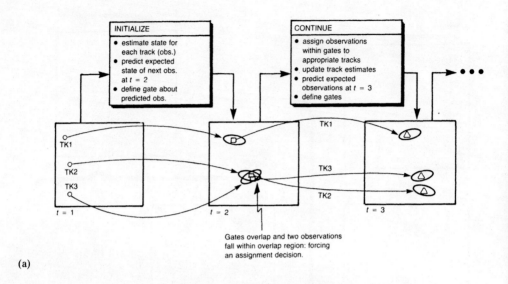

(a)

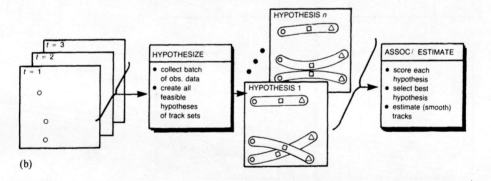

(b)

Figure 6.15 Comparison of recursive (a) and batch (b) association-estimation approaches for tracking.

functions can be based upon spatial parameters or a combination of spatial parameters and target attributes.

Heuristic approaches compute *figure-of-merit* (FOM) values (e.g., weighted sums of variables) based on spatial, temporal (e.g., time since last seen), or attribute variables. The most common applications of such metrics are for coarse gating prior to the application of a more complex scoring function or for association where the attribute or temporal information content is of greater accuracy than the spatial information. An excellent example of an application in which attribute information can supply more information for association than spatial measure-

Table 6.1
Metric and Score Functions

Type	Function	Basis of Value — Spatial Parameters	Basis of Value — Spatial and Attribute Parameters
Observation pairing association metric	Heuristic functions	• Spatial FOM values • Spatial gates	• Weighted spatial and attribute FOMs • Spatial-attribute gates
	Distance functions	• Euclidian distance function • Statistical distance function (variances known)	• Statistical distance functions in which measurement space includes attribute and spatial measurements
Track or scene hypothesis score	Likelihood functions	Likelihood function for each association hypothesis is based on: • Probability of true or false target existences • Probability of track length • Probability of track sequence given length • Probability of residual error given a track sequence	Spatial likelihood function (left) is augmented by: • Probability of attribute measurements given track sequence • Probability of true or false attribute measurement detection
	Posterior probability functions	Bayesian posterior probabilities are computed for each association hypothesis, using prior probabilities for existence of each hypothesis	Spatial and attribute measurement variables are both used to compute probabilities
	Sum of distance functions	Sum of distances for each pairing that makes up each association hypothesis	

ments is the association of secondary surveillance radar data (transponder reply codes) for air traffic control tracking systems. If each aircraft replies with a unique code, the code attribute data can be used to associate data with greater confidence than spatial measurement alone (assuming low error message rates). An example of spatial FOM approaches is provided in Chapter 9.

Statistical distance measures form the basis for most quantitative association metrics. The distance function is measured between a predicted track location and each candidate sensor observation that falls within its association gate. Each possible pairing provides a distance value for comparison and ranking with all others. A large number of distance measures or resemblance coefficients used in cluster analysis may be applicable, depending on the characteristics of the data and the application. The most common statistical measure for spatial distances between predictions and observations is the general (or normalized) measure taking the quadratic form, weighting each observation in inverse proportion to its measurement variance (see Section 6.2).

Hypothesis Scoring

For any collection of existing tracks and new sensor observations (or simply a collection of observations from multiple sensors), a *score* must be developed to rank the possible ways in which the data may be associated into hypotheses. Three general methods for scoring each hypothesis are described in the following paragraphs.

Nonprobabilistic approaches directly compute statistical distances for observation-to-observation or observation-to-track pairings, and the sum of these distances (for the set of all mutually exclusive pairings making up each hypothesis) provides a convenient score.

Non-Bayesian probabilistic approaches compute a likelihood score function based on probabilities of target existence, track length, track sequence, and residual errors (a function of the statistical distances between actual and predicted or smoothed measurements). These approaches are characterized by the assumption that association decisions are correct, and the state estimation process does not account for the possibility that the observations may be incorrectly associated.

Bayesian approaches adopt a score that is an *a posteriori* probability computed from Bayesian inferencing methods, which require *a priori* and conditional probabilities that account for the uncertainty in the association. This characteristic distinguishes the Bayesian approaches; the uncertainty in the origin of observation data (which results from the uncertainty in the association process that assigns observations to tracks) is accounted for in the state update process.

Decision-Making and Hypothesis Maintenance

The number of hypotheses maintained at the end of each recursion can be a single or multiple hypotheses. In single hypothesis association, a hard association decision is made to select a single, unique solution to be carried forward into the next interval. In multiple hypothesis association (also referred to as *branching* or *track splitting*), a soft decision is made by carrying forward more than one possible hypothesis. This defers the hard decision until a future scan when sufficient data is received to delete poor hypotheses and firmly establish good ones.

Use of Observation Data

The observation data used in estimation of the track behavior can be based upon one of several options:

- *Nearest-neighbor*—The single neighboring observation, based upon some "nearness" criteria (e.g., a simple Euclidian distance or statistical distance function) is selected for use in the subsequent recursion of the estimator.
- *All-neighbors, combined*—A combination of the neighboring observations is used to update the next recursion. The combined observation value is a function of the neighboring values, weighted by data that accounts for likelihood that each is the correct contributing observation.
- *All-neighbors, individual*—All neighboring observations are used individually in full multiple hypothesis processes, in which each neighbor is used to create an individual association hypothesis.

The alternative methods for dealing with each of these issues characterize the major approaches developed for data association. Table 6.2 compares the major categories of association algorithms and the design approaches that characterize each.

6.5 TRACK-TRACK DATA ASSOCIATION

Fusion systems that maintain independent sensor-level tracks (i.e., two-level or autonomous systems as described in Section 6.3) require the association of track data from each sensor to merge the multisensor data into a common, central-level track. Sensors are assumed to have individual measurement statistics, described by the covariances of their respective track filters. Because process noise introduced by target behavior is observed by both (all) sensors observing a common track, Bar-Shalom [23] has shown that these covariances may be correlated, and that this correlation must be considered in association metrics. Recursive and batch approaches to this process are discussed next.

6.5.1 Recursive Track-Track Association

Recursive association of track data requires the sequential reporting of track data (state vectors) from each sensor-level tracker to the common association process, as illustrated in Figure 6.16(a). Each sensor-level tracker supplies state vector x_i and covariance P_i data for track i at periodic (scan) intervals. The association and tracking process requires the general functions previously described for recursive report-track association:

- Time and space alignment of the sensor track state vector samples to the central tracks is performed.

Table 6.2
Comparison of Major Data Association Algorithms

Association Algorithm	MAJOR CHARACTERISTICS					REMARKS	MAJOR REFS.
	(1) No. of previous scans used in data assoc.	(2,3) Assoc. metric and hypothesis score	(4) Assoc. decision rule and hypothesis maintenance	(5) Use of neighboring observations in track estimation			
Nearest Neighbor	0 (current scan only)	score is a sum of distance metrics	hard decision	single unique nearest neighbors observation used		• Sequential process • Assoc. matrix contains all pairing metrics	38
Probabilistic Data Association (PDA), Joint PDA (JPDA)		A posteriori probability		all-neighbors (combined) are used		• Tracks assumed to be initiated • PDA for STT, JPDA for MTT • Suitable for dense targets	39 40
Maximum Likelihood	N	likelihood score	soft decision resulting in multiple hypotheses (requiring branching or track splitting)			• Batch process for a set of N scans. In the limit, $N = \infty$ for full scene batch processing • Suitable for initiation	41 42 43
Sequential Bayesian Probabilistic		A posteriori probability or likelihood score		all-neighbors (individually) used in multiple hypotheses each used for independent estimates		• Sequential process with multiple, deferred hypotheses: pruning, combining, clustering is required to limit hypotheses.	44
Optimal Bayesian	∞					• Batch process requires most computation due to consideration of all hypotheses	45

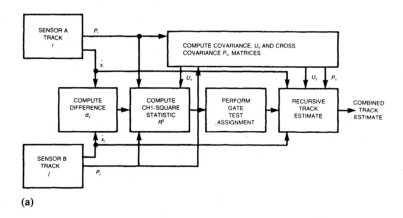

(a)

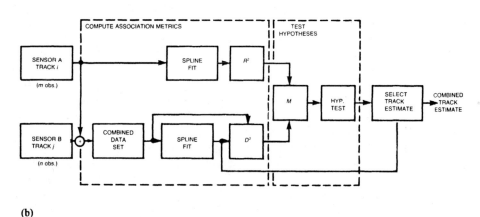

(b)

Figure 6.16 Two track-track association and estimation approaches: a = recursive; b = batch.

- Track-track gating is applied to determine the neighboring sensor tracks to central-level tracks. The metric can be the same statistical distance presented previously, applied to the track state vector differences (rather than observation vector differences) for two tracks, i and j:

$$\mathbf{R}^2 = \mathbf{d}_{ij}^T \mathbf{U}_{ij}^{-1} \mathbf{d}_{ij} \tag{6.6}$$

where

$\mathbf{d}_{ij} = \mathbf{x}_i - \mathbf{y}_i$ = difference vector between track states for i and j
$\mathbf{U}_{ij}$ = covariance matrix for $\mathbf{d}_{ij}$

- Due to the correlation between covariances of the sensor-level tracks, the covariance matrix between two sensors, i and j, must take on a modified form [23] that is a recursive function of the Kalman gain, covariance, and state transition characteristics of each sensor-level track filter:

$$U_{ij} = P_i + P_j - P_{ij} - P_{ij}^T \qquad (6.7)$$

where

P_i, P_j = covariance matrices for x_i, x_j
P_{ij} = cross covariance matrix, a function of the Kalman filter variables K_i, K_j, H_i, H_j, Φ_i, Φ_j

- Association can be performed using any one of the data association algorithms described earlier for report-track processing (Table 6.2). The nearest neighbor approach is generally considered appropriate because the sensor-level tracking process minimizes the false alarm rate (resulting in a low false track alarm rate) and the track state vector dimensionality (position plus velocity terms) minimizes misassociation errors. A track-track assignment matrix is assembled and track pairings are made on the basis of an assignment rule.
- Estimation or track filtering can then be performed using one of three methods. In the track selection approach, one sensor-level track (typically selected from the most accurate sensor) may be chosen to represent the central-level track, directly. In the state vector fusion approach [23], the state estimates from the associated tracks are combined in a linear estimator. This eliminates the need for a complete recursive estimator at the central fusion node and requires that sensors pass only state estimates to the node. Finally, in the measurement fusion approach [24], the observations from sensors are recursively filtered at the central node, once the track association has been made. Roecker and McGillem ([25]; see also the comments by Bar-Shalom in [26]) have quantified the reduction that can be achieved in the covariance of the filtered state vector for measurement fusion, when compared to state vector fusion. Implementation considerations for multiple sensor systems are described further in Section 6.6 for each of these cases.

The central-level tracks may be initiated as soon as sensors confirm tracks by using established sensor-level tracks from the most accurate sensors. Alternatively, central-level tracks may be initiated only after M-of-N sensors' tracks are associated using a selected sensor's track or results of state vector or measurement fusion estimates.

6.5.2 Batch Track-Track Association

Batch association of multiple sensor track data requires that each sensor supply a set of k scans of data in which each track is represented by the observations assigned to the track. The association problem can then be solved by a curve-fitting process that correlates the track curves. The process can consider two categories of association:

- Association of two segments occurring over the same period of time (or, at least, with significant time overlap), so the tracks can be *merged* into a common track.
- Association of two segments occurring over different, but contiguous, periods of time, so they can be *linked* to form a continuous track.

The representative approach by Campbell and Samaan [27] first applies a weighted, least squares spline function to model each sensor-level track segment. The spline function is a polynomial of specified order that represents the segment, providing a high-order model of the segment for developing displacement metrics between the spline and observations at equivalent sample times along the segment. The spline function that minimizes some criteria based on these displacement metrics is determined to best fit the observations. The curve-association process is formed as a hypothesis test to determine the association between two segments, A and B:

1. A spline is fit to segment A and the *root mean square* (rms) value of the residuals (errors between the raw observations for A and the spline fit to A), R_2, is computed.
2. A spline is fit to the combined set of observations from A and B. The rms value of these displacements, D_2, is computed.
3. An association test is performed on the measure M, which relates spline A residuals to the total error in the hypothesis given by $R + D$. Substituting the *root sum square* (rss) measures, $R'_2 = nR$ and $D'_2 = nD$, for rms values results in the following measure:

$$M = \frac{R}{R + D} = \frac{1}{(1 + D'/R')} \tag{6.8}$$

Campbell and Samaan establish the following hypothesis test for track similarity:

h_2 (accept association): $> M - 0.7$

h_1 (uncertain association): $0.5 < M < 0.7$

h_0 (reject association): $< M - 0.5$

Figure 6.16b illustrates the flow of data in this batch method, and the means of selecting the central-level track estimate from the combined data used to compute D_2. Note that alternative linear estimators could also be used to combine the tracks and develop the central-level track.

6.6 STATE ESTIMATORS FOR ASSOCIATION AND TRACKING

The static association processes and the dynamic tracking processes require estimators that convert observation data into accurate estimates of target state. Static location problems generally are solved with batch estimators that collect a set of observations (single or multiple site data) to derive an estimate of state (target location). For dynamic targets, batch estimators may be applied to entire time sequences of data, although recursive estimators are more practical for most real-time applications. In this section, we summarize the principal estimators applied to the data association and tracking problem.

6.6.1 State Estimation

State estimation is the problem of attempting to determine the value ("estimate") of a system's state, $\mathbf{x}$, given redundant observations, $\mathbf{y}$, that may be related to the system state. For example, we seek to determine the geocentric position and velocity of a ballistic object based on observations of the object's topocentric azimuth and elevation. In general an observation equation relates the observed quantities, $\mathbf{y}$, to the object's true (but unknown) state, $\mathbf{x}$:

$$\mathbf{y} = f(\mathbf{x}) + n$$

where n is unknown observational noise.

The estimation problem seeks to determine the value of $\mathbf{x}$ that best fits or explains the observational data $\mathbf{y}$. Generally we have a redundant data set $\mathbf{y}$. That is, there are more data than would be minimally required to determine $\mathbf{x}$. Also, the observation equations can generally be defined to relate $\mathbf{x}$ to $\mathbf{y}$ but not the inverse (i.e., typically we can specify $f(\mathbf{x})$ via analytical and numerical algorithms, but are unable to define $f^{-1}(\mathbf{x})$). Estimation techniques specify how to determine $\mathbf{x}$ given $\mathbf{y}$ subject to the satisfaction of some objective criterion (i.e., optimize $L(\mathbf{B})$, where $\mathbf{B}$ is the estimation matrix). A variety of optimization criteria have been suggested for estimation including: (1) minimizing the sum of the squares of the observation residuals (i.e., the difference between predicted and actual observations); (2) weighted least squares; (3) maximizing a likelihood function; (4) minimizing the error variance; and others. These are described in detail in the following paragraphs.

The practical approach to estimation is based upon the assumption that the model equations can be made linear and that the estimator is linear (see Figure 6.17). Solutions can be extended to nonlinear counterparts for each of the basic methods described here, and the reader is referred to the text by Gelb [28] for detailed derivations of both linear and nonlinear estimators. Table 6.3 compares the optimization criteria and characteristics of the principal linear statistical estimators that are introduced in this location.

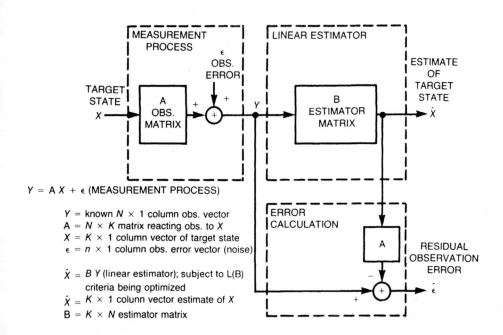

Figure 6.17 State estimation model.

In the estimation of state from multiple, static measurements (Section 6.2) for target localization, the observation matrix, A, is a geometric function that requires nonlinear approximations to the estimation process. Where multiple sensors measure the target state in identical dimensionality and coordinates, a linear estimator may be appropriate to derive a refined estimate from the combined observations, taking into account the measurement characteristics of each sensor. The following paragraphs introduce the principal linear estimators applicable to both static and dynamic applications for fusion of spatial information.

Table 6.3
Linear Statistical Estimators

Estimator	Estimation Criteria and Solution	Properties
Least Squares	• Minimize $Q(x) = \hat{\epsilon}^T \epsilon$, the sum of squares of the components of residual observation errors • Estimate is given by $\hat{x} = (A^T A)^{-1} A^T y$	• Requires no statistical knowledge • Treats all measurements equally, minimizing a function of residual error only
Weighted Least Squares	• Minimize $Q(x) = \epsilon^T W \epsilon$, where **W** is weighting matrix • Estimate is given by $\hat{x} = (A^T W^{-1} A)^{-1} A^T W^{-1} y$	• Applies **W** to weight each measurement individually • For gaussian error statistics, **W** can be set equal to p^{-1} to weight by inverse proportion to covariance
Maximum Likelihood	• Maximize likelihood function, $L(y\|x) \prod_{i=1}^{N} p(y\|x)$, which is a product of conditional probabilities of observation, given state	• Requires knowledge of probability density function of observations, conditioned on state • Knowledge of state statistics not required
Minimum Variance (minimum mean square error)	• Minimize variance of residual error, Q: $Q(x) = E[(\hat{x} - x)]^2$ • Estimate is given by $\hat{x} = (A^{-1} P^{-1} A)^{-1} A^T P^{-1} y$	• Requires covariance matrix, P, of residual error: *a priori* knowledge of state statistics • Effectively requires knowledge of state, measurement second-order statistics

6.6.2 Least Squares Estimator

The basic least squares estimator, which establishes the foundation for most state estimation methods, minimizes the sum of the squares of the components of the residual observation error vector. The estimator is therefore expressed as the value of **x** that minimizes the objective function:

$$Q(\mathbf{x}) = \hat{\epsilon}^T \epsilon \tag{6.9}$$

By setting the partial derivative of Q with respect to **x** to zero and solving for the estimate, we have

$$\hat{\mathbf{x}} = (\mathbf{A}^T \mathbf{A}^{-1}) \mathbf{A} \mathbf{y} \tag{6.10}$$

It is important to note that this estimator does not take into account that the errors for the various observed parameters, **y**, are different; and, further, it does not account for any functional relationships among errors. In other words, all observations are treated equally in the estimate.

6.6.3 Weighted Least Squares

By introducing a weighting matrix, **W**, into the least squares objective function, error contributors to the squaring process can be selectively weighted. This permits the estimator to reduce the square of certain error components more than others, according to the selection of **W**. In effect, the observations with greater importance (usually in terms of accuracy) can make a greater contribution to the solution. The form of the weighted estimator is

$$\hat{\mathbf{x}} = (\mathbf{A}^T\mathbf{W}^{-1}\mathbf{A}^{-1})\mathbf{A}^T\mathbf{W}^{-1}\mathbf{y} \qquad (6.11)$$

where **W** is symmetric and positive definite.

Because the observation residuals are actually samples of a random process, it is reasonable to base the selection of **W** on a statistical criterion. The next two estimators are special forms of this estimator; and, in each case, the weighting function is related to the inverse of the covariance matrix, **P**, which describes the observation noise statistics (under the assumption of Gaussian noise).

6.6.4 Maximum Likelihood Estimator

This estimator maximizes a likelihood function that represents the probability of any value of the target state vector (the variable being estimated) conditioned on the observed value. The likelihood function is therefore the joint probability distribution between the possible values of the random variable and the actual value observed. The estimation process selects the maximum conditional probability of the random variable given the observed sample value.

If the maximum likelihood estimator is applied to a problem of estimating a vector random variable (e.g., the target state vector) that is assumed to be a multivariate Gaussian distribution, the solution is a weighted least square estimator where **W** is the inverse of the covariance matrix of the measurement error. This means that the squares of the error components are weighted by the inverse of the variances of their respective probability distributions.

6.6.5 Maximum Likelihood Estimation for Static Target Localization

A maximum likelihood estimator has been applied to the multiple site localization problem presented earlier in Section 6.2, where multiple measurements and known geometric relationships were used to derive an optimal target position solution. The application requires a nonlinear estimator because of the geometric functions in the location equations. Torrieri's [19] general statistical solution to this class of problems, summarized here, is based upon a maximum likelihood estimator with linearization of the geometric functions. Where the assumption of gaussian measurement noise is not valid, the approach becomes a weighted least squares estimator with the covariance matrix replaced by a set of weighting coefficients.

The state vector, **x**, describing the position of the target is an n-dimensional vector to be estimated in three dimensions, or two dimensions if the problem can be approximated onto a plane. The set of N observations, r_i, $i = 1, 2, \ldots, N$, can be a time sequence from a moving sensor or a set of concurrent measurements from multiple sensor sites. Each individual observation is a geometric function of **x** with additive measurement noise:

$$r_i = f_i(x) + n_i \tag{6.12}$$

The vector function for all i measurements is given by the N measurement equations:

$$\mathbf{r} = \mathbf{f}(\mathbf{x}) + \mathbf{n} \tag{6.13}$$

where

> $\mathbf{f}(\mathbf{x})$ = geometric vector function relating the observation set, **r**, to state **x**;
> $\mathbf{r}$ = observation vector;
> $\mathbf{n}$ = measurement error vector, generally assumed to be random (zero mean) with $N \times N$ covariance matrix, **N**, where $\mathbf{N} = E\{[\mathbf{n} - E(\mathbf{n})][\mathbf{n} - E(\mathbf{n})^T]\}$.

The maximum likelihood estimator is defined as the estimator that maximizes the conditional density function of **r** given a state **x**. This estimate minimizes the quadratic form that computes the scalar distance function:

$$Q(\mathbf{x}) = [\mathbf{r} - \mathbf{f}(\mathbf{x})]^T \mathbf{N}^{-1} [\mathbf{r} - \mathbf{f}(\mathbf{x})] \tag{6.14}$$

Because $f(x)$ is generally a nonlinear geometric function, a linear approximation is developed using a Taylor series expansion [29] to approximate $\mathbf{f}(\mathbf{x})$ as a linear function about a reference state, $\mathbf{x}_0$:

$$\mathbf{f}(\mathbf{x}) \approx \mathbf{f}(\mathbf{x}_0) + \mathbf{G}(\mathbf{x} - \mathbf{x}_0) \tag{6.15}$$

where

> $\mathbf{G}$ = $n \times n$ matrix of partial derivatives of the function f_i, over all elements of state, $\mathbf{x}_i$;
> $\mathbf{x}_0$ = reference state vector, an initial estimate based upon *a priori* data.

$Q(x)$ is then expressed in terms of the approximation, and the estimation expression is obtained by solving for the function that minimizes the scalar $Q(x)$:

$$\hat{x} = \hat{x}_0 + (G^T N^{-1} G)^{-1} G^T N^{-1} [r - f(x_0)] \quad (6.16)$$

This estimator requires an initial estimate of the state, x_0, the gradient matrix, **G**, and the covariance (or weighting, for least squares) matrix, **N**, to process measurement data, **r**. Iterative solutions initialize the process with an initial estimate and add successive observations to converge to more accurate estimates. Foy [29] has suggested validity tests to monitor the performance of the estimator with successive iterations.

6.6.6 Minimum Variance Estimators for Recursive Tracking

Recursive estimators were applied to both report-track and track-track tracking loops in earlier sections. These estimators can be used to perform three functions, depending upon the application:

1. Filters estimate the state, **x**, at time t, given samples of **y** at times previous to and including t.
2. Predictors estimate **x** at some future time $t + K$, given samples of **y** at times previous to and including t.
3. Smoothers estimate **x** at time t, given samples of **y** at times both before and after t.

Recursive dynamic target tracking is a state estimation problem that requires (1) a filter to take the series of observations up to t (or recursively the previous estimate and current observation) and process the data to estimate the state vector at time t, and (2) a predictor to extrapolate the target state to a future sample time. These functions were shown earlier in the tracking feedback loop in Figure 6.7. The prediction function, in this case, performs the temporal alignment of the estimate to the point in time at which the next observation is predicted.

The minimum variance or minimum mean square error estimator reduces the variance of the estimate by minimizing the mean square value of estimation error. In these estimators, statistical models (the first and second moments) of both the random variable to be estimated (the state vector) and the measurement noise processes are required for derivation of the mean square error.

The solution of this problem is the basis of the Weiner-Kolmogoroff theory: the derivation of estimators that are optimal in the linear minimum mean square sense. Two classes of minimum variance filters are well known: the Weiner and Kalman filters. The Wiener filter is a continuous form, frequency domain estimator that is most useful in statistical communication problems. The Kalman filter, on the other hand, is specified in terms of the time domain and is applicable to both stationary and nonstationary state vectors.

6.6.7 The Kalman Filter

The discrete time, recursive solution to the linear, minimum variance estimation problem is provided by the Kalman filter [30], which is the statistical estimator most often applied to dynamic tracking. Nonadaptive or fixed-coefficient filters are also applicable to many tracking applications where statistical estimation is not required. The alpha-beta-(gamma) filters [31, 32] are the classical fixed-coefficient filters applicable to simpler tracking applications where the observation vector includes position data only.

The basic Kalman state equation relates the state at time $k = 1$, $x_{k=1}$, to the state at the previous interval, k, by the following state equation:

$$X_{k+1} = \Phi_k X_k + W_k \tag{6.17}$$

where

Φ_k = state transition matrix;
W_k = plant noise: white, zero mean sequence with covariance, Q_k.

The measurement model relates **y** to the state

$$Y_k = H_k X_k + V_k \tag{6.18}$$

where

H_k = measurement matrix;
V_k = measurement noise; white zero mean sequence with covariance, R_k.

The filter requires a knowledge of the second-order statistics of the plant and measurement noise (**Q** and **R**, respectively, both assumed to be zero mean) to provide the solution that minimizes the mean square error between the true state and the estimate of state. The following four equations form the basis of the filter calculations. The recursive estimator solution is presented in a form in which the estimate of state $x_{k=1}$, is a function of the current measurement, $y_{k=1}$, and the previous estimate, x_k:

$$x_{k(+)} = x_{k(-)} + K_k[Y_k - H_k x_{k(-)}] \tag{6.19}$$

where

$x_{k(+)}$ = estimate of x_k, derived at time $t = k$;
$x_{k(-)}$ = prediction of x_k, derived at time $t = k - 1$.

The Kalman gain matrix, **K**, which minimizes the mean square estimation error is then:

$$\mathbf{K}_k = \mathbf{P}_{k(-)}\mathbf{H}_k^T[\mathbf{H}_k\mathbf{P}_{k(-)}\mathbf{H}_k^T + \mathbf{R}_k]^{-1} \tag{6.20}$$

The covariances of the estimation errors in the filtered and the predicted state estimates are given respectively by

$$\mathbf{P}_{k(+)} = [\mathbf{I} - \mathbf{K}_k\mathbf{H}_k]\mathbf{P}_{k(-)} \tag{6.21}$$

$$\mathbf{P}_{k(-)} = \mathbf{\Phi}_{k-1}\mathbf{P}_{k-1(-)}\mathbf{\Phi}_{k-1} + \mathbf{Q}_{k-1} \tag{6.22}$$

Figure 6.18 is a functional diagram showing the implementation of these equations into a complete filter in a typical tracking system. The inputs are the initial values of the state vector and estimation error covariance matrix and the measurements that are time discrete. The outputs are the state vector estimate and the current value of the estimation error covariance matrix. The form of the filter is that of two dynamic loops: the system model and the estimation error model. The first loop performs the following operations:

1. Residual, or innovations, is computed as the difference between actual and predicted observations.
2. Residual is filtered by the Kalman gain, adding the previous state estimate. This results in the optimal estimate of state, for the current time, which is placed in the track file.
3. The next state is predicted by multiplying the current state (delayed) by the state transition matrix.
4. The next observation is computed by multiplying the predicted state estimate by the measurement matrix.

The estimation error loop performs these operations:

1. Error propagation is simulated using the previous error covariance and the predicted state estimate.
2. Optimal gain is computed, for transfer to the state estimate loop.
3. The estimation error covariance matrix is updated, to provide the current estimate of state errors.

6.6.8 Tracking Filters for Maneuvering Targets

Both the fixed and adaptive coefficient tracking estimators (filters) described earlier are used to provide accurate predictions of future target states for gating and association. The filters' performance is directly related to their ability to predict target behavior *for even one scan beyond the current observations.* Failure to predict the

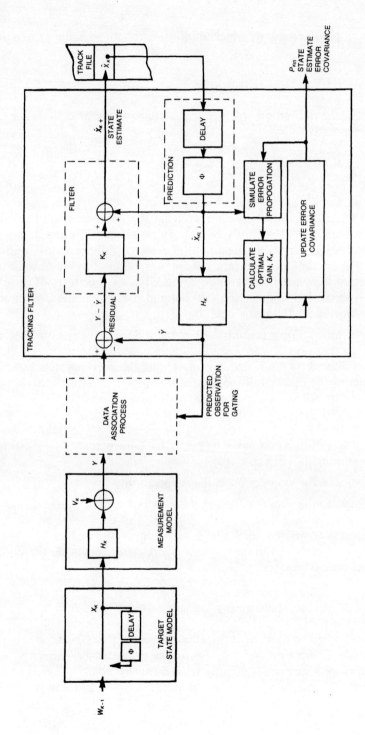

Figure 6.18 Kalman filter in recursive tracking loop.

next target state with the sufficient accuracy to place the gate over the real observation may cause a misassociation or loss of continuity in measurements, resulting in a loss of track. The potential for this occurrence is high when the target behavior changes more rapidly than the steady state model of behavior expected by the filter. The filter designer is faced with two conflicting objectives:

- The need to minimize misassociation with false alarms or nearby targets.
- The need to minimize missed observations that will occur when the target abruptly maneuvers, changing kinematic behavior from that in the tracking predictor's estimate. When this occurs, the next observation may fall outside of the prediction gate and not even be considered a neighbor for association.

These conditions demand two conflicting gate criteria: a tight gate during steady-state behavior and a large gate during the period of maneuver. Adaptation to such maneuvering requires detection of the maneuver and instantaneous adaptation of the estimator to track the maneuver. Maneuver detection can be performed by monitoring the growth in the observation residual or by the use of concentric maneuver gates to detect the increase in the residual. Upon detection of the maneuver, one of several adaptation methods may be used [33–35]:

- Reinitialization of the covariance matrix in the Kalman filter.
- Modification or augmentation of the state variables in the filter.
- Switching between multiple filters, combining multiple filters, or adjustment of a single filter's parameters as a function of maneuver characteristics.

6.6.9 Tracking Filter Implementation Issues

A number of additional Kalman filter implementation issues must be considered in the design of practical tracking systems. Although details and solutions are available in the major text [28], the most noteworthy issues are enumerated here.

Estimate divergence can occur as the estimator reduces the measurement covariance causing the predictions to be weighted more heavily than incoming samples. This may occur following a steady-state period, during which the estimator derives predicted observations with very small variances: as the track deviates (i.e., maneuvers) from the steady-state trajectory, the estimate may diverge unless detection and correction measures are implemented.

Filter initialization is required once a filter is started with new observations that are declared to be a new track. The selection of initial covariance terms is critical to permit the filter to efficiently and rapidly achieve accurate state estimates.

Coordinate selection for the parameters of state has a strong influence upon the performance and complexity of the filter. The use of simple, uncoupled filters for independently tracking line-of-sight coordinates (e.g., range-azimuth-elevation), for example, introduces the need for artificial acceleration terms to account for the nonlinearity of a constant velocity (in x-y-z coordinates) target in those coordinates. Tracking for multiple-sensor systems is generally performed in x-y-z coordinates with fully coupled equations of motion to eliminate these complexities.

Matrix formulation methods of various types have been developed to improve numerical stability and accuracy (e.g., square root and stabilized formulations), to minimize the computational complexity by taking advantage of the diagonal characteristics of the covariance matrix (UDU factorization formulation), and to estimate state when the state functions are nonlinear (extended Kalman filter).

6.6.10 Multiple-Sensor Estimation Considerations

The report-track and track-track association processes described in Sections 6.4 and 6.5, respectively, were integrally related to the estimation functions included in each algorithm. In Section 6.5, three alternative recursive track-track estimation approaches were also described with citations to analyses of relative performance. These descriptions, although applicable to recursive association-estimation with multisensor as well as single-sensor systems, did not highlight the specific considerations required by systems where the estimation process must take into account the different characteristics (e.g., sensor-specific covariances and observation matrices) of multiple sensors contributing observations. In this section, we describe the system architecture alternatives and considerations for multisensor estimation. The general concepts for these fusion alternatives were introduced in Chapter 2, where centralized and distributed architectures were first introduced, and then in Chapter 4, where the centralized and distributed detection problem was addressed. The three general association-estimation approaches available for multiple-sensor systems are summarized in Table 6.4.

Table 6.4
Multiple-Sensor Estimation Alternatives

Association Process	*Multisensor Estimators*
Report-Track Associate multisensor reports (measurements) to central-level tracks	Measurement Fusion: compute estimate using sensor measurements directly, considering two cases: Asynchronous sensors • Sequential computation Synchronous sensors • Parallel computation • Pseudo-sequential • Data compression
Track-Track Associate multisensor-level tracks (estimates of state) to central-level tracks	Measurement Fusion: (same methods as above) State Vector Fusion: compute linear estimate using sensor-level estimates Track Selection: select one sensor track as the best estimate

The two simplest alternatives for track-track estimation, presented earlier in Section 6.5, included *track selection fusion*, in which the "best" sensor-level track estimate is simply selected as the central track estimate and no composite or fused estimate is computed; and *state vector fusion*, in which the associated state vectors are combined in a linear estimator to derive a central-level state estimate. The estimator in this case must consider the sensor-specific parameters and can be implemented in a fashion similar to the Kalman filter to form a minimum mean square error estimate. Both sensor state estimates must be extrapolated to a common time, and the cross-covariance term (6.7) described earlier must be considered. The computation of this term is somewhat complex, and its contribution should be considered relative to implementation considerations [24].

The third and most accurate estimator, referred to as *measurement fusion* [24, 25] uses the sensor measurements (rather than sensor estimates in the case of track-track fusion) to compute a central estimate of target state. Two specific cases of measurement fusion must be considered.

Asynchronous Sensors

When the individual sensor measurement processes are asynchronous, an adaptive (sensor-dependent) estimator is required that adjusts the Kalman measurement (**H**) and measurement covariance (**R**) matrices for the sensors whose data is being processed. The estimator is analogous to the single-sensor case, involving two measurement sets, except that the Kalman process has certain sensor-dependent components (e.g., **H**, **R**, and the measurements, **Y**) that are sequentially applied as each sensor's data is used [36]. The prediction time between estimation and next-observation must also adapt to the asynchronous intervals between the multiple sensor observations.

Synchronous Sensors

When all the sensors sample at the same time period, there are three alternative implementations [37]:

1. The parallel implementation directly modifies the Kalman state update equation (and covariance equation, similarly) to account for all input measurements from I sensors. The modified state equation (compared to eq. (6.14)) becomes

$$\hat{\mathbf{X}}_{k(+)} = \hat{\mathbf{X}}_{k(-)} + \sum_{i=1}^{I} \bar{\mathbf{K}}_{k,i}[\mathbf{Y}_{k,i} - \mathbf{H}_{k,i}\hat{\mathbf{X}}_{k(-)}] \qquad (6.23)$$

Because the parallel solution is computationally demanding, two other alternatives are possible.

2. A pseudo-sequential implementation is similar to the asynchronous sensor solution, except the extrapolated time between sensor observations is set to zero. For each set of i sensor observations, the process is performed recursively i times.
3. A data compression approach requires transforming all of the sensor measurements into a common coordinate system. The same transformations are used to develop a single version of the estimation processes, involving a common state covariance (**P** matrix) and compressed or fused state estimate based on the compressed measurement. The computational requirements of this approach is the lowest of the three approaches and the estimation performance is comparable to the other approaches.

REFERENCES

1. Weiner, Howard L., et. al., "Naval Ocean-Surveillance Correlation Handbook, 1978," Naval Research Laboratory Rep. 8340, Washington, DC, October 1979.
2. Goodman, Irwin R., et al., "Naval Ocean-Surveillance Correlation Handbook, 1979," Naval Research Laboratory Rep. 8402, Washington, DC, September 1980.
3. Bar-Shalom, Y., "Tracking Methods in a Multitarget Environment," *IEEE Trans. on Automatic Control*, Vol. AC-R3, August 1978, pp. 618–626.
4. Blackman, Samuel S., *Multiple-Target Tracking with Radar Applications*, Artech House, Norwood, MA, 1986.
5. Bar-Shalom, Y., and T.E. Fortmann, in *Tracking and Data Association*, ed. W.F. Ames, Academic Press, Orlando, FL, 1987.
6. Hassab, Joseph C., *Underwater Signal and Data Processing*, CRC Press, Boca Raton, FL, 1989.
7. Minor, Lewis G., ed., "Proceedings of Open Sessions of the Workshop on Imaging Trackers and Autonomous Applications for Missile Guidance," GACIAC-PR-80-01, US Joint Service Guidance and Control Committee, November 1979.
8. Woods, E., "Report to Track Asssociation Using a Two-Dimensional Optical Correlator," *Proc. 1987 Tri-Service Data Fusion Symp.*, June 1987, pp. 652–655.
9. McPhillips, K., and R. Fagon, "Data Association and Fusion in an ASW Environment," *Proc. 1988 Tri-Service Data Fusion Symp.*, May 1988, pp. 348–355.
10. Coleman, J.O., "Discriminants for Assigning Passive Bearing Observations to Radar Targets," *Proc. 1980 IEEE Int. Radar Conf.*, Washington, DC, pp. 361–365.
11. Trunk, G.V., and J.D. Wilson, "Association of DF Bearing Measurements with Radar Tracks," *IEEE Trans. on AES*, Vol. AES-23, July 1987, pp. 438–447.
12. Poirot, J.L., and Arbid, G., "Position Location: Triangulation versus Circulation," *IEEE Trans. on AES*, Vol. AES-14, No. 1, January 1978, pp. 48–53.
13. Kelly, R.J., "Global Positioning System Accuracy Improvement Using Ridge Regression," paper presented at AIAA Guidance, Navigation and Control Conf., August 1989.
14. Foy, W.H., "Position Location from Sensors with Position Uncertainty," *IEEE Trans. on AES*, Vol. AES-19, No. 5, September 1983, pp. 658–662.
15. Torrieri, Don J., "Statistical Theory of Passive Location Systems, *IEEE Trans. on AES*, Vol. AES-20, No. 2, March 1984, p.197; see also *Principles of Secure Communication Systems*, Artech House, Norwood, MA, 1985.
16. Spingarn, Karl, "Passive Position Location Estimation Using the Extended Kalman Filter," *IEEE Trans. on AES*, Vol. AES-23, No. 4, July 1987, pp. 558–567.

17. Poirot and Arbid, op. cit.
18. Stansfield, R.G., "Statistical Theory of D.F. Fixing," *J. Inst. Electronic Eng.*, Vol. 94, IIIa, No. 15, 1947, p. 762.
19. Torrieri, Don J., "Statistical Theory of Passive Location Systems," *IEEE Trans. on AES*, Vol. AES-20, No. 2, March 1984, p. 197.
20. Wax, Mati, "Position Location from Sensors with Uncertainty," *IEEE Trans. on AES*, Vol. AES-19, No. 5, September 1983, pp. 658–662.
21. Singer, R.A., and J.J. Stein, "An Optimal Tracking Filter for Processing Sensor Data with Imprecisely Determined Origin in Surveillance Systems," *Proc. 1971 IEEE Conf. on Decision and Control*, December 1971, pp. 171–175.
22. Jaffer A.J., and Y. Bar-Shalom, "On Optimal Tracking in Multiple Target Environments," *Proc. Third Symp. Non-Linear Estimation, Theory and Applications*, September 1972, pp. 112–117.
23. Bar-Shalom, Y., and L. Campo, "The Effect of Common Process Noise on the Two-Sensor Fused-Track Covariance," *IEEE Trans. on AES*, Vol. AES-22, November 1986, pp. 803–805.
24. Willner, D., C.B. Chang, and K.P. Dunn, "Kalman Filter Algorithms for a Multi-Sensor System," *Proc. IEEE Conf. on Decision and Control*, December 1976, pp. 570–574.
25. Roecker, J.A., and C.D. McGillem, "Comparison of Two-Sensor Tracking Methods Based on State Vector Fusion and Measurement Fusion," *IEEE Trans. on AES*, Vol. AES-24, July 1988, pp. 447–449.
26. Bar-Shalom, Y., "Comments on 'Comparison of Two-Sensor Tracking Methods Based on State Vector Fusion and Measurement Fusion,'" *IEEE Trans. on AES*, Vol. AES-24, July 1988, pp. 456–457.
27. Campbell, J.B., and J.E. Samaan, "Algorithm for Decisions on Merging and Linking Target Tracks," *Proc. 1988 Tri-Service Data Fusion Symp.*, May 1988, pp. 414–424.
28. Gelb, A., *Applied Optimal Estimation*, MIT Press, Cambridge, MA, 1974.
29. Foy, Wade H., "Position-Location Solutions by Taylor-Series Estimation," *IEEE Trans. on AES*, Vol. AES-12, No. 2, March 1976, pp. 187–193.
30. Brammer, K., and G. Siffling, *Kalman-Bucy Filters*, Artech House, Norwood, MA, 1989.
31. Fitzgerald, R.J., "Simple Tracking Filters: Steady-State Filtering and Smoothing Performance," *IEEE Trans. on AES*, AES-16, November 1980, pp. 860–864.
32. Fitzgerald, R.J., "Simple Tracking Filters: Position and Velocity Measurements," *IEEE Trans. on AES*, Vol. AES-18, September 1982, pp. 531–537.
33. Ricker, G.G., and J.R. Williams, "Adaptive Tracking Filter for Maneuvering Targets," *IEEE Trans. on AES*, Vol. AES-14, No. 1, January 1978, pp. 185–193.
34. Gholson, N.H., and R.L. Moose, "Maneuvering Target Tracking Using Adaptive State Estimation," *IEEE Trans. on AES*, Vol. AES-13, May 1977, pp. 310–317.
35. Bar-Shalom, Y., and K. Birmiwal, "Variable Dimension Filter for Maneuvering Target Tracking," *IEEE Trans. on AES*, Vol. AES-18, No. 5, September 1982, pp. 621–628.
36. Brandstadt, J., "Track File Fusion," *Proc. 1988 Tri-Service Data Fusion Symp.*, May 1988, pp. 388–395.
37. Willner, D., C. Chang, and K. Dunn, "Kalman Filter Configurations for Multiple Radar Systems," MIT Lincoln Lab TN 1876-21, April 14, 1976.
38. Casnev, P.G., and R.J. Prengaman, "Integration and Automation of Multiple Co-located Radars," *Proc. IEEE EASCON*, 1977, pp. 10-1A–1E.
39. Bar Shalom, Y., and E. Tse, "Tracking in a Cluttered Environment with Probabilistic Data Association," *Automatica*, Vol. II, September 1975, pp. 451–460.
40. Fortmann, T.E., Y., Bar Shalom, and M. Scheffe., "Multi-Target Tracking Using Joint Probabilistic Data Asssociation," *Proc. 1980 IEEE Conf. Decision and Control*, December 1980, pp. 807–812.
41. Sittler, R.W., "An Optimal Data Association Problem in Surveillance Theory," *IEEE Trans. Military Electronics*, Vol. MIL-8, April 1984. pp. 125–139.

42. Stein J.J. and S.S. Blackman, "Generalized Correlation of Multi-Target Track Data," *IEEE Trans. Aerospace and Electronic Systems,* Vol. AES-11, No. 6, November 1975, pp. 1207–1217.
43. Morefield, C.L., "Application of o-i Integer Programming to Multi-Target Tracking Problems," *IEEE Trans. Automatic Control,* Vol. AC-22, June 1977, pp. 302–312.
44. Reid, D.B., "An Algorithm for Tracking Multiple Targets," *IEEE Trans. Automatic Control,* Vol. AC-24, December 1979, pp. 843–854.
45. Singer, R.A., R.G., Sea, and R.B. Housewright. "Derivation and Evaluation of Improved Tracking Filters for Use in Dense Multi-Target Environments," *IEEE Trans. Information Theory*, Vol. IT-20, July 1974, pp. 423–432.

Chapter 7
DATA FUSION FOR OBJECT IDENTIFICATION

In Chapter 2, the data fusion model of the *Joint Directors of Laboratories* (JDL) *Data Fusion Subpanel* (DFS) was introduced; that model defined the notion of the "levels" of fusion products. The Level 1 fusion products of the model were identified as position and identity estimates, and Chapter 6 discussed various aspects of multisensor, multitarget position estimation. Generally speaking, the JDL/DFS model interprets the Level 1 identity estimates to be those associated with single platforms. By way of contrast, the identification of multiplatform *sets* is usually attributed to Level 2 or Level 3 processing, which is more symbolic than numeric and where the target set identification estimate is usually considered to result from some type of template-matching process that considers the deployment patterns of the single target estimates provided by Level 1 processing, in addition to a variety of contextual information. However, this is somewhat of an ideal because many factors (sensor characteristics, observation geometries, *et cetera*) affect whether the first available identity estimates in any real system are associated to single targets or target clusters. Thus, the Level 1 identity product in a real system is that which can be derived from the first available observation sets and may relate to either "point" targets or target groups. At the present time, the identification of the class of a target group, based on "grouped observables" from sensors is not only a concept that has not been fully clarified but no R&D has been conducted to develop computer-based methods to estimate "group identity" from such observables.

As a result, the focus here is on automated methods to operate on "first available" multisensor observables, which are assumed to have sufficient resolution to identify single objects, in order to derive estimates of either target class (e.g., friend or foe), membership within a class (e.g., F-15, MIG), or specific "serial number" within the subset of a class.

Section 7.1 provides an overview of the many methods used for identity estimation. Section 7.2 focuses on and contrasts the Bayesian and Dempster-Shafer methods in some detail, the rationale being that these are the methods most frequently applied or discussed in the literature. The material herein has been graciously contributed by two close colleagues, Dr. David L. Hall (Section 7.1), and Dr. Dennis M. Buede (Section 7.2).

7.1 OVERVIEW OF DATA FUSION ALGORITHMS FOR IDENTITY ESTIMATION*

7.1.1 Introduction

Object identity estimation is a much broader problem than object position estimation because identity is a much broader concept than position, involving a larger number of variables. In multisensor classification processing, we may require several observational models or extensive pretesting to characterize the sensing process. We can require complex alignment algorithms; we may need to postulate an order of battle; and so on. Moreover, the nonmetric aspect of identity permits numerous types of *identity representation* techniques to be employed, each of which may be operated on in different ways. These complexities make the identity declaration process more challenging in breadth than that of positional fusion. As a result, a number of techniques have been developed to address fusion of identity data. This section provides an overview of algorithms that have been utilized for classification fusion. A taxonomy and algorithm overview is presented first in Section 7.1.2. Details are provided for each method in Section 7.1.3.

In what follows, it is important to recall the architectures of Chapter 2. Some of the methods described are those that *directly* estimate identity, whereas others combine or fuse declarations of identity that are presumed to derive from still other (e.g., feature-based) algorithms.

7.1.2 Taxonomy of Identity Fusion Algorithms

Precise and unique categories of algorithms utilized for identity fusion do not exist. Techniques successfully used for identity fusion range from well-known, statistically based algorithms, such as classical inference and Bayesian methods, to ad hoc methods such as templating, voting, and evolving techniques such as adaptive neural networks. Comparison of the complete range of techniques smacks a bit of an "apples and oranges" comparison, and any partitioning of algorithms into categories may be arguably arbitrary. Nevertheless, Figure 7.1 presents a conceptual taxonomy of identity fusion algorithms. Three major categories are shown: (1) physi-

---

*As noted above, much of this section has been graciously contributed by Dr. David L. Hall, a close colleague of the authors.

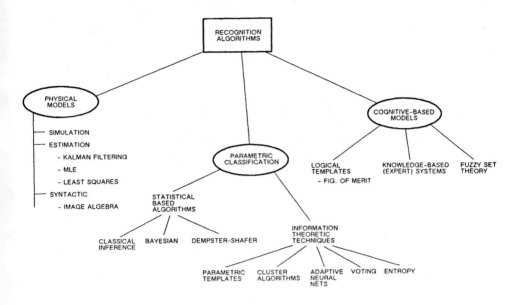

Figure 7.1 Taxonomy of identity-classification algorithms.

cal models, (2) parametric classification techniques, and (3) cognitive-based models.

Physical models attempt to model accurately observable or calculable data (e.g., RCS, IR spectra, etc.) and estimate identity by matching predicted (modeled) observations with actual data. Techniques in this category include simulation and estimation methods such as Kalman filtering. Although modeling identity may be difficult, it is conceptually possible to estimate identity utilizing classical techniques. The process for such estimation is completely analogous to positional data fusion. In a sense, this category is included for completeness, as relatively few operational systems employ such techniques. However, such methods are employed in basic research efforts (e.g., [1]). However, we also include in this category those syntactic (or "model-based") methods that develop a grammar and syntax of physical object descriptive components (or "primitives") individually derived from sensor data, thereby permitting "assembly" of an object's identity from the postulated structural relationships.

Parametric classification seeks to make an identity declaration based on parametric data, without utilizing physical models. A direct mapping is made between parametric data (e.g., features) and a declaration of identity. We may further subdivide these into statistically based techniques such as classical inference, Bayesian inference, the Dempster-Shafer method, and information theoretic techniques such as templating, cluster algorithms, adaptive neural nets, voting methods, and entropy methods. The information theoretic methods generically attempt to induce natural groupings in the data which are then associated with object classes or types.

A third major category of identity fusion algorithms is cognitive-based models. These methods seek to mimic the inference processes of human analysts in recognizing identity. Techniques in this category include logical templates, knowledge-based (expert) systems, and fuzzy set theory. These methods in one way or another are based on a perception of how humans process information to arrive at conclusions regarding identity of entities.

An overview of the techniques identified in Figure 7.1 is provided in Figure 7.2. That figure lists each algorithm and summarizes the kernel process, the character of the required input, and the character of the output. Further details on each of these algorithms are given in Section 7.1.3 of this chapter. Because of the strong analogy between positional data fusion and the use of physical models for identity fusion, no further details are provided on the use of physical models for identity fusion.

7.1.3 Algorithm Descriptions

A description of several identity fusion algorithms is provided in this section; space prevents an exhaustive review of all methods (e.g., "feature-based" algorithms and the associated methods from statistical pattern recognition). The level of presentation varies depending upon the extent to which these algorithms are well known in the literature. For example, only a brief description is provided for classical inference, as it has such a rich history in mathematical statistics. By contrast, entropy methods have not yet been adequately studied for identity estimation and are relatively rare in the literature—hence, some additional detail is provided.

Statistically Based Algorithms

Statistically based algorithms include classical inference, Bayesian inference, and the Dempster-Shafer method. These techniques utilize *a priori* knowledge about the observation process to make inferences about identity. Each technique will be described.

Classical Inference

Classical inference techniques [2] compute the probability of an observation given the assumption of an *a priori* hypothesis. Hence, classical inference describes the

METHOD	KERNEL PROCESS	REQUIRED INPUT CHAR.	CHAR. OF OUTPUT
CLASSICAL	Pr (OBSERVATION/H)	EMPIRICAL PROBABILITY, POPULATION DISTRIBUTION FOR STATISTIC	Pr (ERROR/DEC. ON H)
BAYESIAN	"A POSTERIORI" Pr (H /EVIDENCE) (UPDATES BELIEF ON H /GIVEN NEW DATA)	EMPIR/SUBJ. PROBABILITY EXHAUSTIVE DEFINITION OF CAUSES "A PRIORI" PR (CAUSES) MUTUALLY EXCLUSIVE CAUSES	UPDATED LIKELIHOOD OF THE OCCURRENCE OF AN EVENT
DEMPSTER SHAFER	"Pr (H/MULT. EVIDENCE) AND PR (ANY H TRUE)" — GEN'L UNCERTAINTY	EMPIR/SUBJ. "PROBABILITY" EXHAUSTIVE (INCL. DISJ) Pr (Hj EVIDENCE)	UPDATED LIKELIHOOD OF THE OCCURRENCE OF AN EVENT AND LEVEL OF IGNORANCE
FUZZY SET THEORY	SET ALGEBRA WHERE SET ELEMENTS HAVE MEMBERSHIP FUNCTION	SUBJECTIVE MEMBERSHIP FUNCTIONS FOR ALL SET ELEMENTS	PROFILE OF GOAL SET ELEMENTS AND MEMBERSHIP FUNCTION
CLUSTER ANALYSIS	SORTING OF OBSVNS. INTO "NATURAL GROUPS" BASED ON "SIMILARITY MEASURE"	PARAMETRIC OR SUBJECTIVE DATA	CLUSTER ELEMENTS AND SIMILARITY MEASURES
ESTIMATION THEORY	"BEST STATE (IDENTITY) ESTIMATE FOR GIVEN OBSVNS. (LEAST SQUARES)	QUANTITATIVE OBSVNS.. STATE/ OBSVN MODEL	STATE (IDENTITY) VECTOR
ENTROPY	COMPUTES MEASURE OF INFORMATION CONTENT	EMPIRICAL OR SUBJECTIVE PROBABILITY	IDENTITY ESTIMATE WHICH MAXIMIZES INFO CONTENT
FIGURE OF MERIT	COMPUTES DEGREE OF SIMILARITY BETWEEN PARAMETERS AND ENTITIES	ATTRIBUTE VECTORS FROM SENSOR OBSVNS	NUMERICAL VALUE OF SIMILARITY OF CANDIDATE OBJECT TO POSTULATED CLASS
EXPERT SYSTEMS	COMPUTER PROGRAM TO MIMIC HUMAN INFERENCE PROCESS	OBSERVATION DATA TO SUPPORT INFERENCES	DECLARATION OF INFERENCE (OBJECT IDENTITY)THAT BEST SATISFIES LOGIC CONDITIONS
TEMPLATES	PATTERN MATCHING TECH- NIQUE FOR COMPLEX ASSOCIATIONS	OBSERVED DATA RECORDS	DECLARATION THAT DATA SUPPORTS (MATCHES) A TEMPLATE — HENCE. OBJECT ID
ADAPTIVE NEURAL NETS	NON-LINEAR TRANS- FORMATION FROM OBSERVATION TO RECOGNITION SPACE	OBSERVED PARAMETRIC DATA (FEATURES)	MATCH OF INPUT DATA TO PRE-SPECIFIED CATEGORIES

Figure 7.2 Overview of analytical methods for identity fusion.

probability of observed identity-related data, given a hypothesis on the existence of the object. For example, suppose that two different types of radars exhibited agility in *pulse repetition interval* (PRI) as illustrated in Figure 7.3. The figure shows the probability density function *versus* PRI for each radar, where the two radars (type 1, denoted E1, and type 2, denoted E2) exhibit overlapping ranges of PRI. The probability that radar type 2 will emit at a pulse repetition interval of $PRI_N \leq PRI \leq PRI_{N+1}$ is given by the cross-hatched area under the probability density function in Figure 7.3. Classical inference seeks to confirm or refute a proposed hypothesis

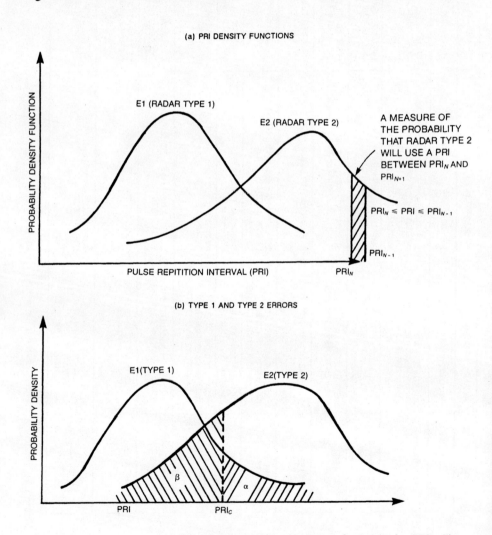

Figure 7.3 Hypothetical example of classical inference to identify classes of radars having PRI agility.

about identity. Thus, if we have observed a pulse repetition interval of PRI_{obs} we seek to determine whether the associated radar is of type 1 (E1) or of type 2 (E2); that is, we seek to assert whether the observations support the existence of an object of given (hypothesized) identity.

Classical inference proceeds as follows. A critical value of PRI (e.g., PRI_c) is chosen. If the observed value of PRI is greater than PRI_c (i.e., if $PRI_{obs} > PRI_c$), then we say that the evidence fails to refute the hypothesis that the radar is of type 2. Alternatively, if $PRI_{obs} < PRI_c$, we say that the evidence fails to refute the hypothesis that the radar is of type 1. In our example, because of the overlap of PRI (due to sensor agility) the decision based on PRI_c may still result in an erroneous identity declaration. In particular, the lower half of Figure 7.3 illustrates that there is a finite probability, α, that the observed PRI would be greater than PRI_c for radar of type E1, and a finite probability of β that the observed PRI would be less than PRI_c for a radar of type E2. These misidentification errors are termed a *type 1 error* and *type 2 error*, respectively. The choice of the value of PRI_c is up to the analyst and can be influenced by many factors of the overall problem. Although it may be changed to minimize either type 1 or type 2 errors, in general, there still remains a finite probability of misidentification.

In the preceding, the two PRI distributions could have come from different sensors, and in such a case the fusion process is represented by the decision-making strategy discussed earlier. Alternative fusion strategies using classical inference usually seen in the literature result from either a generalization of the approach to the multivariate case (where the [measured] random variables come from multiple sensors) or from logical combinations of sensor-specific decisions that are individually based on a classical inference approach.

The advantages of classical inference techniques are that they are very well known, they exploit the sampling distributions (these can be difficult to obtain, however), and they provide a measure of the decision error probability. If there is a requirement to extend the approach to the case of multivariate statistics (as can be required in real-world situations), this requires the *a priori* knowledge and computation of multidimensional probability density functions. This is a serious disadvantage for realistic applications. Additional disadvantages include the following:

1. They can assess only two hypotheses at a time, namely, the hypotheses H_0 *versus* an alternate hypothesis, H_1.
2. Complexities arise for multivariate data.
3. Classical inference does not take direct advantage of *a priori* likelihood assessments.

Moreover, this method requires the availability *(a priori)* of the density (i.e., sampling) functions as well, often a serious limitation because the acquisition of such data can require significant investments in peacetime surveillance and may be politically provocative. Combined, these disadvantages are formidable; in real systems, such methods are only used if the necessary data are readily available.

Bayesian Inference

The Bayesian inference technique (e.g., [3]) resolves some of the difficulties with classical inference methodology. Bayesian inference updates the likelihood of a hypothesis given a previous likelihood estimate and additional evidence (observations). Bayesian inference takes its name from the English clergyman, Thomas Bayes, who died in 1760. A paper by Bayes, published in 1763, contains the inequality that is known today as Bayes theorem [4]. The technique may be based on either classical probabilities, or subjective probabilities (i.e., it does not necessarily require probability density functions, a potentially major simplification).

Suppose $H_1, H_2, \ldots, H_j$, represent mutually exclusive and exhaustive hypotheses (i.e., the existence of an object of identity i) that can "explain" an event E (or datum, observable, *et cetera*), which has just occurred. Then,

$$P(H_j|E) = \frac{P(E|H_j)P(H_j)}{\sum_j P(E|H_j)P(H_j)} \tag{7.1}$$

and

$$\sum_j P(H_j) = 1$$

where

$P(H_j|E)$ = the *a posteriori* probability of hypothesis H_j being true (object j existing) given the evidence, E;

$P(H_j)$ = *a priori* (and unconditional) probability of hypothesis H_j being true;

$P(E|H_j)$ = the probability of observing evidence E given that H_j is true (sometimes called the *forward conditional probability* or *likelihood* of H_j).

This is a pedagogically pleasing formulation, in that new evidence, E, is used to improve prior hypotheses about events.

The process of using a Bayesian formulation for fused identity estimation is illustrated in Figure 7.4. Multiple sensors observe parametric data (e.g., RCS, PRI, IR spectra) about an entity whose identity is unknown. The sensors each provide an identity declaration; that is, a hypothesis about the object's identity, based on the observations and a sensor-specific classifier algorithm (unspecified but frequently is a feature-based classifier using pattern recognition methods). For each sensor, the previously established uncertainty characteristics in classifier performance (developed either experimentally or theoretically) provide estimates of the

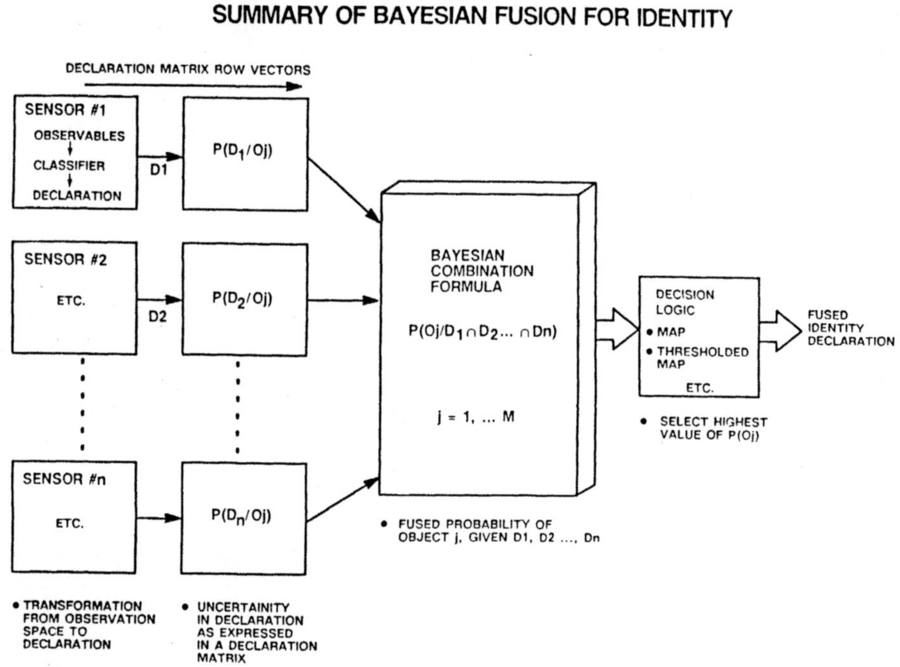

SUMMARY OF BAYESIAN FUSION FOR IDENTITY

Figure 7.4 Summary of Bayesian fusion.

probability that the sensor would declare the object to be the declared type, given that the true object is object j; that is, P (Declaration $/O_j$). These declarations are then combined via a generalization of equation (7.1). This provides an updated, joint probability for each possible entity, O_j, on the basis of the multisensor declarations; that is,

$$P(O_j/D1 \cap D2 \cap \ldots \cap D_n), j = 1, \ldots, M$$

is the probability of having observed object j (of the object set M) given declaration (evidence) $D1$ from sensor 1, declaration $D2$ from sensor 2, and so on. We can then apply a decision logic and select a joint declaration of identity by choosing the object whose joint probability $P(O_j/D1 \cap D2 \cap \ldots)$ is the greatest; this type decision rule is typically called *maximum a posteriori probability* (MAP). Thus, the Bayes formulation provides a means to combine identity declarations from multiple sensors to obtain a new (and, we hope, improved) identity declaration. Required input to the Bayes formulation includes the ability to compute, that is, model, $P(E/H_j)$ and the *a priori* probabilities of the hypothesis, $P(H_j)$, being true. When no *a priori* information exists concerning the relative likelihood of H_j, the "principle of indifference" is used in which $P(H_j)$ for all j are set equal. Disadvantages of Bayesian inference include

1. Difficulty in defining prior likelihoods.
2. Complexity when there are multiple potential hypotheses and multiple conditionally dependent events.
3. Requirement that competing hypotheses be mutually exclusive.
4. Lack of an ability to assign general uncertainty.

The Dempster-Shafer Method

Shafer and Dempster created a generalization of Bayesian theory that allows for a general level of uncertainty (see, e.g., [5, 6, 27]). The two methods produce identical results when all the basic or singleton hypotheses considered are mutually exclusive (viz., H_i does not overlap H_j for all $i = j$) and there is no general level of uncertainty.

The D-S approach derives in part from a behavioral model of how humans distribute their belief over propositions: the model, attributable to Shafer, asserts that this distribution occurs in a fragmentary way. In particular, the model argues that belief in a proposition (existence of a target of some particular type) is allocated not only to the single proposition under consideration but to disjunctions that include it. Because of this feature, the D-S approach is distinguished in that it is (1) two-valued—that is, each proposition is assigned two values of an uncertainty measure (analogous to but not equal to probability)—and (2) there is a general level of uncertainty in the way that evidence is attributable to a proposition, which results in the proposition being plausible but not directly supported or refuted by the evidence. Thus, we have for each proposition, P, an "evidential interval" bounded by an uncertainty measure called *support*, $S(p)$, and by a measure called *plausibility*, $Pl(p)$ as shown in Figure 7.5.

The values of $S(p)$ and $Pl(p)$ derive from what are called *probability masses*, which represent the distributed belief in the proposition. Notionally, we can consider each sensor in a multisensor system to produce propositions regarding identity (based on the evidence of observables), each proposition having a probability mass and $S(p)$ and $Pl(p)$ values. The final distinguishing characteristic of this approach, attributable to Dempster [6], are his "rules of combination" or formulas by which multiple mass distributions can be combined. Combined mass distributions are used to derive the joint or fused $S(p)$ and $Pl(p)$ values for any given proposition (object identity), completing the process. The decision criterion would be analogous to the Bayesian MAP, but here looking for maximum values of $S(p)$ and $Pl(p)$, indicating maximally supportive joint evidence and minimally refuting joint evidence.

The concept of using a Dempster-Shafer approach to fuse multisensor identity data is illustrated in Figure 7.6. Analogous to the Bayesian approach, individual

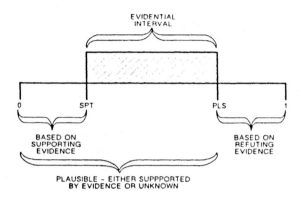

Figure 7.5 Evidential interval and uncertainty.

sensors collect parametric data from which probability mass values for each type of possible object are derived (the formalisms of how to compute such mass volumes are still incomplete). Such mass distributions are analogous to the row vectors

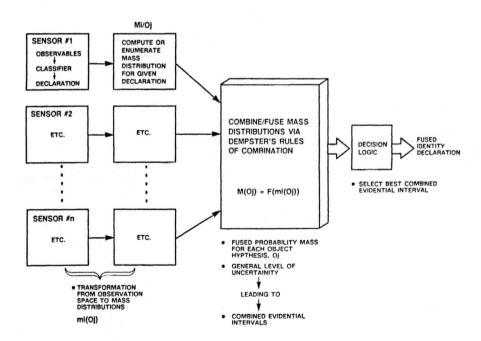

Figure 7.6 Summary of Dempster-Shafer fusion.

of conditional probabilities in a Bayesian sensor probability matrix (the declaration matrix). That is, these distributions also reflect the uncertainty in the sensor-specific classification process. Dempster's rules of combination provide a prescription for combining these masses, leading to a joint evidential interval, $[(H_i), \text{Pls}(H_i)]$, for each possible hypothesis (i.e., possible identity of observed entities). Some additional details on the Dempster-Shafer method are provided in Section 7.2.

At this time considerable controversy remains concerning the use of a Dempster-Shafer *versus* a Bayesian approach to identity fusion. (Section 7.2 expands on a comparison of these methods.) Key issues involve the performance of these approaches in a tactical environment involving electronic warfare, the computational complexity of Dempster's rules of combination, and finally the process by which a probability mass is devised for the case of real-world sensors. Numerous prototype systems are under development, with mixed results.

Information Theoretic Algorithms

A number of identity fusion methods are grouped under the category of information theoretic algorithms. These techniques are not based on a statistical approach, but instead make use of a transformation or mapping between parametric data and a resultant identity declaration. The overall philosophy is similar, i.e., similarity in identity is said to be reflected in similarity of observable parameters, but no attempts are made to directly model the stochastic aspects of the observables. The techniques include parametric templates, cluster algorithms, adaptive neural networks, voting, and entropy methods. With the exception of parametric templates, each of these techniques is described below. The discussion of parametric templates is combined with logical templates in the next subsection.

Cluster Analysis Methods

Cluster analysis methods employ a set of heuristic algorithms that have become popular in the biological and social sciences. Cluster analysis is a generic name for a wide variety of procedures that can be used to group data into natural sets or clusters. These clusters of data are interpreted by an analyst as representing a meaningful object category. For example, parameters such as pulse repetition interval and radio frequency might be used to separate radars by function or activity (e.g., tracking *versus* acquisition *versus* search). The concept is illustrated in Figure 7.7. One or more sensors observe a phenomenon (e.g., various radars of different types). The sensor data are selected and put into a cluster analysis algorithm. The algorithm groups the data into clusters that may be interpreted as implying object-class membership, such as radar by type or activity.

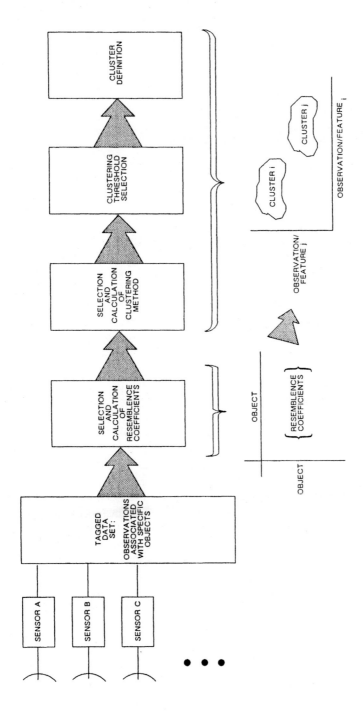

Figure 7.7 Concept of cluster analysis.

Five basic steps are required for any type of cluster algorithm:

1. Select the sample data.
2. Define the set of variables (or features) with which to characterize the entities in the sample.
3. Compute the similarities among the data.
4. Use a cluster analysis method to create groups of similar entities on the basis of data similarities.
5. Validate the resulting cluster solution.

All these steps are important in cluster analysis. Definition of the set of variables by which to characterize the entities in the sample entails selection of observable features that (we hope) will allow separation of data into identifiable groups. Numerous other features could have been selected with potentially different resulting clusters. Features also may be created by transformational processes on basic observables, such as Fourier transforms (resulting in Fourier coefficients). Ultimate selection of useful features depends on knowledge of the underlying physical processes or simply on trial and error. The features in turn may be scaled or weighted by a variety of means and normalized to permit integration independent of scale.

All cluster algorithms require the definition of a similarity metric or association measure that provides a numerical value representing the degree of "closeness," or, alternately, dissimilarity, between any two feature vectors Y_i and Y_j. Four major types of similarity are described by Aldenderfer and Blashfield [7]. These include (1) correlation coefficients such as Pearson's product-moment correlation coefficient, (2) distance measures such as Euclidean distance and Mahalanobis's distance, (3) association coefficients for binary variables (e.g., Jaccard's coefficient and Gower's coefficient), and (4) probabilistic similarity coefficients. Summaries of these measures are given by Sneath and Sokal in [8].

Given a scaled data set and selected similarity metric, a number of algorithms have been developed to search for natural groupings of clusters in the feature space. Aldenderfer and Blashfield [7] identify seven families of clustering methods summarized in Figure 7.8. The figure lists each family of clustering methods (i.e., hierarchical agglomerative methods, iterative partitioning, hierarchical decisive, density search, factor analytic, clumping, and graph theoretic methods). For each family of methods, a brief description of the algorithm strategy is provided along with references describing the details of the algorithms. One of the difficulties in studying cluster methods is the plethora of techniques derived in diverse applications such as biology, sociology, and psychology. Many of these algorithms having different names are in fact identical mathematical procedures. (Note that these statistically nonparametric methods are similar but are in addition to so-called unsupervised or feature-based techniques of pattern recognition; see, e.g., [9].)

Family of Clustering Methods	Algorithm Approach	References
Hierarchical Agglomerative Methods	For each observation pair, compute the similarity measure; use linkage rules to cluster observation pairs that are most similar; continue hierarchically comparing and clustering observations to observations, observations to clusters, clusters to cluster until no additional clustering is feasible via the clustering rules. Types of rules include single linkage, complete linkage, and average linkage.	Sneath, P., and R. Sokal Numerical Taxonomy, San Francisco, W. H. Freeman, (1973). Dubes, R., and A. Jain "Clustering Methodologies in Exploratory Data Analysis", Advances in Computers Vol. 19, 1980, pp. 113-228.
Iterative Partitioning Methods	1) Begin with an initial partition of data set into a specific number of clusters, compute cluster centroids. 2) Allocate each data point to the cluster with the nearest centroid. 3) Compute new centroids of the clusters. 4) Alternate steps 2 and 3 until no data points change clusters.	Anderberg, M. Cluster Analysis for Applications, New York, Academic Press, (1973).
Hierarchical Devisive Methods	Logical opposite approach to hierarchical aggomerative methods. Initially all data are assigned to a single cluster. This cluster is divided into successively smaller chunks using either a monothetic or polythetic strategy. Monothetic clusters are defined by certain variables on which certain scores are necessary for membership. Polythetic clusters are groups of entities in which subsets of the variables are sufficient for cluster membership.	Everitt, B. Cluster Analysis, New York, Halsted, (1980).
Density Search Methods	Algorithms search the feature space for natural modes in the data that represent volumes of high density. Strategies include a variant of single linkage clustering and methods based on multivariant probability distributions.	Everitt, B., D. Wishart, "Supplement CLUSTAN User Manual," 3d ed., Program Library Unit, Edinburgh University, 1980, 1982.
Factor Analytic Methods	Also known as factor analysis variants, inverse factor analysis, and Q-type factoring. Methods form a correlation matrix of similarities among cases. Classical factor analysis is then performed on the NxN correlation matrix. Data are assigned to clusters based on their factor loadings.	Skinner, H. "Dimensions and Clusters: A Hybrid Approach to Classification," Applied Psychological Measurement, Vol. 3, (1979), pp. 327-341.
Clumping Methods	Clumping methods allow case membership in more than one cluster. Cases are iteratively assigned to one or more clusters to optimize the value of a criterion referred to as a cohesion function.	Cole, A. J., and D. Wishart, "An Improved Algorithm for the Jardine-Sibson Method of Generating Overlapping Clusters," Computer Journal, Vol. 13, (1970), pp. 156-163.
Graph Theoretic Methods	Use of graph theory to develop a hierarchical tree linking data into clusters (analogous to hierarchical aglomerative methods).	Dubes and Jain (1980). Ling, R., "An Exact Probability Distribution of the Connectivity of Random Graphs," Math Psychology, Vol. 12, (1975), pp.90-98.

Figure 7.8 Summary of cluster algorithm families.

Cluster analysis methods provide a valuable tool for exploring new relationships in data that may lead to identification paradigms. However, because of the heuristic nature of cluster algorithms, their application is fraught with potential biases. In general, data scaling, selection of a similarity metric, choice of clustering algorithms, and (sometimes) even the order of the input data may substantially affect the resulting clusters. Aldenderfer and Blashfield [7] provide a dramatic example of these effects. Hence, use of cluster methods must be judged on their effectiveness and repeatable ability to form meaningful identity clusters. The user must be the sole judge of these results.

Adaptive Neural Nets

Artificial neural net models, or adaptive neural systems, are hardware or software systems that seek to emulate the processes some have postulated for biological nervous systems. A neural network consists of layers of processing elements, or nodes, that may be interconnected in a variety of ways. Figure 7.9 illustrates a three-layer network with each layer having four processing elements. Data vectors are entered on the left-hand side of the network and the neural net performs a nonlinear transformation, resulting in an output vector on the right-hand side of the network. Such a transformation could produce the type of mapping from data to identity categories as is performed by cluster analysis techniques. Thus neural nets can be used to

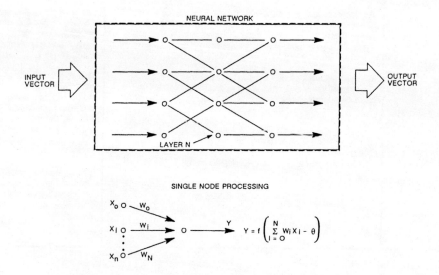

Figure 7.9 Overview of adaptive neural systems.

transform multisensor data into a joint declaration of identity for an entity. Recent applications of neural nets to data fusion have included hull-to-emitter or "Hultec" association (identifying ships from observed parametric radar data: PRI, RF, *et cetera*) by Priebe and Marchette in [10] and pattern recognition for sonar applications [11]. A discussion of other defense and fusion applications is provided by R. North in [12].

The processing concept for a single node of a neural net is shown in the lower half of Figure 7.9. Each node in a layer receives inputs from up to N nodes in the preceding layer. These inputs, $X_0, X_1, \ldots, X_N$ are combined in the processing element to produce an output, Y, which in turn acts as the input to nodes in subsequent layers of the net. The combination of X_i to produce Y is performed using a nonlinear function of the weighted values of x; that is,

$$Y = f\left(\sum_{i=0}^{N} w_i x_i - \theta\right) \tag{7.2}$$

where the function f may assume a variety of forms, including, for example, step functions or sigmoid functions.

A number of variations are possible in formulating a neural network; for example, the number of layers in a network, the number of nodes per layer, the interconnectivity among nodes, the selection of the nonlinear and weighting functions, and other details. Lippmann provides a description of these variations in [13] and shows a taxonomy of commonly used networks for classification applications:

- *Binary input nets*
  Supervised (Hopfield net, Hamming net)
  Unsupervised (Carpenter-Grossberg classifier)
- *Continuous-valued input nets*
  Supervised (perceptron, Multilayer perceptron)
  Unsupervised (Kohonen self-organizing features maps).

The terms *supervised* and *unsupervised* refer to the training mode used. A neural net can be adapted or trained to perform correct classifications (i.e., mapping the input data from known entities into their associated known identity classes) by systematically adjusting the weights, w_i, in the equation for Y. This is typically performed using a sample or training data set (in which object identities are known). Nets trained with supervision (e.g., the Hopfield net and the perceptron) are provided with information or labels that specify the correct mapping between input data and output classifications. Unsupervised training approaches are trained simply to form output data clusters without specifying the "correct" input-output mapping. Nets also may be trained adaptively, allowing the weights to vary as is done in adaptive signal processing.

At this time a number of experiments have been performed using neural nets for fusion of identity data. Preliminary results such as in [14] suggest that neural nets are superior to traditional cluster methods for identity fusion, especially when the input data are noisy and when data are missing. However, at this time much groundwork remains to develop a theoretical framework to understand basic issues such as

1. Selection of a network model,
2. Choice of number of layers and nodes,
3. Development of a training strategy,
4. Incorporation of neural nets with traditional classifiers.

Voting Methods

Perhaps the most conceptually simple technique for combining identity declarations from multiple sensors is voting (including Boolean AND/OR processing). In this process, each sensor's ("hard") declaration is treated as a vote in a democracy in which the majority or plurality rules or, in more complex structures, decision trees are employed. Each sensor provides an input declaration of identity of an observed entity (e.g., $D1$ from sensor 1, $D2$ from sensor 2). The voting algorithm searches the declarations to find a majority declaration (or other simple decision rule) and declares the joint declaration to be that of the majority, plurality, *et cetera*. Additional complexities may, of course, be introduced via weighting the sensor input, application of thresholds, and other decision logic. Although conceptually simple, the voting technique can be valuable, especially for real-time techniques, when accurate *a priori* statistics are not available, or may be attractive from an overall cost-benefit point of view.

Entropy Methods

In the late 1940s Shannon [15] and others began to develop the concept of information entropy (analogous to the concept of thermodynamic entropy in physics) to describe the extent to which a communication system is ordered *versus* disordered or random.

The entropy measure reflects and quantifies the information in a generalized "message" on the basis of its probability of occurrence. The basic philosophy is that frequently occurring "messages" or data are of low value, and "surprising" or "rare" messages (e.g., "World War III is about to occur") bear greater value. Presuming we adhere to this philosophy, we seek a function that decreases with increasing message probability.

The "self-information" of message m is defined as follows (see, e.g., [16]):

$$S_m = -\log_2 Pr(m) \sim \text{bits} \tag{7.3}$$

where the base of the logarithm defines the units, here bits, corresponding to a base of 2. Extending this concept to a series of messages or symbols (or, for instance, identity-bearing observations or features), we have the average message information as the entropy, H_m (see also [16] here):

$$H_m = \sum_{i=1}^{j} P_i(-\log_2 P_i) \tag{7.4}$$

where there are j symbols or components of the message. Note, of course, that this is an expected value in a statistical sense. Various books on information theory discuss the many interesting properties of this parameter. For example, note that H_m is a maximum when all P_i are equal.

In [17], this concept is brought to bear on an example involving aircraft identification using a rule-based (i.e., knowledge-based) approach. In this paper, Kelley and Simpson first derive (following the work of Dretsky [18]) the informational value of a rule, then its entropy. This expression is given by

$$H(\text{rule}) = -\wedge P(\text{true}) \log_2 P(\text{true}) \\ - P(\text{false}) \log_2(\text{false}) \tag{7.5}$$

where $P(\ ) = P(\text{antecedent evaluation})$, and $\wedge$ is a term called *information leverage* and given by

$$\wedge = 1 + \frac{\log_2 p(\text{consequent true})}{\log_2 p(\text{antecedent true})} \tag{7.6}$$

They extend their analysis to rule chains:

$a \rightarrow b$
$b \rightarrow c$
$c \rightarrow d$

as

$$H(\text{chain}) = -\underset{a\ b\ c}{\wedge \wedge \wedge} P(a = \text{true}) \log_2 P(a = \text{true}) \\ - P(a = \text{false}) \log_2 P(a = \text{false}) \tag{7.7}$$

where terms are as described earlier. These expressions can be used in rule-selection strategies in a knowledge-based approach to object identification. Kelley and Simpson point out however that single rule selection to maximize entropy does not necessarily assure maximum progress toward a goal or solution. Because of this constraint, they derive the rule chain entropy, where maximization of this entropy would better assure concurrent progress toward a solution. A final concern is that entropy, as a measure within any global rule set construct, will exhibit specific variance characteristics that should, ideally, be studied or estimated. Kelley and Simpson suggest a variance-modified rule chain entropy to compensate for the variance of unknown outcomes.

Hence, we see that this measure can be applied, with a creative approach, to identification problems. It would appear, intuitively, that entropy measures could be applied both effectively and more straightforwardly in strictly numeric (e.g., feature-based) approaches to identity estimation than in symbolic-based approaches.

Cognitive-Based Models

The third class of identity algorithms identified in the taxonomy in Figure 7.1 is the class of cognitive-based models. These are models that seek to mimic the human cognitive processes in identifying entities from multiple sensor data. Three basic models are described here: logical templating methods, knowledge-based or expert systems, and fuzzy set theory. The use of figures of merit for templating is also described. The former two techniques are especially designed for making high-level inferences about the existence and intent of complex entities such as battlefield units or force structures.

Logical Templating Methods

Templating is the name given to logic-based pattern recognition techniques used in multisensor data fusion processing for event detection or situation assessment, but the method has application to single-object identity estimation as well. This class of techniques has been successfully used in data fusion systems since the mid-1970s. The term *templating* comes from the concept of matching a predetermined pattern (or template) against observed data to determine whether conditions have been satisfied, thereby allowing an inference to be made (see also Section 9.2 for additional discussion regarding templating).

This pattern matching concept can be generalized for complex patterns involving logical conditions (e.g., Boolean relationships), fuzzy concepts, and uncertainty in both the observational data and the logical relationships used to define a pattern. Templating represents a means of combining parametric pattern matching with logical relationships. For example, a template could compare the observed value of the pulse repetition interval of an emitter with *a priori* thresholds, and it could determine the temporal and spatial relationship of the observed emitter to other possible entities. Templating is one way to bridge the gap between algorithmic approaches, such as cluster analysis, and figure-of-merit approaches and heuristic methods based on artificial intelligence. Moreover, attribute fusion techniques such as Bayesian inference, Dempster-Shafer inference, and fuzzy logic can be incorporated into template algorithms.

The basic templating algorithm can be used to match multisensor observational data against prespecified conditions to determine if the observations provide evidence to identify an entity. The input to the templating process is one or more observations, which may include parametric and nonparametric data, over a period of time; hence, temporal variation (not normally included in other methods mentioned so far) can be included in these structures. The output is a declaration of whether the observations match a predetermined profile. The output may also include an associated confidence level or probability of the object declared by the template matching process.

Figure 7.10 shows a generic template. A generic template is illustrated because every application requires information specific to the application domain. However, some types of information are similar on all templates. The domain-independent information includes template identification, threat type, acceptance threshold, rejection threshold, necessary conditions, sufficient conditions, and the components that describe or make up the object.

The template-based identification process must also provide a means of retrieving the template from a data base or naming the template for system users. Threat type describes the category of tactical threat addressed by the template. Acceptance and rejection thresholds are user-specified numerical or logical criteria utilized in the automatic template process. Necessary and sufficient conditions pre-

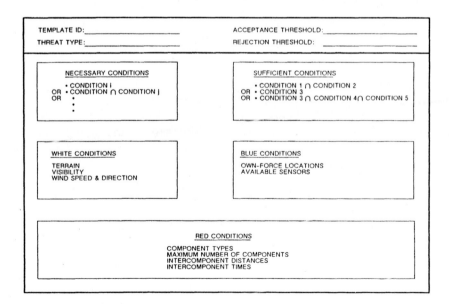

Figure 7.10 Example of a generic template.

scribe the required observations, logical relationships, and data patterns that cause the template to be accepted as a candidate in the automatic template process.

In Figure 7.10, *white conditions* refers to those environmental factors that influence both enemy activities and friendly or neutral force activities. These may include terrain, weather, or other effects. *Blue conditions* refers to predetermined conditions concerning friendly forces, such as location and available sensors. *Red conditions* refers to observations made about enemy units or entities. Thus, templating techniques provide ways to represent the context of the observational process. Subsection 9.2.2 provides an example of the basic template processing flow for event recognition, but the logic is basically the same for identity estimation.

Templates or various aspects of templates have been used with varying degrees of success in several data fusion systems. Recent research is described by Hall and Linn in [19], Gabriel and Gabriel in [20], and Schroeder and Wright in [21].

The robustness of templating methods has also led to their use in situation assessment (Level 2) processing. A discussion of these applications, which bear strong similarity to those for identity estimation, is included in much of Chapter 9.

Figures of Merit

Figures of merit (FOMs) are metrics that are usually defined on the unit scale and derived from any plausible basis to establish a degree of association between observations and object identity. In turn, *measures of correlation* (MOC) are weighted combinations of FOMs. Creative use of this very general method provides a practical and reasonable approach to identity estimation.

An FOM can represent the degree to which observations relate to a hypothesized object identity (reflecting "how close" observation O_i is to a "reference" observation O_r expected for a known object), or it may reflect object-object associations based on distances or other measures. In essence, the FOMs and MOCs are sets of "templates" that reflect expected observations, behaviors, logical relationships, or any other basis for profiling an object's (or set of objects) identity. An overview of FOM and MOC processing is described in Subsection 9.2.3, including representative logic identity declaration.

Knowledge-Based (Expert) Systems

In recent years, a large number of expert, or knowledge-based, systems have been developed successfully, with commercial, industrial, and military applications. These systems have evolved from experimental prototypes used to illustrate principles of expert system research to tools used daily in industry. It is only natural that the expert systems approach should be applied to multisensor data fusion; see [22] for comments on the technologies for fusion.

Figure 7.11 illustrates the concept of an expert system computer program. The structure of the program comprises four logical parts:
- A knowledge base that contains facts, algorithms, and a representation of heuristics,
- A global data base that contains dynamic data,
- A control structure, or inference engine,
- A human-machine interface.

The control structure uses input data and facts and attempts to search through the knowledge base to reach conclusions.

The knowledge base for an expert system contains the information that constitutes the system's "expertise." Knowledge may be represented by several mechanisms:
- Production rules in the form, If (evidence exists for X) then do Y, where Y may involve performing a computation, updating a data base, *et cetera*.
- Semantic nets, which are graphical representations that describe relationships between general objects and more specific instances of that object (e.g., the general object, aircraft, and specific instances of *aircraft*, such as B-1, 747, or F-15).
- Frames, which represent objects and classes of objects via structured data records.
- Scripts, which use the concept of a play (e.g., acts, scenes, settings, actions) to describe situations and events.

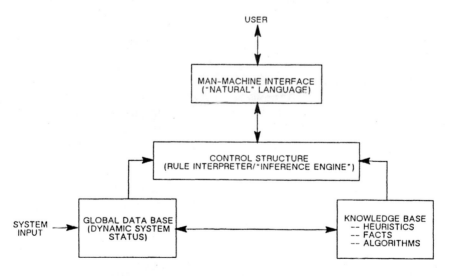

Figure 7.11 Basic structure of an expert system.

Each of these types of knowledge representations has advantages and disadvantages for representing information for specific applications.

The primary differences between expert systems and ordinary computer programs are twofold. First, an expert system deliberately separates the control process from algorithms in the knowledge base. Ordinary computer programs tend to embed the program control (e.g., logic) with the problem-specific algorithms (knowledge base). This is an efficient design when all of the logical paths are known and fixed, but breaks down when the complete set of branches is unknown or very large (potentially infinite). The advantage of an expert system is that the knowledge base can be readily modified without reprogramming the system. Cases can be added or paths modified as experience grows, building up a complex knowledge base.

A second difference between ordinary computer programs and expert systems is implied by the dynamic and inexhaustive search through the knowledge base. The expert system program searches for a supportable inference, but does not exhaustively search the knowledge base. Hence, an expert system obtains an answer, but not necessarily the only answer. Indeed, the conclusion may differ, depending on the order in which the data is entered. Certainly, the order of the chain of reasoning depends upon the data order. This aspect makes it difficult to evaluate expert systems.

Depending on the knowledge representation technique used, these approaches permit varying degrees of flexibility in representing an object's identity through both numeric-type characterizations and symbolic or reason-based characterizations, which can be a distinctive advantage in cases where parametric profiling is insufficient. However, the successful design and development of expert systems is still a difficult process. Even today, the overall "knowledge engineering" process is not fully understood and frequently costly; test and evaluation are difficult from both a philosophical and a practical point of view because, in many cases, the underlying truth (e.g., true object identity) may never be available so that the applied knowledge cannot be formally tested; finally, achieving efficient processing can still be quite difficult for meaningful military applications (see Section 2.6 regarding parallel processing methods for knowledge-based systems).

Fuzzy Set Theory

The last cognitive-based inference model introduced is fuzzy set theory. Fuzzy set theory was developed by Zadeh [23, 24]. The basic concept is that people frequently deal with concepts that are imprecise because of indistinct boundaries of definition. Proponents argue that commonly used terms such as *tall, short, attractive,* or *ugly* are imprecise not because the human thought processes utilizing these terms do not understand them, but rather because these terms refer to attributes that are inherently imprecise. Fuzzy set proponents argue that such imprecision can be addressed mathematically via an extension of Boolean set theory.

Fuzzy sets are defined as follows. A set A has members $X_0, X_1, \ldots, X_n$. Each element, X_1, of set A has an associated value, $\mu A(X_1)$, which indicates the degree to which X_1 belongs to set A. The function is called the *membership function*, $\mu(X)$, and has a value between 0 and 1 with $\mu A(X) = 0$ indicating that element X is *not* a member of set A, and $\mu A(X) = 1$ indicating that element X is *completely* a member of set A. The values of $\mu(X)$, within this range, must be provided by the person defining the fuzzy sets. Hence, in fuzzy set theory, sets are defined by ordered pairs, $[X, \mu(X)]$ in which X is an identified set element, and $\mu(X)$ is the associated membership value for element X. By contrast, Boolean sets are defined by identifying the elements that completely belong to a set (hence $\mu(X)$ is either 1 or 0).

The real value of fuzzy set theory to data fusion is its extension to fuzzy logic; because of this, we classify this method in the "cognitive" group. Fuzzy logic deals with approximate modes of reasoning; see [24]. In classical two-valued logic, a proposition, p, is either true or false. Classical logic uses truth tables and manipulation rules to follow a chain of reasoning to determine the truth (or falseness) of a proposition. By contrast, in fuzzy logic, a proposition has a membership value ranging from 0 (completely false) to 1 (completely true), representing the membership of the proposition in the truth value set. In a sense, a fuzzy truth value may be viewed as an imprecise characterization of a numerical truth value.

Fuzzy logic is well defined, with a means of representing fuzzy propositions, combination rules to create syllogisms, and inference utilizing fuzzy probabilities. Basic rules of inference are defined, analogous to classic logic (e.g., conjunctive rule, Cartesian product, projection rule). Currently, commercial software tools that support fuzzy inference are beginning to appear.

The value of fuzzy set theory and logic for data fusion is still being researched. Nevertheless, it is intuitively appealing to allow fuzzy reasoning in describing the identity of battlefield objects or object sets. Statements about the approximate coposition of enemy units, enemy intent, and operational objectives cannot realistically be made without utilizing approximate reasoning. Perhaps the largest hurdle in gaining acceptance of fuzzy logic in data fusion systems may be the unfortunate choice of the name *fuzzy*. Much work remains in this area to define membership functions in real-world situations and comparison of fuzzy reasoning with methods utilizing probability, evidential intervals, or ad hoc confidence intervals.

7.2 COMPARISONS BETWEEN BAYESIAN AND DEMPSTER-SHAFER TECHNIQUES*

Of all the methods presented, the Bayesian and Dempster-Shafer techniques have been the ones most actively researched and employed, and because of this they

---

*The remainder of this section has been graciously contributed by Dr. Dennis M. Buede, a close colleague of the authors.

warrant special attention; hence the existence of this section. As the various methods presented in the preceding section employ representative "calculi" available for representation and processing uncertainty, the reader interested in uncertainty representation may want to refer to Chapter 12, which gives some additional perspectives on uncertainty in the context of AI applications.

7.2.1 Perspectives on the Formalisms

This section concentrates on some of the formalisms of probability theory with (1) Bayes' rule and (2) Dempster's rule applied to belief functions. Shafer and Tversky [25] provide good example-based comparisons of these two approaches. Each of these approaches is defined by a set of axioms. The differences between these axiom sets are discussed in the sections which follow. Both of these calculi are commutative and associative, guaranteeing that the order in which independent sets of information are received will not affect the result. More detailed, theoretical comparisons and generalizations of these approaches can be found in Kyburg [26], Shafer [27], and Lindley [28].

As mentioned in Section 7.1, a major difference of these two axiom structures is how they deal with the conjunction of an event and its negation; that is, A union (not A). Probability defines this conjunction to be the universal event and assigns it a probability of 1.0. Evidence theory operates on propositions about events and defines a proposition, called *uncommitted* or *don't know* to which a belief measure can be assigned. As a result, the belief assigned to the conjunction of a proposition and its negation can be, and usually is, less than 1.0.

The requirement by the Bayesian approach for prior probabilities that is troublesome to some people can be circumvented in the Dempster-Shafer formalism. All initial beliefs can be assigned to the "uncommitted" state and redistributed by Dempster's rule as new evidence arrives.

A distinct advantage that probability theory is its incorporation into an axiom-based *decision logic*. This rational decision logic is presented in detail in [29]. Schum [30] and Henrion [31] deal with some of the practical advantages of this logic; namely, that probabilities can be used to guide decisions. Formalisms for decision-making using belief functions and the other measures of uncertainty are currently being developed.

Yet another modification of a traditional Bayesian approach involves so-called belief networks. The development of belief networks (directed acyclic graphs) began in about 1980. These networks accommodate dependencies in uncertain situations and facilitate the computation of uncertain measures in large problems. Pearl [32] collects much of this work and presents a unified approach to fusing information of varying levels of abstraction using probability theory. The structure of these networks is independent of the uncertainty calculus used for computation.

If the network can be shown to be singly connected (there is at most one path between any two nodes), then the probabilistic updating can be accomplished by local propagation in an isomorphic network of parallel autonomous processors. Gordon and Shortliffe [33] and Shenoy and Shafer [34] have developed much of this theory for the Dempster-Shafer approach. Shafer [35] points out that the bodies of evidence being combined via Dempster's rule must be independent. Shachter [36] has consolidated and advanced the theory behind influence diagrams that include probabilities within the decision analysis process.

The body of literature on the Bayesian and D-S techniques is quite large and so it is wise to avoid strong, authoritative assertions regarding either mathematical formalism or applications. Because of this, in what follows, an attempt has been made to remain as unbiased and objective as possible. Nevertheless, important distinctions do exist and it is hoped that the discussion expands the reader's knowledge and perspectives about these two important methods. Pearl [32] quotes Shafer as follows: "Probability is not really about numbers; it is about the structure of reasoning." Consistent with this thought, this section describes the two approaches to uncertainty management and applies the associated reasoning processes to aircraft identification problems.

7.2.2 Bayesian Probability Theory

In this subsection the framework for probability theory and Bayesian revision is developed. Then, a discussion of the meaning and interpretation of probabilities is presented. Finally, a data fusion example for the identification of aircraft is illustrated.

Mathematical Properties

An important property of an uncertainty management system is whether it is based on a set of axioms. The subject of uncertainty is so complex that any system not resting firmly on a set of axioms is subject to considerable internal incoherence for any application. Probability theory has been studied so extensively that several axiom sets have been developed, each having the same result.

The nomenclature used here is as follows:
- A, B, and C represent events that may or may not occur,
- $p(A)$ is the probability that event A will occur,
- $p(A/B)$ is the probability that event A will occur given that event B has occurred (or information B is known),
- The intersection of two events, A and B, defines the event C common to A and B, and is designated by $C = AB$,

- Two events are called *mutually exclusive* if their intersection is the null (or empty) event, ϕ,
- A set of events, $\{A_n\}$, is called *collectively exhaustive* if their union defines the universal (or certain) event, I.

Probability Axioms

1. $p(A/B)$ is a real, nonnegative number: $p(A/B) \geq 0$.
2. The probability of the certain event, I, is normalized: $p(A/A) = p(I) = 1$.
3. If the events in $\{A_n\}$ are mutually exclusive given B, then

$$p\left(\sum_n A_n | B\right) = \sum_n p(A_n | B) \tag{7.8}$$

that is, the probability of the summation of events is the sum of probabilities when there is no overlap between the events.

4. The conditional probability of the intersection of two events is

$$p(BC|A) = p(C|AB)p(B|A) \tag{7.9}$$

that is, the probability of B and C is the probability of C given that B has occurred times the probability of B, everything conditioned on A.

From these axioms the following results can be obtained:

$$p(\phi) = 0 \tag{7.10}$$
$$p(\text{not } A) = 1 - p(A) \tag{7.11}$$
$$p(A \text{ union } B) = p(A) + p(B) - p(AB) \tag{7.12}$$

An important concept in probability theory is probabilistic independence. Two events, A and B, are considered probabilistically independent of each other if their joint probability is equal to the product of their marginal (or unconditioned) probabilities:

$$p(AB) = p(A)p(B) \tag{7.13}$$

which is to say that $p(A|B) = p(A)$ and $p(B|A) = p(B)$.

Similarly, two events, A and B, are considered conditionally independent of each other if there is an event C such that

$$p(AB|C) = p(A|C)p(B|C) \tag{7.14}$$

The joint probability of several events can be found through several forms of the chain rule:

$$p(ABC) = p(A)p(B|A)p(C|BA) \qquad (7.15)$$
$$= p(B)p(C|B)p(A|BC), \text{ and so on.}$$

An approach to computing the marginal probability of an event using conditional probabilities is called *expansion:*

$$p(A) = p(AI) = p[A(B + \text{not } B)] \qquad (7.16)$$
$$= p(A|B)p(B) + p(A|\text{not } B)p(\text{not } B)$$

The general expression for (7.16) using $\{B_n\}$, where B_1 through B_n are mutually exclusive and collectively exhaustive, is

$$p(A) = \sum_{i=1}^{n} p(A|B_i)p(B_i) \qquad (7.17)$$

The Bayesian Form

Reverend Thomas Bayes discussed an approach to the updating of uncertainty based upon new evidence (see [4]). Rewriting (7.9) yields

$$p(C|AB) = \frac{p(BC|A)}{p(B|A)} \qquad (7.18)$$

Now, let us interpret C as one element of a mutually exclusive and collectively exhaustive set of potential outcomes and write it C_i, interpret B as a set of data that has been collected, and expand (using (7.17)) B as a function of C:

$$p(C_i|AB) = \frac{p(BC_i|A)}{\sum_j p(B|C_jA)p(C_j|A)} \qquad (7.19)$$

Finally, we can rewrite the numerator of (7.19) using (7.9):

$$p(C_i|AB) = \frac{p(B|C_iA)p(C_i|A)}{\sum_j p(B|C_jA)p(C_j|A)} \qquad (7.20)$$

where

$p(C_i/A)$ is an *a priori* (or prior) probability of C_i, based upon the state of information A;

$p(C_i/AB)$ is the *a posteriori* (or posterior) probability of C_i given the data B and the prior state A;

$p(B/C_iA)$ is called the *likelihood function*, that is, it is the likelihood of observing the data B given C_i and the prior state of information;

$\Sigma_j p(B/C_jA)p(C_j/A)$ is the "preposterior" or probability of the data occurring, given the prior state of information, but conditioned on all possible outcomes C_j.

Equation (7.20) is Bayes's rule, which gives the posterior probability of C_i given that data B is observed and a prior state of information A. This posterior probability is equal to the likelihood function (probability that the data would be observed conditioned on C_i and A) times the prior probability that C_i would occur, divided by the prior probability of observing the data.

It is possible to aggregate probability statements from a lower level of abstraction to a higher one using either equation (7.8) or (7.12). Although this presentation has concentrated on discrete events, there is a similarly complete development of probability for continuous valued variables, including both scalars and vectors.

The likelihood expressions represent how solid or subject to change a given probability statement is. These functions must be developed prior to collecting the data by thinking through the problem. Note that the preposterior is simply the combination of the likelihood functions and the prior distributions.

An often-voiced concern of potential users of Bayes's rule is the need for a prior distribution over the events of concern. In the real world, this necessitates a subjective (or personal) interpretation of probability (see the next subsection) or arbitrarily setting the probabilities for each outcome equal, sometimes called the *principle of indifference*. Too often the large amount of information that is available for justifying estimates of prior probabilities is ignored due to concern about formal correctness and validity.

Interpretation and Assessment of Probabilities

Along with the vast mathematical developments associated with probability theory, there are many interpretations or definitions of the concept of probability. Generally speaking, three types of probabilities typically are described in the literature:

- *Empirical probability*—Wherein probability values are derived from "long-run" interpretations of sampling distributions. In such cases, probabilities are assigned *a posteriori;* that is, with the sampling distribution in hand.

- *Classical probability*—Wherein probability is defined by the potential for the event of interest to occur as compared to other equally probable outcomes. This is the interpretation first taught to students of probability and typically couched in the context of simple experiments (e.g., tossing coins). When calculable, this probability can be assigned *a priori* (prior to the experiment).
- *Subjective probability*—Wherein probability is assigned on the basis of an experience, based on the interpretation of the available information. Although there is a body of formal mathematical literature on subjective probability, specific values derive from individual, personal judgment.

For military or defense applications, the appropriate probability often will be of a subjective type because frequently it is the only way that the occurrence of so-called singular or rare events (as many military or defense events are) can be probabilistically estimated.

A Data Fusion Example

The Bayesian approach to data fusion, based on updating probabilities, is shown as an influence diagram in Figure 7.12. This process first converts all sensor reports for a specified track in the given time period to likelihood functions based on type. For this example, there will be two sensors, an *identification-friend-foe-neutral* (IFFN) sensor that responds with a "Friend" declaration, $p_{IFFN}(\text{data}|\text{FRD})$ when it detects and receives a valid response to its interrogation, and an electronic support measures sensor that can identify aircraft at the type level, $P_{ESM}(\text{data}|T_k)$. An air-

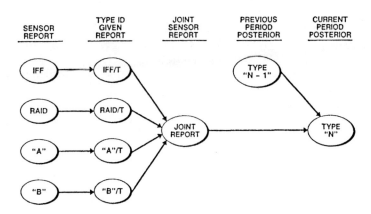

Figure 7.12 Influence diagram for Bayesian type identification fusion.

craft of type k is represented as T_k. For the IFFN sensor the expansion equations must be used to produce probabilities conditioned on type as follows:

$$P_{\text{IFFN}}(\text{data}|T_k) = P_{\text{IFFN}}(\text{data}|\text{FRD}) \cdot p(\text{FRD}|T_k)$$
$$+ P_{\text{IFFN}}(\text{data}|\text{FOE}) \cdot p(\text{FOE}|T_k) \quad (7.21\text{a})$$
$$+ P_{\text{IFFN}}(\text{data}|\text{NEU}) \cdot p(\text{NEU}|T_k)$$

or

$$P_{\text{IFFN}}(\text{data}|T_K) = P_{\text{IFFN}}(\text{data}|\text{FRD}) \cdot p(\text{FRD}|T_k)$$
$$+ P_{\text{IFFN}}(\text{data}|\text{not FRD}) \cdot p(\text{not FRD}|T_k) \quad (7.21\text{b})$$

for each target type k. The first formulation (7.21a) is appropriate if the foe aircraft have some capability of mimicking the proper response to the IFF interrogation; the second form (7.21b) is appropriate if they do not. The likelihood functions, $p_{\text{IFFN}}(\text{data}|\quad)$, are based upon some type of *a priori* test measurements. Note, of course, that the second term nominally requires access to a FOE's "mimicking" or deception capability. In most cases this is unlikely unless such deception-generating equipment is obtained covertly—thus, surrogate equipment is often used to obtain such data. The values for the $p(\text{FRD}|T_k)$, $p(\text{not FRD}|T_k)$, $p(\text{FOE}|T_k)$, and $P(\text{NEU}|T_k)$ for a scenario in which only one side has a given type of aircraft are either 1 or 0. For scenarios in which the same aircraft is employed by both sides, an additional type aircraft will be defined (e.g., friendly k and enemy k). As long as one sensor (e.g., the IFFN sensor) can distinguish between these two types, the Bayesian approach will function. The set of arrows from the "sensor reports" to the "type identification given report" in Figure 7.12 depicts this first operation (i.e., per equation (7.21a) or (7.21b)).

The second step is to compute the joint sensor report based on the reports of all sensors during a given time period. Because in this case the individual identification reports rely upon different phenomenologies (so called "non commensurate sensors), these probabilities can be considered independent and the joint likelihood values are

$$p(\text{data}|T_k) = \prod_i p_i(\text{data}|T_k) \quad (7.22)$$

for all k.

Finally, Bayes's rule is used for the last link of the influence diagram in Figure 7.12:

$$p(T_k|\text{data}) = \frac{p(\text{data}|T_k) \cdot qT_k}{p(\text{data})} \quad (7.23)$$

where

qT_k = the previous period's value of $p(T_k|\text{data})$;
$p(\text{data}) = \sum_k p(\text{data}|T_k) \cdot qT_k$.

$p(\text{data})$, the preposterior, is the probability of seeing the data that was collected based upon the priors, qT_k. The larger $p(\text{data})$ is, the more predictive the priors are of the situation as it evolves. When no identification sensors report detections for the given time period, there is no updating to do and the values of $p(T_k/\text{data})$ for the current period are set equal to those from the previous period.

Having completed the discussion of type identification using the Bayesian approach, the computation of friend, foe, and neutral identification as well as class identification can be addressed. The means for computing these posterior probabilities is to use the posterior probabilities for type and the scenario defined (i.e., *a priori*) values of $p(\text{FRD}|T_k)$, $p(\text{FOE}|T_k)$, and $p(\text{NEU}|T_k)$:

$$p(\text{FRD}|\text{data}) = \sum_k [p(T_k|\text{data}) \cdot p(\text{FRD}|T_k)]$$

$$p(\text{FOE}|\text{data}) = \sum_k [p(T_k|\text{data}) \cdot p(\text{FOE}|T_k)] \qquad (7.24)$$

$$p(\text{NEU}|\text{data}) = \sum_k [p(T_k|\text{data}) \cdot p(\text{NEU}|T_k)]$$

The posterior probabilities for aircraft class (e.g., bomber, small fighter, large fighter, and commercial jetliner) can be found in the analogous manner; for example, $P(\text{BOM}/\text{data}) = \Sigma_k[p(T_k|\text{data}) \cdot p(\text{BOM}|T_k)]$. This approach even permits employment of kinematic attributes such as altitude, speed, and maneuvers to produce probabilities of class membership through the inclusion of such terms via an expansion similar to that in equation (7.21a).

7.2.3 Dempster-Shafer Evidence Theory

This subsection follows the structure of Subsection 7.2.2; mathematical properties are discussed first, then interpretation of beliefs, and finally a data fusion example.

Mathematical Properties

The development of evidence theory began in the 1960s when Dempster [6, 37] developed the mathematical foundations of a two-valued uncertainty mapping, upper and lower uncertainty measures, between two spaces. A result of this work is Dempster's rule of combination, which operates on belief or mass functions as Bayes's rule does on probability functions. Shafer [27], a student of Dempster, has extended the development of belief functions and is the major proponent of evidence theory.

Evidence theorists talk of a "frame of discernment," Θ, a finite set of the single possible answers or hypotheses about a problem. The power set of Θ, traditionally labeled as 2^θ, is the set of all subsets of Θ. (If Θ has N elements, then 2^θ has 2^{N-1} elements.) Whereas probability theory operates on the frame of discernment and provides logic for computing elements of the power set, Dempster's rule operates on the power set; then the computation of upper and lower uncertainties (support and plausibility as described in Section 7.1) is completed using the results. The advantage of this is that Dempster's rule operates directly on all belief measures placed on specific elements of the power set. Probability measures can also be placed on specific members of the power set, but Bayes's rule must be used in concert with other probability operations (e.g., equation (7.16)), as shown in equation (7.21a,b). Because 2^θ is considerably larger than Θ, Dempster's rule will generally be much more computationally inefficient than Bayes's rule, but this is not the only basis for comparison. An entire set of uncertainty logic for each approach must be compared. Results on this topic are presented in the next subsection.

Belief measures are defined to be the basic metrics of uncertainty declarations within evidence theory. These measures span the power set and are operated on by Dempster's rule to pool two (or more) separate opinions or sources of data, which correspondingly lead to a pooled set of belief measures. Belief measures are also the values from which upper and lower measures of uncertainty are computed. Shafer's axioms for belief functions follow.

Belief Axioms

1. $b(\phi) = 0$, belief in the impossible is zero.
2. $b(\Theta) = 1$, normalization or the belief in the universal event is unity.
3. For every positive integer n and every collection of subsets of Θ $\{A_1, \ldots, A_n\}$

$$b(\cup A_n) \geq \sum_i b(A_i) - \sum_{i<j} b(A_i A_j) + \cdots + (-1)^{n+1} b(A_1 A_2 \cdots A_n) \qquad (7.25)$$

where $\cup A_n$ = union of A_1 through A_n.

The first two axioms are similar to those for probability. The third is similar to the third for probability, but deals with sets that are not mutually exclusive, an important distinction. A similar axiom for probability theory would have an equal sign rather than an inequality.

There is no comparable belief axiom for the fourth axiom of probability theory, conditional probability. One of Shafer's main arguments with probability theory is that conditional uncertainty is defined in the axiom structure rather than derived from the axioms; see [35].

One of the unique attributes of evidence theory is the ability to assign uncommitted belief to the entire frame of discernment (in effect, this assigns belief to a level of admitted ignorance regarding the meaning of the evidence). This has the effect of producing results like

$$b(A) + b(\text{not } A) \leq 1 \tag{7.26}$$

Dempster's Rule of Combination

Dempster's rule is used to combine probabilistically independent sets of evidence. Suppose two independent belief or mass functions, b_1 and b_2, exist over a common frame of discernment. Also suppose Θ has been divided into different subsets, $\{A_n\}$ and $\{B_m\}$, for these two belief functions, respectively. Then Dempster's rule of combination states

$$b(A_i B_j) = \frac{b_1(A_i) b_2(B_j)}{1 - Q} \tag{7.27}$$

where $Q = \sum_r \sum_s b_1(A_r) b_2(B_s)$ such that $A_r B_r = \phi$, the null event.

In effect, the Q term accounts for conflicts in the belief distributions from the sources b_1 and b_2 and assures that the combined belief is normalized to the unit interval. This formulation of Dempster's rule also is valid when $\{A_n\}$ and $\{B_m\}$ are identical.

When there is no uncommitted belief (i.e., ignorance) in either belief function, Dempster's rule is identical to Bayes's rule. The two rules are also identical when the frames of discernment for the belief functions being operated upon contain the same singleton hypotheses and their conjunctions; that is, $\{A_n\} = \{B_m\} = \{A, \text{not } A\}$.

For the sensor that computes belief measures b_k for K elements of the power set and an uncommitted belief, u, the following equation represents the belief for each element k at the time n resulting from n applications of Dempster's rule:

$$b_k(n) = \begin{cases} b_k, \text{ if } n = 1, \text{ or} \\ \dfrac{b_k \left[(b_k + u)^{n-1} + \sum_{j=1}^{n-1} u^j \right]}{\sum_{k=1}^{K} \left\{ b_k \left[(b_k + u)^{n-1} + \sum_{j=1}^{n-1} u^j \right] \right\} + u^n}, \text{ if } n > 1 \end{cases} \tag{7.28}$$

The denominator in this equation for $n > 1$ normalizes the belief assigned to all elements of the power set to sum to 1.0. Note that the uncommitted belief at time n is u^n, a simple function of the number of sensor reports. Thus, equation (7.28) provides a means to calculate belief over several observational cycles or scans.

Dillard [38] develops combination rules for multiple sensor fusion that are similar to the single-sensor cumulative (n-scan) belief function in (7.28). The generalization of (7.28) to the case of i multiple sensors is as follows:

$$b'_k(n) = \frac{\Sigma_{K_1 K_2 \cdots K_i = K} [b_1(n) \cdot b_2(n) \cdots b_i(n)]}{[1 - Q]} \tag{7.29}$$

as the nth scan, multisensor (joint or fused) belief in proposition or object-class k in the frame of discernment. Define also

$$U'(n) = \frac{\prod_{j=1}^{i} U_j(n)}{[1 - Q]} \tag{7.30}$$

as the joint or fused uncommitted belief at the nth scan. In both of these,

$$Q = \Sigma_{K_1 K_2 \cdots K_i = \phi} [b_1(n) \cdot b_2(n) \cdots b_i(n)] \tag{7.31}$$

is the belief attributed to conflicting declarations and used in normalizing (7.28) and (7.29). Finally, the cumulative belief across all sensors through time period n for proposition/target k is given by

$$b_k(n) = \begin{cases} b'_k(1) & n = 1 \\ \dfrac{b'_k(1) \prod_{j=2}^{n} [b'_k(j) + u'(j)] + \sum_{j=1}^{n-1} \left[\prod_{r=1}^{j} u'(r) \right]}{\sum_{K=1}^{K} \left\{ b'_k(1) \prod_{j=2}^{n} [b'_k(j) + u'(j)] + \sum_{j=1}^{n-1} \left[\prod_{r=1}^{j} u'(r) \right] \right\} + \prod_{j=1}^{n} u'(j)} & n > 1 \end{cases} \tag{7.32}$$

In the preceding expressions,

$b_k(n)$ is the belief associated with element k of the frame of discernment by sensor i in time period n;

$u_i(n)$ is the uncommitted belief declared by sensor i in time period n.

The D-S technique has other aspects that add to its complexity. For example, Shafer [27] introduces different types of support functions.

Finally, the upper and lower uncertainties must be computed from the belief function. Suppose that after applying Dempster's rule the power set is represented by $\{D_m\}$. The lower bound of uncertainty is called the support, s, and is the sum of all beliefs assigned the element itself and any elements that are subsets of it:

$$s(D_j) = \sum_r b(D_r) \qquad (7.33)$$

where each D_r must be a subset of D_j.

The upper bound of uncertainty is the plausibility, pl, which is defined to be 1 minus the support of (not D_j), which is the union of all elements whose intersection with D_j is the null vector:

$$pl(D_j) = 1 - s(\text{not } D_j) \qquad (7.34)$$

where (not D_j) = $\bigcup_k D_k$ such that $D_k D_j = \phi$.

Interpretation and Assessment of Beliefs

The meaning of a belief measure has been defined in the context of the reliability of receiving the information and the meaning of the information. Prade [39] defines the belief measure $b(p)$, to be "the probability that the evidence is *exactly and completely described* by p, i.e., the weight of evidence in favor of p, but not a probability measure on P," where p is a proposition about the state of the world and P is the set all propositions. Implicit in this discussion is a major difference between the two approaches: probability is assigned to events or states of the world; belief is assigned to propositions about events or states of the world.

A Data Fusion Example

The Dempster-Shafer approach uses Dempster's rule to proceed directly to a joint sensor report, bypassing the second column of actions in Figure 7.12. Table 7.1

Table 7.1
Example of Dempster's Rule of Combination

IFFN		
.4	.12	.28
(1111111111)	(1000000000)	(1111111111)
.6	.18	.42
(1111000000)	(1000000000)	(1111000000)
	.3	.7
	(100000000)	(1111111111)
	ESM	

illustrates this procedure for the IFFN and ESM sensors. Each entry of the table has two elements: a belief value and a 10-bit vector that represents the proposition to which belief is attached. This example assumes there are ten types of aircraft, the first four (in the indexing scheme) of which are friendly, the next five of which are foes, and the last is a commercial airliner. The power set for this example has $2^{10} - 1$, or 1023 elements. If a bit is set to 0, this reflects that the sensor declares that aircraft in question is not of the type associated with that bit. If the bit is set to 1, the aircraft may be of that type:

(1000000000) = target type 1, a friend

(1111000000) = any of the first four target types, a friend

(0000100000) = target type 5, a foe

(0000111110) = a foe

(0000000001) = a commercial airliner

(1111111111) = the uncommitted state (target unknown)

For the conceptual scenario being developed here, there are ten aircraft types, four aircraft classes (bomber, small fighter, large fighter, and commercial airliner), and the "nature" (friend, foe, or neutral) of the aircraft. Considering all intersections of these categories, only 19 (e.g., friendly bomber, type 3) of the 1023 power set elements are interesting. Clearly these 19 elements can be represented by a bit vector with less than 10 elements, but the 10-bit vector can be interpreted immediately and will be used throughout this example.

The row-column intersections in Table 7.1 result in an unnormalized belief value that is the product of the row and column beliefs; this matrix of belief values is the outer product of the two belief vectors.

$$[b_D(B)] = \mathbf{b}_{\text{ESM}}(C)\mathbf{b}_{\text{IFF}}^T(D) \qquad (7.35)$$

where

$[b_D(B)]$ is the matrix of belief values that result from Dempster's rule on the frame of discernment, B.

$\mathbf{b}_{\text{ESM}}(C)$ is the column vector of beliefs from the ESM sensor on its frame of discernment, C.

$\mathbf{b}_{\text{IFF}}^T(D)$ is the transpose of the column vector of beliefs from the IFF sensor on its frame of discernment, D.

$B_{ij} = C_i D_j$.

The bit vector that each belief element is to be associated with results from the intersection of the row bit vector and the column bit vector. Whenever an intersection results in the null vector (0000000000), the other results of the operation must be normalized to sum to 1.0. Refer to (7.27).

For the Dempster-Shafer representation of the ESM sensor, we assert a representative processing rule wherein positive belief is assigned to the most likely target type and any other target type that has a $p_{ESM}(data | T_k)$ greater than 50% of the value of the most likely type. The ESM sensor assigns the remainder of the belief to the uncommitted state. So

$$b_{ESM}(B''T_k'') = \frac{p_{ESM}(data | T_k)}{\sum_k p_{ESM}(data | T_k)}$$

$$\text{if } p_{ESM}(data | T_k) > .5 p_{max} \quad (7.36)$$

$$= 0, \text{ otherwise}$$

$$b_{ESM}[B(1111111111)] = 1 - \sum_k b_{ESM}(B''T_k'')$$

where $B''T_k''$ represents the bit vector associated with the kth target type, and $p_{max} = p_{ESM}(data | T_k)$ for the most likely T_k.

This (special) version of the ESM sensor will be called the *truncated Dempster-Shafer ESM sensor,* as it truncates low belief values and assigns that belief to the uncommitted state.

If additional sensor reports had been associated with these two reports, then Dempster's rule would be repeated; the result from Table 7.1 would form one of the entries and each new sensor the second entry, in turn. Ultimately, multiple belief assignments to the same bit vector would have to be aggregated; for example, from Table 7.1, $b(1000000000) = .12 + .18 = .3$.

Dempster's rule is again used to combine the joint sensor report with the beliefs of the previous period. Dempster's rule requires that some prior beliefs be established for each new track, or estimate of aircraft position and velocity, that is to be formed (based on one or more associated sensor reports). Unlike the Bayesian approach, the Dempster-Shafer approach permits all belief to be assigned to the uncommitted state:

$$b(1111111111) = 1.0 \quad (7.37)$$

This belief function is called *vacuous* because the combination of any other belief function, b_1, with the uncommitted state via Dempster's rule is simply b_1.

Finally the lower and upper bounds on belief (support and plausibility functions, respectively) are computed using (7.33) and (7.34).

7.2.4 Representative Comparison between Bayes and Dempster-Shafer

Continuing with the conceptual scenario and notation of the example described above, the first part of our comparison examines the data required by each method. The Bayesian approach requires the following additional identification data: a vector of prior probabilities, a matrix of *friend-foe-neutral* (FFN) to type probabilities, and matrix of class to type probabilities. In the example of 10 types and 4 classes this is a total 80 numbers (see Table 7.2). The effect of the priors becomes insignificant in the computation of the posteriors as time goes on. So these probabilities can be set equal, although a number of elicitation techniques make the assignment of these numbers meaningful, see Watson and Buede [29]. All of the other probabilities, FFN-given-type, and class-given-type, are ones and zeros. When both sides employ the same aircraft, additional aircraft are defined (as discussed earlier) and these probabilities are again ones and zeros.

Table 7.2
Additional Identification Data Required by the Bayesian Approach

Variable	Data		
Prior	q_{T_1} q_{T_2}	...	$q_{T_{10}}$
$p(\text{FRD} \mid T_k)$	$p(\text{FRD} \mid T_1)$	...	$p(\text{FRD} \mid T_{10})$
$p(\text{FOE} \mid T_k)$	$p(\text{FOE} \mid T_1)$	...	$p(\text{FOE} \mid T_{10})$
$p(\text{NEU} \mid T_k)$	$p(\text{NEU} \mid T_1)$	...	$p(\text{NEU} \mid T_{10})$
$p(\text{BOMBER} \mid T_k)$	$p(\text{BOMBER} \mid T_1)$	...	$p(\text{BOMBER} \mid T_{10})$
$p(\text{S FTR} \mid T_k)$	$p(\text{S FTR} \mid T_1)$	...	$p(\text{S FTR} \mid T_{10})$
$p(\text{L FTR} \mid T_k)$	$p(\text{L FTR} \mid T_1)$	...	$p(\text{L FTR} \mid T_{10})$
$p(\text{COMM AIR} \mid T_k)$	$p(\text{COMM AIR} \mid T_1)$	...	$p(\text{COMM AIR} \mid T_{10})$

We continue this comparison using a computer-based simulation of a simple one-on-one tactical engagement scenario. In this engagement the friendly and enemy aircraft are flying parallel to each other in opposite directions. The parallel paths begin when the aircraft are over 100 nautical miles (nmi) apart; the paths are separated by 5 nmi. The friendly aircraft has a radar for position tracking, an IFF sensor for friend identification, and an ESM sensor for type identification. The capabilities of the ESM sensor are varied throughout this comparison to illustrate the relative performance of the Bayesian and Dempster-Shafer methods. The ESM sensor initially truncates beliefs as was described in (7.36). In general this ESM sensor is better able to distinguish between specific enemy types than friendly types.

Using the simulation, the first comparison of the Dempster-Shafer and Bayesian fusion approaches focuses on target identification when (1) an existing enemy aircraft under radar track is detected by the ESM sensor and (2) a friendly aircraft is simultaneously detected by the ESM and IFF sensors. Figures 7.13 and 7.14 show the belief and probability distributions over time for these two cases. Note that the time difference for reaching a high confidence (e.g., 0.9) identification in these two situations is a result of the ESM sensor being less diagnostic (discriminating) of friendly aircraft than of enemy aircraft. Both of these graphs dispell the notion that the plausibility and support functions are upper and lower bounds of probability in this process of dynamically updating uncertainty. In both cases the Bayesian probability is more predictive after 3 seconds. The Bayesian approach takes half the

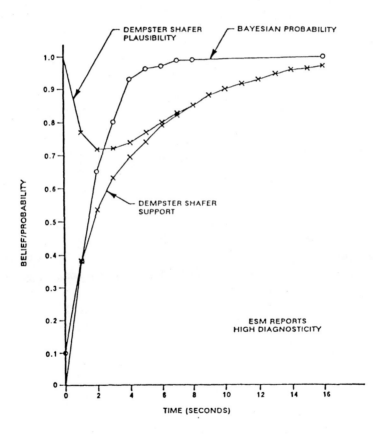

Figure 7.13 ESM confidence on type identification.

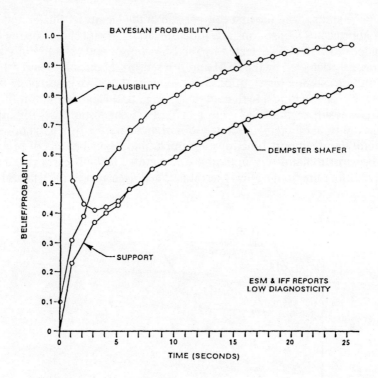

Figure 7.14 ESM and IFF confidence on type identification.

time of the Dempster-Shafer approach in both cases. These times are not significant for some tactical situations, but may be for split-second decisions.

These results suggest several areas that require additional investigation:

1. To what degree are the results determined by the way in which the Dempster-Shafer ESM sensor was defined?
2. What is the impact of having more or less uncommitted belief in a sensor report?
3. How does each approach deal with conflicting sensor reports?

To answer the first question, a Dempster-Shafer ESM sensor that did not truncate the uncertainties for the most unlikely types of aircraft was defined. This sensor assigned the same amount of belief (0.39) to "uncommitted" as the sensor that truncated small beliefs; but it did so by normalizing the belief assigned to aircraft types to 0.61 (1 minus that assigned to "uncommitted"). Figure 7.15 shows that the sensor using truncation proceeds more quickly to high-confidence declarations than the sensor that normalizes the beliefs after setting aside the "uncom-

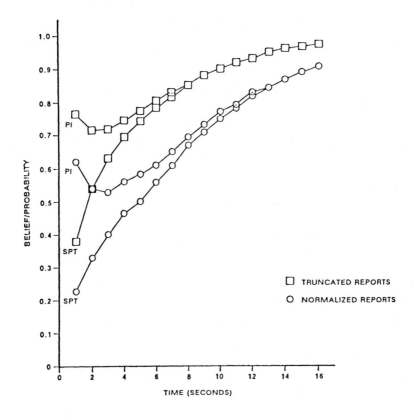

Figure 7.15 Alternate Dempster-Shafer sensors.

mitted" belief; thus the earlier results suggest that the Bayesian approach converges three to four times as fast as the normalized Dempster-Shafer sensor with almost 40% of its belief uncommitted.

Figure 7.16 demonstrates how belief accumulates for the normalized sensor as the "uncommitted" belief is varied. When the uncommitted belief is set to zero, the Dempster-Shafer process is identical to the Bayesian approach. The normalized Dempster-Shafer sensor with an uncommitted belief of 0.2 is about the same as the truncated Dempster-Shafer sensor with uncommitted belief of 0.39. The differences between the Bayesian probability and the support generated by these three Dempster-Shafer sensors is shown in Figure 7.17, illustrating that the uncommitted value in the sensor report has a significant effect; as dictated by (7.28).

Another graphic illustration of the effect that the uncommitted term has on Dempster-Shafer interference is shown in Figure 7.18. In this case the ESM sensor

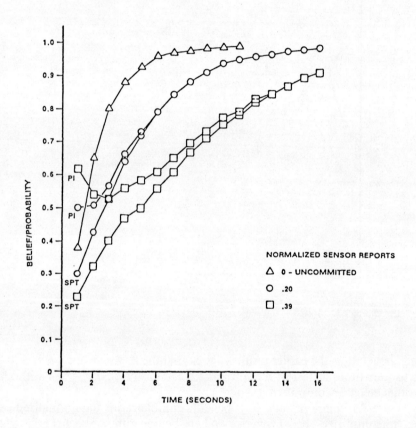

Figure 7.16 Dempster-Shafer kinematics.

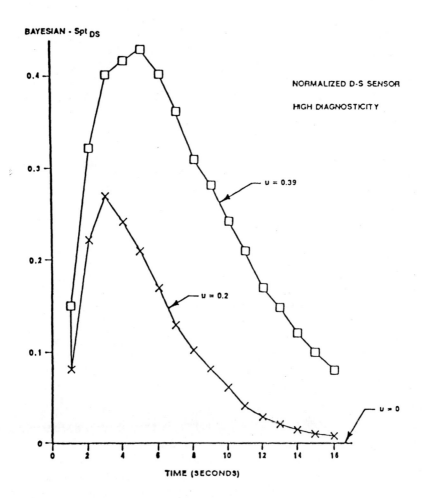

Figure 7.17 Dynamic effect of the Dempster-Shafer uncommitted belief.

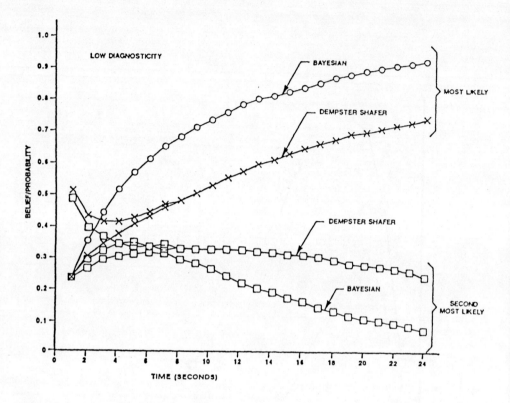

Figure 7.18 Confidence on type identification.

is not very discriminating. The Bayesian and Dempster-Shafer results are shown for both the most likely and second most likely target types. After 8 sec the uncommitted belief, which started at 0.27, has dwindled to zero; this value is u^n for time period n and independent of any of the sensor reports. Essentially this uncommitted belief is split between the hypotheses and allocated disproportionately to the less likely hypotheses, thus slowing the rise of support and plausibility compared to the probability.

The third question was addressed by substituting a slightly conflicting sensor report in the third of every four reports over 20 time periods. The $p(\text{data} | T_k)$ follow for both reports:

T_k	1	2	3	4	5	6	7	8	9	10
Main Report	.06	.09	.03	.03	.50	.30	.06	.12	.03	.10
Conflicting Report	.06	.09	.03	.03	.30	.50	.06	.12	.03	.10

The Dempster-Shafer sensor utilized a truncated report (only types 5 and 6 were reported) and had an uncommitted belief of 0.39. Figure 7.19 shows the results of a conflicting report at times 3, 7, 11, 15, and 19. Both approaches exhibit similar behavior. The gap between the plausibility and support functions of the Dempster-Shafer approach is sometimes called *ignorance;* but this value is not affected by the presence of the conflicting report.

The computational load for a single sensor report and update concerning identification is shown in Table 7.3. Logical operations were not counted; note the Dempster-Shafer computations require significant logical operations. The two soft sensor approaches are driven by the number of basic elements. In this example the Bayesian approach focused on 10 target types; then it generalized to FFN and class.

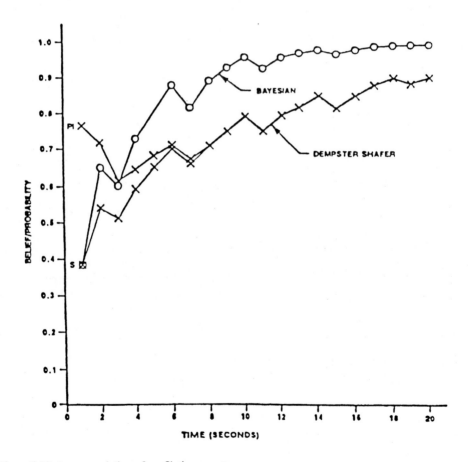

Figure 7.19 Accommodation of conflicting reports.

Table 7.3
(U) Single Report Processing Load

	Floating Point Operations			
	×	/	+	−
Deterministic	0	0	0	0
Bayes (IFF)	223	14	196	0
Dempster-Shafer	361	41	801	19

The IFF sensor was chosen for the Bayesian benchmark because it requires more processing than the ESM sensor. The Dempster-Shafer approach permits 19 unique levels of abstraction, down noticeably from the 1023 in the power set. For a realistic scenario involving on the order of 30 aircraft types, the Dempster-Shafer approach would have to deal with over 40 levels. Table 7.3 demonstrates that the computation time associated with the Dempster-Shafer approach (which is driven by the square of the number of unique levels of abstraction) exceeds that of the Bayesian approach, both of which exceed the deterministic approach.

REFERENCES

1. Swerling, P., "Probability of Detection for Fluctuating Targets," *IRE Trans.*, IT-6, April 1960.
2. Henkei, R.E., *Tests of Significance*, Sage Press, Beverly Hills, CA, 1976.
3. Berger, J., *Statistical Decision Theory: Foundations, Concepts, and Methods*, Springer-Verlag, New York, 1980.
4. Bayes, T., *Essay toward Solving a Problem in the Doctrine of Changes*, Philosophical Transactions of the Royal Society, London, 1763.
5. Lowrance, J.D., and T.O. Garvey, "Evidential Reasoning: A Developing Concept," *Proc. Int. Conf. Cybernetics and Society*, October 1982.
6. Dempster, A.P., "Upper and Lower Probabilities Induced by a Multivalued Mapping," *Ann. Math. Statistics*, Vol. 38, 1967.
7. Aldenderfer, M.S., and R.K. Blashfield, *Cluster Analysis*, Quantitative Applications in the Social Sciences Series, Paper 07-044, Sage Publications, London, 1984.
8. Sneath, P., and R. Sokal, *Numerical Taxonomy*, W.H. Freeman Publisher, San Francisco, 1973.
9. Duda, R.O., and P.E. Hart, *Pattern Classification and Scene Analysis*, Wiley-Interscience, New York, 1973.
10. Priebe, C., and D. Marchette, "An Application of Neural Networks to a Data Fusion Problem," *Proc. 1987 Tri-Service Data Fusion Symp.*, June 1987.
11. Widrow, B., and R. Winter, "Neural Nets for Adaptive Filtering and Adaptive Pattern Recognition," *IEEE Computer*, March 1988.
12. North, R., "Neurocomputing: Its Impact on the Future of Defense Systems," *Defense Computing*, January 1988.
13. Lippmann, A., "An Introduction to Computing Neural Nets," *ASSP Magazine*, Vol. 3, No. 4, 1987.
14. Bowman, C., "Artificial Neural Network Adaptive Systems Applied to Multisensor ID," *Proc. 1988 Tri-Service Data Fusion Symp.*, May 1988.

15. Shannon, L.E., "Mathematical Theory of Communication," *Bell System Tech. J.*, Vol. 27, 1948.
16. Ingels, F.M., *Information and Coding Theory*, Intext Educational Publishers, Scranton, PA, 1971.
17. Kelley, B.A., and W.R. Simpson, "The Use of Information Theory in Propositional Calculus," *Proc. 1987 Tri-Service Symp. Data Fusion*, June 1987.
18. Dretsky, F.I., *Knowledge and the Flow of Information*, MIT Press, Cambridge, MA, 1982.
19. Hall, D.L., and R.J. Linn, "Comments on the Use of Templating for Multisensor Data Fusion," *Proc. 1989 Tri-Service Data Fusion Symp.*, May 1989.
20. Gabriel, J.R., and M.H. Gabriel, "Data Representation and Matching for Events and Templates," *Proc. 1988 Tri-Service Data Fusion Symp.*, May 1988.
21. Schroeder, F., and L. Wright, "Automatic Correlation of Multiple Intelligence Sources," *Proc. 1988 Tri-Service Data Fusion Symp.*, May 1988.
22. Llinas, J., D.L. Hall, and E. Waltz, "Data Fusion Technology Forecast for C3/MIS," *Proc. Third Int. Conf. Command, Control, Communications and MIS*, Bournemouth, England, March 1989.
23. Zadeh, L.A., *Fuzzy Sets and Systems*, North-Holland Press, Amsterdam, 1978.
24. Zadeh, L.A., "Fuzzy Logic," *IEEE Computer*, April 1988.
25. Shafer, G., and A. Tversky, "Languages and Designs for Probability Judgment," *Cognitive Science*, Vol. 9, 1985, pp. 309-339.
26. Kyburg, H.E., Jr., "Bayesian and Non-Bayesian Evidential Updating," *Artificial Intelligence*, Vol. 31, 1987, pp. 271-293.
27. Shafer, G., *A Mathematical Theory of Evidence*, Princeton University Press, Princeton, NJ, 1976.
28. Lindley, D.V., "Scoring Rules and the Inevitability of Probability," *Int. Statistical Rev.*, Vol. 50, 1982, pp. 1-26.
29. Watson, S.R., and D.M. Buede, *Decision Synthesis: The Principles and Practice of Decision Analysis*, Cambridge University Press, Cambridge, England, 1987.
30. Schum, D.A., "Probability and the Processes of Discovery, Proof, and Choice," *Boston University Law Rev.*, Vol. 66, 1986, pp. 825-876.
31. Henrion, M., "Uncertainty in Artificial Intelligence: Is Probability Epistemologically and Heuristically Adequate?" in J. Mumpower (ed.), *Expert Systems and Expert Judgment*, Springer-Verlag, New York, 1987.
32. Pearl, J., *Probabilistic Reasoning in Intelligent Systems: Networks of Plausible Inference*, Morgan Kaufmann Publishers, San Mateo, CA, 1988.
33. Gordon, J., and E.H. Shortliffe, "A Method for Managing Evidential Reasoning in a Hierarchical Hypothesis Space," *Artificial Intelligence*, Vol. 26, 1985, pp. 323-357.
34. Shenoy, P.P., and G. Shafer, "Propagating Belief Functions with Local Computations," *IEEE Expert*, Fall 1986, pp. 43-52.
35. Shafer, G., "Lindley's Paradox," *J. Amer. Statistical Assoc.*, Vol. 77 (Theory and Methods Section), No. 378, June 1982, pp. 325-351.
36. Shachter, R.D., "Evaluating Influence Diagrams," *Operations Research*, Vol. 34, 1986, pp. 871-882.
37. Dempster, A.P., "A Generalization of Bayesian Inference," *J. Royal Statistical Soc.*, Series B, Vol. 30, 1968.
38. Dillard, R.A., "Computing Probability Masses in Rule-Based Systems," NOSC Technical Document 545, Naval Ocean Systems Center, September 1982.
39. Prade, H., "A Computational Approach to Approximate and Plausible Reasoning with Applications to Expert Systems," *IEEE Trans. Pattern Analysis and Machine Intelligence*, Vol. PAM-7, May 1985, pp. 260-283.

Chapter 8
MILITARY CONCEPTS OF SITUATION AND THREAT ASSESSMENT

The military theories laid down in history are the frameworks within which military commanders decide and act. Each theory presents a somewhat different view of how warfare should be conducted, so military commanders following different theories may react differently when presented with similar situations or threats (although many factors other than guidance from such theories influence a commander's judgment).

In general, the goal of combat operations is to control the power distribution on the battlefield. To do so requires an understanding of the battlefield environment, and the application of an effective problem-solving strategy. In a sense, the guidelines associated with problem-solving approaches represent the application of military theories. Each theory suggests that certain emphases be placed on the specific aspects of the problem of controlling power distribution; over time, military theory has evolved in response to improved weaponry, capability for maneuver and covertness, and in communication and data processing, among other technologies.

For example, Sun Tzu [1] concentrates upon complicating the enemy's problems, emphasizing deception. His goal is to present a sudden unsolvable problem to the enemy so that surrender occurs without actual battle. Clausewitz [2] concentrates on the enemy's center of gravity and disrupting the enemy's problem-solving process by disrupting the ability to act. Beaufre [3] emphasizes political interaction and constraints with special attention to goal structure conflicts to determine political maneuvering and strategy to manipulate constraints. The current American approach to war attempts to focus on the enemy's decision process (deterrence), complicate the enemy's problem-solving structure (triad of forces), and manage risk in conventional conflicts (attrition-based strategies).

American theorists recently have criticized such attrition-based strategies in favor of maneuver-based warfare. Most discussions of maneuver warfare deal with basic concepts rather than implementation, yet the ultimate value of maneuver warfare depends on the practicality and effectiveness of the implementation. At the tactical level, maneuver warfare concentrates on disrupting the enemy's ability to generate and execute plans, depending mainly on the rapid shifting of problems through fast transient maneuvers. The feasibility of maintaining the required tempo over long periods of time and against an enemy expecting such a technique can be questioned; however, successful implementation of maneuver warfare strategy depends much more on the ability to rapidly exploit such opportunities than the prospect of totally disrupting the enemy's decision process.

The capability required by the maneuver warfare approach does fit well into a power distribution control approach. Mobility and maneuverability make the commander's control of the power distribution much easier, especially if combined with the ability to dig rapidly into prepared positions. Maneuver forces also have the potential for disrupting the enemy's attempt to control risk with an attrition-based strategy. Recent studies of military war games [4, 5] suggest that the frequently used attrition models represented by Lanchester's mathematical framework [6] do not correlate well with historical battle results, but reveal instead that out-maneuvering the enemy appears to be the dominant factor that most influences the potential for victory. The debate and thought incited by the maneuver warfare advocates is good, but it is dangerous to blindly accept their concepts without carefully examining the implementation and required trade-offs. Thus, Lanchester's equations remain useful and instructive (we use them in Chapter 11) but the implications of the results should be developed carefully and with the realization that there are other points of view.

Which theories provide the optimum guidance for the conduct of war of course is arguable but there is no doubt regarding the correctness of the goal: to control battlefield power distribution. (Such theories, of course, only attempt to suggest frameworks for understanding very complex phenomena and cannot be evaluated in the fashion of theories for physical phenomena.) In an actual situation, there are two opposed commanders with different specific goals. Hence, combat becomes a race to formulate and solve the friendly force problem while blocking the solution of the enemy's problem.

This chapter presents a top-down hierarchical description of the military problem-solving and decision processes within which the specific analyses of situation and threat assessment (which each exploit data fusion techniques) are conducted. This knowledge is crucial to the design of semiautomated decision-analysis support systems because such systems attempt to replicate or at least approximate human reasoning patterns based on this knowledge. In fact, one of the distinctions between Level 1 and Level 2 or 3 processing is that in Levels 2 and 3 symbolic reasoning methods, as opposed to numeric approaches at Level 1, are employed in the system design to accomplish this goal. One crucial aspect of successfully assessing the situation and threat is understanding that such assessments optimally derive from examining the data from multiple contextual viewpoints. Situation or threat analysis is not a concatenation-type process; it is a multiple-perspective, contextually based process.

Additionally, because a unified, consistent definition of the exact nature of such analyses does not exist in the U.S. military literature, it is considered important that the architect of a data fusion process understand the problem-solution context within which the fusion process is conducted. In an effort to provide this view, the chapter sequentially describes the overall military problem-solving process, offers definition of the situation and threat assessment elements and processes, discusses the important aspects of countermeasure techniques, and the necessity for "multiple-perspective analysis." Two levels and examples of decision-making processes are described in Section 8.1. One, at the strategic and more abstract and generalized level, closely follows Orr's work [7] in C^3I concepts. The other, at the tactical and more concrete and specific level, closely follows Wohl's work [8, 9] on the SHOR paradigm.

Section 8.2 describes both the concepts and component elements of a situation assessment. In Section 8.3, some of the major elements of the "fog of war," concealment-cover-deception techniques, and their effects on situation and threat processing are discussed. Space prevents an equivalent discussion on the use and effects of electronic countermeasures but analogies can be drawn to the discussion of Section 8.3. Summary comments on situation assessment are made in Section 8.4, followed by a discussion on threat assessment in Section 8.5, which parallels that for situation assessment in Section 8.2. Finally, the importance of so-called multiple perspective analysis for both situation and threat assessment is discussed in Section 8.6, and summary comments on threat assessment are made in Section 8.7.

8.1 THE MILITARY PROBLEM-SOLVING PROCESS

Several researchers [10] have approached the military problem-solving and decision-making processes in terms of a state transition model. The state transition model approach is useful because military engagements may be thought of as discrete (but complex) configurations of entities that go through successive realignments as the battle proceeds. The transitions can be identified as actions of single entities or entity groups. Hence, state transition models are useful because we can define states and transitions and map these to real objects and activities on the field. Moreover, the level of abstraction of such approaches can be selected to fit the focus and purpose of the analysis.

Dieterly [11] describes problem solving and decision making in terms of the state transition model proposed by Reitman [10]. According to this model, also described in Orr [7], the basic decision-problem condition involves state A, state B, and a transition from state *A* to state *B*, as in Figure 8.1. Dieterly points out that things are not usually so simple, however, because the states and transitions are not always known. Eight basic problem models are identified in Figure 8.2, where solid lines indicate known states or transitions and broken lines unknown states or transitions.

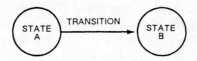

Figure 8.1 Basic decision-problem condition.

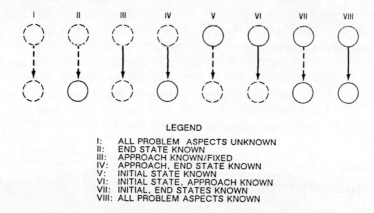

Figure 8.2 Decision-problem models (after [11]).

Each of these situations presents different difficulties. Dieterly calls model VIII the *trivial problem* (both states and transition known), and the problem solver's goal is to reduce all other models to this trivial case. Model I, called the *intuitive model,* is most difficult because there is no information initially. In general the problem-solving process starts with one of the cases and uncovers information to proceed to other cases, terminating in the trivial case, VIII.

Real situations are typically more complicated. Dieterly discusses five "classes of conditions," as indicated in Figure 8.3. A multiple-end state results when an action can have two or more results, as in random processes. A multiple transition

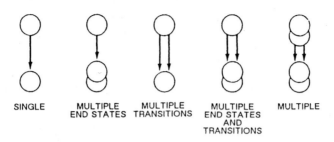

Figure 8.3 Dieterly's classes of conditions (after [11]).

occurs when two or more transitions or actions from an initial state result in the same end. Multiple initial conditions occur when two or more initial states are transformed to the same final state. Each of the (eight) models in Figure 8.2 could have any of the (five) conditions in Figure 8.3. Hence, Dieterly considers 40 possible problem situations, and somewhat different problem-solving techniques are indicated for each.

Reidelhuber [12] presents a model of tactical decision making that integrates various C^3 functions with battlefield state estimates. His model, as shown in Figure 8.4, is generally hierarchical and multileveled (reflecting coordination among multiple military units), but it can be considered an elaboration of the concepts of Figures 8.1–8.3. In Figure 8.4, *mission* can be associated to a desired end or power distribution, whereas *actual state* can be associated to current state (or situation) of each combatant unit.

The military commander, charged with employing the forces at his command, uses the abstract problem-solving process depicted in Figure 8.5 [11]. Notionally, at the start of the process the commander is not sure of his initial state (situation), the desired state, or the desired transition. By working through several problem-solving steps the commander proceeds to the extreme right side and the desired final state. The first step consists of a parallel refinement of the initial state, desired state, and possible transitions. This involves clarification of the desired final state, clarification of the situation, and the generation and evaluation of possible plans of action.

Determination of the current situation is a key function of the C^3I system. Relevant functions include surveillance to locate and identify enemy units, intelligence gathering, analysis of data to determine the meaning of data patterns, assessment of enemy capability and intent, and presentation of the situation in a useful form. Attempts to disrupt this step involve concealment, creation of ambiguity, and deception. Concealment involves using various means to prevent sensors from obtaining information about actual deployments, capability, or intention. Ambiguity involves generating situations that can be interpreted in contradictory ways

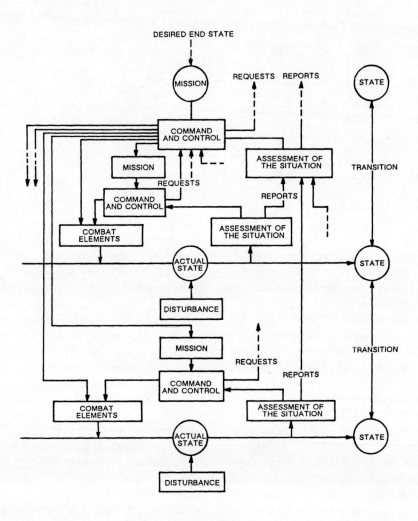

Figure 8.4 Hierarchical decision-making model (after [12]).

by the enemy. This is most common with regard to concealing intent but can also be used to confuse interpretations about capability and deployment. A good knowledge of the enemy surveillance, intelligence gathering, and analysis process is extremely valuable in this respect. Finally, deception involves creating a wrong conclusion by the enemy as to one's intent, capability, or deployment. This may involve controlling the data the enemy is allowed to collect or, more commonly, injecting false or misleading data into the enemy C^3I system. All of these techniques

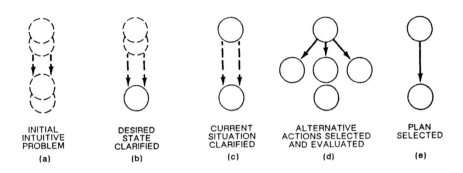

Figure 8.5 Military problem-solving process (after [11]).

can lead to misinterpretation of the actual problem, and hence to inappropriate actions on the part of the enemy commander; more is said on this in Section 8.3.

Thus, much of modern warfare is information warfare. Situation assessment plays a key role in a general five-step problem-solving process (Figure 8.5), and, in turn, data fusion plays a role in improving the reliability and specificity of the component information.

The final stage in solving a military problem involves executing the selected plan. This should be a dynamic process with minor alterations made in real-time according to the actual engagement outcomes and detected shifts in the power distribution. Good C^3I systems are essential at this point. The resulting real-time control lessens the dependence of combat operations on precise estimates of the situation and probable outcome. This control is the key to effectively managing battles in the chaotic and uncertain conditions of combat. Commanders therefore attempt to ensure the smooth operation of the combat operations process for their forces, while attempting to disrupt the enemy's process.

Note that this overall process is goal driven; that is, driven by the intent to achieve a desired final power distribution. The process of problem solving is set in motion by the specification of a military (or political) objective.

8.1.1 The SHOR Model of Military Decision Processing

Following up on and extending Wason's work in behavioral response [13], Wohl formulated the so-called SHOR paradigm [8, 9] as a mechanism to describe the salient features of the military decision-making process. *Stimulus-Hypothesis-Option-Response* (SHOR) was a model devised to deal explicitly with both information input uncertainty and consequence-of-action uncertainty in military problem solving. In contrast to Orr's description of the process as goal driven (wherein the first step is determining desired power distribution or end state), Wohl's model describes a data-driven (stimulus-driven) or reactive approach to problem solving and decision making. Alternatively, Orr's model could be considered strategic in nature, dealing with both political and military goals, whereas Wohl's model could be considered tactical in nature, dealing with the immediacy of combat situations within the context of strategic goals. Some of the following description of the SHOR paradigm is taken from Wohl's article in [9].

The SHOR paradigm, summarized in a top-level flowchart in Figure 8.6, provides a useful mechanism for describing these salient features of an individual's decision tasks. It was devised to deal explicitly with two realms of uncertainty in the decision-making process:

- Information input uncertainty, which creates the need for hypothesis generation and evaluation.
- Consequence-of-action uncertainty, which creates the need for option generation and evaluation.

Figure 8.7 is an elaboration of the SHOR paradigm to include the concept of the commander's mental model. Besides providing the decision maker with an

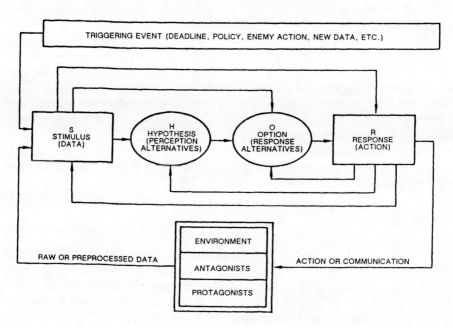

Figure 8.6 Dynamics of tactical decision process—the SHOR paradigm (after [9]).

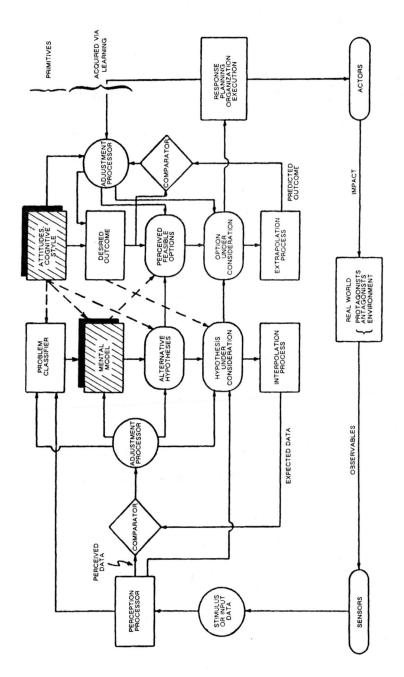

Figure 8.7 The SHOR paradigm elaborated (after [9]).

internal representation of the problem, a mental model functions as a theory or framework from which to generate hypotheses. Wohl asserts that a hypothesis set derives directly from the interaction of input information with a commander's mental model or internalized representation of the problem situation.

Another key aspect of the elaborated SHOR paradigm is the introduction of the psychological concepts of attitude and cognitive style. Attitudes have a direct impact on the desired outcome, and an indirect impact on option availability and selection, whereas cognitive styles act through differential weights with an impact on both the hypotheses and option portions of the paradigm.

The elaborated SHOR paradigm provides a means for describing the decision task in general terms. What is necessary to provide analytic substance to the SHOR paradigm is an overlay that unmasks the operator-centered elements amenable to analysis: information processing, data processing, hypothesis or option generation, and action selection. Figure 8.8 shows the relationships between the SHOR paradigm and three primary human decision-making elements or subtasks.

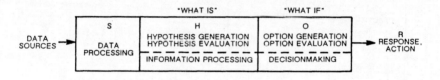

Figure 8.8 SHOR paradigm in terms of task elements (after [9]).

According to Wohl, some of the tasks and subtasks that are specific to data processing include, but are not necessarily limited to, the following.

8.1.2 Data Processing Tasks in the SHOR Model

Data Reduction, Compression, and Aggregation

The task of reducing data to a form commensurate with human information processing involves eliminating unwanted data, filtering corrupting data, and organizing and categorizing the remaining useful data. Clearly, a form of data fusion occurs at this step. Of course, determining which data are useful is highly situation dependent, as are the categories into which the data are classified. The compression and aggregation of useful data involves their transformation into a form that reduces the (human's) fractional capacity necessary to consider it (and thereby affects operator workload).

Data Search and Scan

These subtasks are associated most often with the human interaction with a large data base that may be common to many users. On the other hand, the battlefield itself is dynamic, which necessitates repetitive scanning of the data base; thus, *search* refers to what data to seek, whereas *scan* refers to how often to sample the data. Maintenance of the focus of attention is an aspect of cognition; such capability is typically achieved using knowledge-based systems in a decision-aiding approach.

Attentional Allocation

A primary issue associated with data processing is the attentional allocation among and within the various data types. The object of attentional allocation is to maximize (through judicious sequencing of data source monitoring), the information content gleaned from the data base. The data fusion system designer must consider this aspect of information processing when selecting strategies for aggregation of typically asynchronous scan intervals from different sensors. Selection of fusion time epochs creates what is, in effect, the overall scan interval visible to the user (i.e., sensor-specific scans are blended into a fused scan cycle).

Event Detection, Recognition, and Confirmation

The task of event detection is concerned with recognizing that a problem exists or that something is happening (e.g., a missile launch or change in enemy objective). An event is detected by either a discrepancy or correlation between the operator's estimated (or forecasted) state of affairs and the observed state. The detection and recognition of an event is often difficult, as it may involve estimating subtle differences between two sets of highly uncertain variables. The objective of detection is to correctly recognize the event occurrence as quickly as possible. Clearly, this presents the operator with a trade-off among late or missed detections, confirmations, and false alarms. This trichotomy spawns what is perhaps the most fertile area for aiding the operator. For example, performance improvements may be obtained in several ways: by use of better alarms and alerting displays; by simplifying the format in displays; by providing relational displays that explicitly present the observed data in terms of simplified relationships; or by providing predictive displays that aid the operator to anticipate future events. Automatic event detection based on fused multisensor observations represents yet another aiding mechanism. Event detection in C^3I systems is frequently achieved through data fusion processions employing templating techniques; see Chapter 9 for details.

8.1.3 Information Processing Tasks in the SHOR Model

Wohl distinguishes between data and information processing by asserting that data processing encompasses those tasks within which the operator interacts with the data-gathering function (e.g., by directing the sensors). In contrast, information processing encompasses those tasks in which the operator is essentially passive to the external environment and focused more on analysis-type processes. According to Wohl [9], the information-processing elements of an operator task are associated with functions of data correlation, hypothesis generation and evaluation, status assessment, *et cetera*. Figure 8.9 shows that typical information-processing subtasks include the following.

Data Association

Data associated with a target, battlefield event, *et cetera*, will often be reported independently via a multiplicity of sensors each differing in coverage area, spectrum, resolution, response time, and observables sensed. Data association is necessary to determine, for example, whether target data, which have been reported by different sources, represent the same target or different targets. This is essentially the same association function defined in Chapter 1 and described for tracking problems in Chapter 6. It is a basic data fusion process that has to do with determining which measurements or observations associate to which objects or object groups in the field of view.

State and Parameter Estimation

This subtask provides continual estimation of the dynamic variables (states) pertinent to the military engagement. Typical states that change continuously with time include troop strength, positions, movements, and platform locations. Variables that are constant, or change much more slowly in time than states, are defined by Wohl as parameters. These include aircraft capabilities, missile ranges, and radar frequencies. States are generally correlated with each other, so that good estimation of certain states is likely to yield good estimates of other states, provided the operator is cognizant of the underlying correlations. Such correlation typically results from knowledge of doctrinal force patterns, *et cetera*. Understandably then, achieving the objective of state estimation with minimal uncertainty will depend strongly on the quality of the information base and the operator's perception of the engagement dynamics.

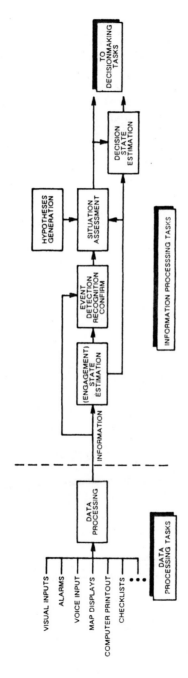

Figure 8.9 A taxonomy of display and information-processing tasks (after [9]).

Hypothesis Generation

The hypothesis generation subtask is perhaps the most critical part in a complex information-processing task. Hypothesis generation addresses the problem of how people generate a reasonable set of hypotheses *(situation perceptions)* and modify the set when the need arises. It involves what is commonly termed *creative thinking*. The DFS model labels this task *situation abstraction*.

The fundamental purpose of hypothesis generation is to extract meaning from data. This definition then gives rise to a measure of performance for the hypotheses, because the extraction of meaning from data can be measured in terms of hypothesis uncertainty and its rate of change. The task of hypothesis generation and testing thus is really a task of *uncertainty reduction*. How human decision makers accomplish hypothesis generation is a critical issue in fusion system design, because the design of information processing and decision aids must assist them in "natural" ways.

Situation or Threat Assessment

Situation assessment involves identifying the probable situation causing the observed data and events. In general terms it is the probabilistic categorization of the observed data over a set of alternative hypotheses (i.e., pattern recognition). The inputs to this task include the event detections, the state estimates, and a set of generated hypotheses for evaluation. The outputs of the task are, at least theoretically, the conditional probabilities of the various hypotheses being considered. The task objective is to maximize the probability associated with the (most) correct hypothesis; diagnostic accuracy as well as timeliness are the dominant concerns in the assessment process.

Decision State Estimation

Given the assessed situation and the engagement (input) state estimates, this task estimates the pertinent variables for subsequent decision making. Thus, this task forms the interface between information processing and decision making. The theoretical decision states are transformations and compressions of the states of the engagement. Clearly, not all engagement information is necessary for decision making; however, these states typically include the variables important for dynamic decision making, such as the time available for action.

Wohl's description of the SHOR model in [9] continues on to the option generation and response functions. The description here is truncated at the hypothesis function because we assert that sensor- or observation-based data fusion processes do not directly contribute to the option generation and response planning functions. Rather, as depicted in Chapter 2, data fusion is characterized as a *hypothesis-producing process;* that is, a process that aids in the formulation of a perspective of the battlefield but not one intrinsic to option or response development.

8.2 DEFINING SITUATION ASSESSMENT

Extended efforts to determine whether formal definitions of the concept of situation assessment exist in military references (e.g., [14]) revealed only imprecise and somewhat disparate interpretations in different works. The Data Fusion Subpanel of the JDL/TPC[3] (see Chapter 2), in an effort to foster some clarification of terms, has suggested that the notions of situation assessment and threat assessment be made more distinct. In this taxonomy, situation assessment is considered the process by which the distributions of fixed and moving entities are associated with environmental, doctrinal, and what are called *performance* data (e.g., hostile vehicle trafficability performance, sensor line-of-sight performance, *et cetera*—see Section 9.4). What generally distinguishes this process from so-called Level 1 analysis as described in Chapter 2 is that the Level 1 position-track and identity estimates are analyzed here in the context of prescribed event and activity sequences, estimated organizational force structures, and the overall battle environmental factors. This analysis raises the level of abstraction within which the analyst perceives the data. Another distinguishing feature of this process is that it typically requires a significant amount of *a priori* database information to support the component analyses.

This situation assessment process is made distinct from threat assessment by separating from it both the multiperspective (i.e., "red, white, and blue" perspectives—see Section 8.6) and quantitative enemy force assessments required to properly estimate the enemy's course of action and force lethality as part of threat analysis. In the context of multiperspective interpretation, situation assessment seems closest to establishing the "white" perspective in which force deployments are considered in the context of environmental effects (e.g., terrain and weather). However, situation assessment can also be thought of as a means to estimate the enemy battle plan; that is, what the enemy is doing (activities) and attempting to achieve (intent, goals). This viewpoint is closer to the "red" perspective of the battlefield.

Further, the task of situation assessment involves identifying the probable situation causing the observed events and activities. In general terms, this task involves categorizing hypotheses according to some type of probability or confidence scale. In this sense, as mentioned previously, it is a pattern-recognition process. This separation of concepts and definitions is not reflected in either formal military references or the professional and popular defense industry literature. In fact, quite the opposite cases exist; most of the literature mixes the terms and processes of situation and threat assessment such that distinctions are fuzzy.

Reference [15] contains an enumeration of the elements of a military situation according to one Soviet view. This work is very helpful in establishing a comprehensive list of such elements. From these elements, we can deduce component functions of a situation assessment (see Chapter 2, Figures 2.9 and 2.10). The force estimation function builds on the Level 1 data fusion products of unit positions and identities to develop order of battle estimates and estimated event-activity patterns. Additionally, with the aid of yet other data bases, the force patterns are analyzed, as part of this Level 2 process, in the context of environmental and sociopolitical factors.

8.3 DEALING WITH CONCEALMENT, COVER, AND DECEPTION

There is one critical aspect of the situation assessment task: it is clouded with uncertainty, and generally that uncertainty includes the difficulty of confidently estimating the enemy commander's intent. We can enumerate a class of situation-estimating "countermeasures," which include

- *Concealment*—Methods to prevent sensors from observing deployments, capabilities, intentions.
- *Cover*—Methods (e.g., camouflage and avoidance) to deny the adversary the intelligence data needed for carrying out operations.
- *Deception*—Defined as usual, this includes the injection of false or misleading data.
- *Ambiguity*—Generating situations that have multiple interpretations (especially concerning intent).

The first three countermeasures, often called *CC&D techniques,* are important components of the information war. The following discussion, derived in part from [16], elaborates on the important aspects of countermeasure techniques such as CC&D. As noted previously, parallels can be drawn to the use of electronic countermeasure techniques and their effect on situation or threat analysis.

A skillful user of CC&D seeks to provide an adversary with pieces of information that appear genuine in themselves and that fit a course of action that the adversary would find reasonable. In this, the CC&D practitioner attempts to exploit the "anchoring bias" of the cognitive process [17], by presenting the strongest indications of the deception story first. If the intended victim (friendly situation analyst) has already formed an estimate of the most likely course of action, the practitioner need only take those actions necessary to provide substantiating evidence. Once the victim has focused on a single most likely course of action, receipt of later information will be evaluated in terms of whether it matches the current hypothesis. The victim may then ignore contradictory evidence, fit ambiguous evidence to match the hypothesis as if no ambiguity existed, and accept deceptive activity with little scrutiny [16].

The situation and threat assessment processes are vulnerable to CC&D at each step. Figure 8.10 decomposes the process and shows the opportunities for a skillful opponent to employ CC&D. To begin with, an adversary can control the timing and type of force activity to manage the opposing force's perception of the observable features of such action. Assuming that an analyst began with an accurate baseline of enemy locations and activity, this type of deception would cause errors in threat-situation monitoring. The opponent's CC&D plan attempts to orchestrate observable activity so that collection distortions and interpretation errors are propagated through the higher levels of intelligence analysis [16].

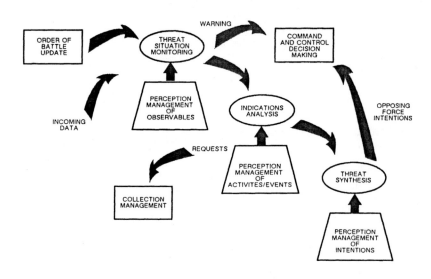

Figure 8.10 Threat assessment vulnerabilities to CC&D (after [16]).

Indications analysis, relying on an incorrect statement of the current situation, will misdirect requests for additional collection. Key activities will be missed, and others will be assessed as having occurred when they have been only simulated. The final step in the process, threat synthesis, matches key activities with the hypothesized set of courses of actions. If those key activities are not correctly identified, or if the set of courses of action is incomplete, an incorrect assessment of the opposing decision maker's intentions will be given to the blue force commanders. Unfortunately, many of the developments in intelligence systems in recent years have increased our vulnerability to CC&D at the same time that they have increased our ability to collect data and to analyze activity. Together with a greatly increased ability to collect, we have developed systems to help the analyst exploit that capability by focusing attention on items that analysts have identified as "keys" [18]. Thus, each enemy course of action can be broken down into indicators—steps that must be taken to realize that action, indicators into key activities, activities to observables. The result is a system with great power for focusing attention on significant pieces of information and leading to conclusions of intent based upon a clear path of reasoning. However, the weakness of this system is that the discriminators at each step become high-value targets for an opponent's CC&D activities [16].

Soviet doctrine for cover and deception, for example, derives from the doctrinal requirement for surprise, one of the basic principles of what the Soviets call *operational art*. These basic principles, in the order assigned in [19], include

- Mobility or tempo
- Concentration of efforts
- Surprise
- Activeness of combat
- Preservation of effectiveness
- Conformity of goals to conditions
- Interworking

For example, in the discussion of [19] describing methods to achieve surprise, six general types of CC&D activities are mentioned:

- Leading the enemy astray
- Secrecy of preparation
- Unexpected use of nuclear weapons
- Delivery of attacks at unexpected place or times
- New means or methods of warfare
- Avoiding repetition of methods

The first of these methods, leading the enemy astray with regard to one's intended course of action, is the doctrinal basis for Soviet deception operations, just as secrecy of preparation is the basis for the widespread Soviet use of cover and camouflage for offensive purposes. The last two methods are useful in understanding how the Soviets have been able to continue to surprise their opponents in intervention actions over the past 30 years. Use of such methods puts our intelligence system on notice that hypotheses limited to past patterns of Soviet actions not only can fail as aids to detection of new patterns, but also increase the probability that the Soviets will exploit our tendency to correlate the elements of a new course of action with an old one.

According to [16], countering deception is a two-step process. The first step is to identify the targets for deception; the second step is to identify how CC&D directed against those targets can be recognized and then exploited. These techniques must then be integrated into both sensor-related improvements and the development of expert systems and other automated tools for intelligence analysis.

Identifying areas of CC&D vulnerability can benefit from the great effort already expended in the development of structured indications and warning systems. These systems break down a range of courses of action into the steps (indicators) required to achieve them, decompose these steps into their key activities, and then identify the observables associated with each key activity. These observables are the high-value targets for CC&D operations, as by managing an opponent's collection of these observables, the deceiver exerts control on the basis of the victim's perceptions. Identifying the observational-category targets for CC&D involves a "reverse engineering" process. We know how individual collection sources and disciplines can be exploited by an opponent in a general way and we know which disciplines are employed to collect given observables. By matching collection means and CC&D method pairs to the association of collection means and observables, we can construct a CC&D matrix for each key activity. Table 8.1, after [16], presents a sample CC&D means matrix for one possible key activity—the deployment forward of a technical unit (such as a bridging unit). Each row in the matrix summarizes how the observable within a particular collection discipline could be simulated through specific CC&D techniques [16]. The more elaborate the deception, the greater number of these methods would be employed and the larger the number of units simulated.

Table 8.1
Representative CC&D Techniques (after [16])

Activity to be Concealed or Simulated: Deploy Technical Unit Forward	
Collection Discipline	Cover & Deception Means
ACINT	• Simulate sound of vehicles
COMINT	• Maintain normal levels of communications at garrison locations to mask movement of units • Maintain strict communications silence by the moving units
ELINT	• Stagger radar check-out before deployment to simulate normal activity • Maintain normal levels and types of radar activity in garrison areas
HUMINT	• Release covering explanation for activity (e.g., exercise announcement, troop rotation)
IRINT	• Deploy through areas with high levels of background heat (urban areas, major roads) • Deploy through areas which absorb IR emissions • Simulate normal garrison activity with nonessential vehicles
OPINT	• Camouflage deploying vehicles as non-military • Simulate essential vehicles in garrison with nonessential ones • Move at night
PHOTINT	• Simulate essential vehicles with dummies in garrisons
VISINT	• Divert foreign observers from deployment routes • Allow observers to see staged activities elsewhere

Uncovering CC&D depends in part on discovery of inconsistency. The hostile force would be likely to apply their military doctrine to CC&D operations, in that they will strive to achieve consistency with the least effort necessary and with integration ("interworking") of CC&D operations in all domains. Counter CC&D therefore means to progress from line items in the means matrix to correlation of line items in the matrix, and then to the higher levels of the situation and threat assessment processes. This search for inconsistency takes place on three levels, each demanding a higher level of human (or machine) intelligence [16]:

- Single collection discipline
- Multiple collection disciplines
- Analytical procedures involving one or more intelligence disciplines

Each technique, properly employed, should further stretch the deceiver's web of consistency in the observables, until finally it gives way and is revealed.

Single discipline techniques address the weaknesses in the collection and interpretation chain and can be divided into two types:

- Bringing the target into the field of view
- Increasing target discrimination

Success in either of these can be achieved by improving either the sensor capabilities, the data fusion and the exploitation processing, or even by alerting interpreters to the likelihood of a particular CC&D method.

Multiple discipline techniques seek to break down inconsistencies between two or more observables associated with the same key activity and are important components of an overall data fusion process for situation and threat assessment. The first step in applying these techniques is to take advantage of the means matrix to identify the opportunities for multiple discipline correlation. There follows a determination of whether the current collection schedules for the sensors involved allow simultaneous coverage. Planning for such coverage increases the burden of activity required to maintain a deception, especially if the sensors are multispectral. Finally, the analyst's ability to make effective use of multisource coverage requires that the interpretation of the collection be organized by activity.

An analyst can also uncover a CC&D operation by comparing current activity with the knowledge base of an opponent's capability and options. These comparisons are intelligence cross-discipline consistency checks, in which current intelligence is matched with basic intelligence, on the one hand, and threat assessment, on the other.

Basic intelligence provides the analyst with a reference of what an opponent can do. This includes, for example, the physical capability of equipment. In addition, it provides an organizational and doctrinal reference for current activity. Discrepancies in such factors become the basis for requests for additional collection and for expanding the scope of the analytical evaluation.

Once the time and space relationships between indicators and an opponent's likely courses of action have been established, these can also help uncover a CC&D operation and help the analyst recognize when the actual course of action does not match any of the hypotheses. One of the major benefits of the structured warning systems is that the analyst can be alerted to the inconsistent absence of activity. This absence could occur under any of the following conditions:

- The activity is present, but is being covered.
- The activity is not present, and other key activities are being staged.
- The activity is not present, and other activities are part of a course of action outside the current range of hypotheses.

The analyst can identify which explanation applies by increasing collection and exploitation efforts to uncover activities if they exist. If the activity is not found, solutions must be sought along both collection and analysis paths.

To support such analyses, increased effort would be applied to determine if some of the observed activities are actually simulations (i.e., CC&D activities). At the same time, the threat assessment process needs to reevaluate whether the absent activity is a necessary part of a course of action, and whether a new hypothesis would be consistent with the current combination of active and inactive indicators. The discovery of CC&D operations during this process has an additional value in that their existence is itself an indication of an opponent's course of action.

It should be clear that the determination of the existence of CC&D activities can and should be intrinsic to the overall data fusion process for situation and threat assessment. Note, however, that two of the three methods for CC&D detection suggested in [16] require an agile, adaptable multisensor collection capability, which cannot always be realized in many real-world applications. Failing this capability, the detection of CC&D activity relies primarily on correlation, inconsistency assessment, and absence-of-indicators (missing data) analysis, which must be carefully installed into the data fusion process.

Finally, it should be clear upon reading Chapter 9 that the methods being examined in the community to conduct situation and threat assessments are consistent with, and should include, the methods for CC&D detection described here.

8.4 SUMMARY COMMENTS ON SITUATION ASSESSMENT

From all the preceding discussions, it can be appreciated that no single method, algorithm, or procedure performs data fusion to support situation assessment. As asserted at the outset, there is not even a clear definition of the term *situation assessment* in the military or professional literature. There are, however, some salient points that apply to this function:

- Situation assessment is the result of a hierarchical hypothesis formulation and evaluation process, and each candidate hypothesis (situation) has an associated value of uncertainty.
- Optimality is achieved by formulating the least uncertain hypothesis.
- The processes of situation abstraction, or data analysis, by which the situation is depicted in terms of the current values of its elements, is essentially equal to the data fusion process; that is, they are synonymous.
- Situation assessment is a dynamic, chronologically ordered process in which the levels of consolidation (fusion) and abstraction usually increase over time. As a result, the fusion processes exhibit these same qualities over time.

Table 8.2 shows how data fusion can be applied in the processing of a subset of the situation elements toward the accumulation of a situation assessment. The table is only representative of the processing possibilities as a full set of possible combinations of information (situation elements) is clearly much too large and complex to depict diagrammatically.

8.5 THE CHARACTER AND COMPOSITION OF THE THREAT

As for the notion of situation assessment, there are somewhat varying definitions of what constitutes a "threat." Across several documents, it is not unusual to observe definitions of these terms that overlap to varying degrees and that are sometimes considered equivalent by different authors. In our approach, the distinctive aspect of the definition of the threat is that such definition quantifies the *destructive power* as well as the *vulnerability* of the hostile force and the *intent* of the hostile force. Thus, whereas situation assessment establishes a view of activities, events, maneuvers, locations, and organizational aspects of force elements and from it estimates what is happening or going to happen, threat assessment estimates the degree or severity with which engagement events will occur. Brentor [20] contends that every force element is characterized by such destructive power and vulnerability and that the balance of power depends on complex interrelationships between the destructive power of one side and the vulnerability of the other.

There are various interesting perspectives on threat analysis. For example, it is frequently asserted that the significance of the threat is in proportion to the perceived capability of the enemy to carry out that threat. To assess this significance then, the combination of *intent* and *capability* has to be determined. Said another way, the perception of the threat is in proportion to the level of danger it implies. The danger level, in turn, is the potential for harm if the threat is carried out; this view assesses the threat in terms of the potential losses to friendly forces. Both views provide a basis for quantification of the threat, in terms of either capability or projected losses of friendly forces. We prefer the former view because it does not interpose an engagement model from which loss estimates are calculated. We believe

Table 8.2
Segment of Fusion Processing for Situation Assessment

Situation Element	Specific Data (Sensed, Observed, Fused)	Information Processing		
		Primary Processing Product	Consolidation (Fusion) Product	Affected Blue Decision
• Estimate of Enemy	Composition of hostile forces (unit type) Individual unit locations Unit-Location-Activity Associations	Forward edges Strong, weak points Type of combat formation Nuclear delivery systems	Main enemy grouping Nuclear weapons employment concepts	Preferred combat sequence High value targets Nature of maneuvers
• Estimate of Own Troops	Individual unit deployments (type, location) Deployments patterns by function (artillery, tanks, troops, etc.) Unit-Location-Activity-Associations Qualitative Data: Degree of training Morale Quality of Commanders	Forward edge Strong, weak points Nuclear delivery systems	Compatibility of assigned mission and current deployments (regrouping requirements)	Preferred main thrust Structure of combat formations
• Estimate of Adjacent Units	General positions/deployment Nature of operations/missions	Locus of adjoining forces Strong, weak points Nuclear delivery systems		Possible joint efforts

Input Sources:
Multi-Sensors
Prisoners
Local populace
HUMINT
Other Blue Staff
Reference Info

Fusion Processing Functions

• Level 1 Fusion Processing, Followed by:

• Step 1 - Level 2 Fusion
  - Activity estimation
  - Key events
  - Maneuver detection
  - Critical Node Nets

• Step 2 - Level 2 Fusion
  - Hierarchical Event Trees
  - I&W Elements
  - Preliminary Nuclear Assessment
  - Battle Locus Estimation

• Step 3 - Level 2 Fusion
  - Red mission estimates
  - Estimated Red courses of action

• Step 4 - Level 2 Fusion
  - Fusion of Step 3 Fusion Results for Multiple Situation Elements
  - Optimal Enemy Mission. Course of Action Estimate
  - Increasing Consolodation, Level of Abstraction

that the commander is better prepared to couple the estimated capability with estimated intent to derive a threat estimate, although the use of flexible engagement models as decision or processing aids could certainly help derive improved decisions as well.

8.5.1 Elements of the Threat

In our definition, characterizing the threat thus amounts to (1) quantitatively portraying its force capability, and (2) coupling this picture with an estimate of intent. This view is not unlike that of the *intelligence preparation of the battlefield* (IPB) process defined by [21], in which "threat evaluation" consists of a detailed study of enemy forces: their composition and organization, their tactical doctrine, their weapons and equipment, and their supporting battlefield functional systems. The focus of this step of the IPB process is to determine an opponent's capability and intent.

Quantifying Force Capability

The enumeration of force capability in the practical, operational case would be a hierarchical process, governed or bounded by specific areas of interest at each command echelon. A useful, generic template of threat elements that can be used for force capability estimation follows:

1. *Strength*—Enumeration of the number and size of enemy units committed and number and size of enemy reinforcements available for use in the area of operations. Ground strength, air power, naval forces, nuclear and CB weapons, electronic warfare, unconventional warfare, surveillance potential, and all other forms of strength that might be significant to consider.
2. *Composition*—Structure of enemy forces (order of battle) with description of unusual organizational features, identity, armament, and weapons systems.
3. *Location and disposition*—Description of the geographical location of enemy forces in the area, including fire support elements, command and control facilities, air, naval, missile forces, and bases where appropriate.
4. *Availability of reinforcements*—Description of enemy reinforcement capability in terms of ground, air, naval, missile, nuclear, and CB forces and weapons; terrain, weather, road and rail nets, transportation, replacements, labor forces, prisoner of war policy, and possible aid from sympathetic or participating neighbors.
5. *Movements and activities*—Description of the latest known enemy activities in the area.

6. *Logistics*—Description of levels of supply, resupply ability, and capacity of beaches, ports, roads, railways, airfields, and other facilities to support supply and resupply; hospitalization and evacuation; military construction, labor resources, and maintenance of combat equipment.
7. *Operational capability to launch missiles*—Description of the total missile capability that can be brought to bear on forces operating in the area, to include characteristics of missile systems, location and capacity of launch or delivery units, initial and sustained launch rates, size and location of stockpiles, and other pertinent factors.
8. *Serviceability and operational rates of aircraft*—Description of the total aircraft inventory by type, performance characteristics of operational aircraft, initial and sustained sortie rates of aircraft by type, and other pertinent factors.
9. *Operational capabilities of combatant vessels*—Description of the number, type, and operational characteristics of ships, boats, and craft in the naval inventory, base location, and capacity for support.
10. *Technical characteristics of equipment*—Description of the technical characteristics of major items of equipment in the enemy inventory where not already considered.
11. *Electronics intelligence*—Description of the enemy intelligence-gathering capability using electronics devices.
12. *Nuclear and CB weapons*—Description of the types and characteristics of nuclear and CB weapons in the enemy inventory, stockpile data, delivery capabilities, nuclear and CB employment policies and techniques, and other pertinent factors.
13. *Significant strengths and weaknesses*—An estimate of the significant enemy strengths and weaknesses can be developed from the facts presented in the preceding list.

Assessing Hostile Intent

In addition to some type of enumeration of force capability as just listed, the definition includes the assessment of intent. Although some military analysts argue that true intent can be determined only from COMINT (i.e., deriving intent on the basis of intercepted message traffic, presumably revealing force intentions in orders, *et cetera),* there are arguably other ways to conduct an intent analysis. Such methods may include

- Estimation of courses of action from patterns of movement, events, activities.
- Assessing operational readiness of key force elements.
- Critical node analysis as described in Section 9.2.2 wherein key indicators are monitored as crucial to the execution of a particular activity.

Figure 2.11 of Chapter 2 shows, analogous to Figure 2.9 for situation elements, a representative depiction of elements constituting a threat, consistent with the preceding discussion. Note again our insertion of countermeasure elements, a product of situation assessment, as also important in correctly understanding the available data for a threat analysis.

8.5.2 Threat Assessment Functions

Similar to the deduction of situation assessment functions from the enumerated elements, Figure 2.12 of Chapter 2 offers a representative set of threat assessment functions and techniques that would be involved in such analysis. Note that there are various Level 2–Level 3 data fusion interfaces in the overall process.

Note that force capability is derived from force counting functions, supported by Level 1 data fusion products and by Level 2 assessments of deployment patterns. This function is primarily attempting to estimate force lethality, measured by the quantified potential of hostile forces to inflict blue losses. Such estimates can be supported by the use of so-called fast-time engagement models, which can estimate the results of a hypothesized conflict.

The intent estimation function employs a variety of tools to assess intent; the analysis of CC&D effects, provided by an interface to the Level 2 situation assessment function, is crucial to this analysis.

8.6 MULTIPLE PERSPECTIVES OF THREAT ELEMENTS

In some training materials being developed for intelligence analysts, an interesting concept, called the *concept of shifting perspectives,* has been posed as the proper framework for intelligence data analysis, including situation and threat analysis. In this approach, any given set of threat elements (i.e., any threat element set instantiation) should be viewed from "white," "red," and "blue" perspectives, and the perspectives integrated to fully develop the model of the threat. More is said on all this in Chapter 9 but let us define the basic ideas:
- The white perspective looks at the data in the context of time and spatial references. Analyses from this perspective result in models of threat element deployments and movements and aid in formulating notions of intent. (Note the commonalities in approach between the analysis associated with the white perspective and the process of situation assessment described in Section 8.2.)
- The red perspective looks at the data in the context of the estimated red war plan. This analysis requires that the blue analyst "think red"; that is, transpose blue's perspective to that of the enemy commander being confronted. In this way, it is felt that optimum estimates can be determined of red intent and

probable courses of action. Developing such views ideally requires understanding of the enemy's (1) ideology, (2) history, (3) culture, (4) political structure, (5) education and training. In addition, estimates must be made of the red war plan that defines why, where, and when the enemy will enter into combat and with what force structures, schedules, and operations.
- The blue perspective looks at the data in the context of the blue mission goals and objectives. Ideally, this analysis considers the threat element data in the context of each of the nine principles of war (see Chapter 9).

The principles of war define a structure for blue thinking in which each principle defines a potential use of threat model information.

This approach helps ensure that the analysis of threat element data is more comprehensive than that usually applied from just the blue perspective. This general approach, combined with the threat element taxonomy listed in Subsection 8.5.1, provides the framework for an optimum, robust characterization of the threat.

8.7 SUMMARY COMMENTS ON THREAT ASSESSMENT

Although the concept of shifting perspectives is useful in expanding upon possible implications of the existence or measure of given threat elements, data fusion is used in threat assessment to help accomplish the blue mission. This occurs in two ways: by a fusion process that helps develop the correct intelligence product, and by a fusion process that aids in developing a complete threat model.

It is instructive to recall that data fusion is part of an intelligence cycle that culminates in an intelligence *product*. Further, part of the definition of data fusion claims that, to be useful, fusion should contribute to product comprehensiveness; that is, completeness. There is an analog to the classical systems engineering process here: after all the analysis and partitioning to fully understand the input (here the observations, assertions, and hypotheses), a synthesis establishes a single, unified estimate of reality.

Data fusion to develop the product entails the aggregation of information consistent with the intended user's view (or an estimate of it). Often, this is an echelon-specific view, although the "one-up, two-down" multiechelon perspective has been well-trained in the U.S. services and may be a better representation of the expected norm. This data fusion is more than just proper aggregation; it may involve estimation, interpolation, or other analytical methods to assemble the best product or it may involve the use of knowledge-based techniques for this purpose. Data fusion to assemble the most complete threat model generally derives from two thought processes: those associated with *expectations* regarding the various aspects of the threat, including the effects of various countermeasure techniques, and those associated with estimating the *uncertainties* in the threat model.

In Chapter 9, when we look at the various ways that researchers have attempted to estimate battlefield situations and threats, we will emphasize that all such methods derive from the broad notion of "expectation templates." Expectation templates, which form the basis for data fusion for situation and threat assessment, are essentially *a priori* conceptual models of the threat. They result from the cumulative training and experience in any individual, and are difficult to change (hence the difficulty in implementing the concept of shifting perspectives, especially in establishing the red view). These conceptual models have frailties: they include various biases (both cultural and cognitive biases), and they are affected by human emotions (especially stress).

Nevertheless, the fusion processes applied to derive maximally complete threat estimates are founded on these (imperfect) expectation templates. The expectation templates (or models), depending on their structure, can be used to fuse both multisource capability estimates and estimates of intent derived from multisource data. As mentioned earlier, some argue that estimates of intent can be derived only from communication intelligence, but analyses and fusion of maneuver characteristics, critical node, or functional network behavior (e.g., resupply networks) can be most instructive in estimating intent.

It is very difficult to deal with the issue of uncertainty estimation at this level of abstraction. Techniques for fusing multiple uncertain estimates whose uncertainty scales may be mixed (statistical, ranked, or ad hoc) usually must resort to some type of scale normalization. The only alternative would seem to be the presentation of multiple sets of uncertainty measures to the decision maker, who would have to quickly assess the composite meaning of such metrics. Conversely, the scale normalization option usually involves normalizing all measures to the scale of the least specific metric. Neither choice seems attractive; but another choice, which balances these, may be to fuse the measures from a set of "key" indicators. In such cases, a composite measure that is easily understandable and reasonably specific may result. The next chapter will examine how these concepts of data fusion for threat and situation assessment have been implemented in either laboratory prototype or near-operational systems.

REFERENCES

1. Tzu, S., *The Art of War*, trans. S.B. Griffith, Oxford at the Clarendon Press, London, 1963.
2. Von Clausewitz, C., *On War*, Princeton University Press, Princeton, NJ, 1976.
3. Beaufre, A., *Deterrence and Strategy*, Praeger Publishers, New York, 1966.
4. Dupuy, T.N., "Can We Rely upon Computer Combat Simulations?" *Armed Forces Journal Int.*, August 1987.
5. McQuie, R., "Battle Outcomes: Casualty Rates as a Measure of Defeat," *Army Magazine*, November 1987.
6. Gye, R., and T. Lewis, "Lanchester's Equations: Mathematics and the Art of War—A Historical Survey and Some New Results," *Mathematical Science*, Vol. 1, 1976.

7. Orr, G.E., *Combat Operations C³I: Fundamentals and Interactions,* Air University Press, Maxwell Air Force Base, Alabama, July 1983.
8. Wohl, J.G., "Force Management Requirements for Air Force Tactical Command and Control," *IEEE Trans. Systems, Man, and Cybernetics,* Vol. SMC-11, pp 618–639, 1981b.
9. Wohl, J.G., et al., "Human Decision Processes in Military Command and Control," *Advances in Man-Machine Systems Research,* Vol. 1, JAI Press, 1984, pp. 261–307.
10. Reitman, W.R., *Cognition and Thought: An Information Processing Approach,* John Wiley and Sons, New York, 1965.
11. Dieterly, D.L., "Problem Solving and Decisionmaking: An Integration," NASA Tech. Memo. 81191, Ames Research Center, CA, April 1980.
12. Reidelhuber, O., "Modeling of Tactical Decision Processes for Division-Level Combat Simulations," in *Systems Analysis and Modeling in Defense,* R.K. Hubert (ed.), Plenum Press, New York, 1984.
13. Wason, P., "The Psychology of Deceptive Problems," *New Scientist,* Vol. 63, 1974, pp. 382–385.
14. *Dictionary of Military and Associated Terms,* JCS Pub. 1, US Government, Superintendent of Documents.
15. Ivanon, D.A., et al., *Fundamentals of Tactical Command and Control—A Soviet View,* translated under auspices of the USAF, U.S. Government Printing Office, Washington, DC, 1977.
16. Goldsmith, R.P., and R.F. Gerenz, "Techniques for Detecting Cover and Deception," *Proc. 6th MIT/ONR Workshop on C³ Systems,* MIT, July 1983.
17. Heuer, R.J., Jr., "Strategic Deception: A Psychological Perspective," presented at the 21st Annual Conv. Int. Studies Assoc., Los Angeles, March 19–22, 1980.
18. Gerenz, R.F., "A Methodology for Improving the Strategic Warning Process," *J. Defense Research,* Crisis Management Edition, April 1977.
19. Saukin, V.Y. *The Basic Principles of Operational Art and Tactics (A Soviet View),* Moscow, 1972. USAF translation, Chapter 3, pp. 167–277.
20. Brentor, R., *Decisive Warfare—A Study in Military Theory,* Stackpole Books, Harrisburg, PA, 1969.
21. *IPB-Intelligence Preparation of the Battlefield,* U.S. Government Printing Office, Washington, DC, Sup R 66000-A, June 1983.

Chapter 9
IMPLEMENTATION APPROACHES FOR SITUATION AND THREAT ASSESSMENT

The processes of situation assessment and threat assessment have much in common with characteristics of command and control systems. First, situation and threat assessment, like command and control, are multilevel activities; almost invariably, information flows across the levels of a command hierarchy (both up and down), command takes place at several levels, decisions are made at several levels, and decisions at a given level entail coordination with those above and below. This suggests that the analytical processes for *situation and threat assessment* (STA) must be sensitive to command cycles and communications processes within and across levels and that information input requirements and decision-aiding techniques and methodologies will vary appreciably as one ascends or descends the hierarchy [1].

Second, C^2 and STA serve mutiple and manifold functions, including perception and information management, decision making and direction giving, and implementation. Both planning (preliminary and detailed) and operations are central to these processes, and intelligence, fire and air support, and personnel and logistics all feed into the C^2 and STA analytical and decision process. Central to command and control and STA is the inference making process, with option identification and selection peculiar to C^2 (recall that fusion is a hypothesis-generating process that concludes before option development). These processes correspond at least in a very general way with intelligence and operations; both require a wide range of aiding systems and methodology bases to yield accurate, useful inferences and comprehensive but manageable and easily communicated option sets.

Third, both C^2 and the assessment processes apply to multiple military decision occasions. Included are peacetime, crisis, and wartime (conventional and nuclear) situations operating at multiple levels. Overlaid on these tiers are strategic C^2, theater C^2, allied C^2, and tactical C^2, which injects even more complexity into the overall C^2-STA milieu, reinforcing the realization that, by their very nature, C^2 decision making and assessment processes are both complex and distributed.

Finally, according to Hopple [1], from which these comments are derived, the key shared characteristic of all C^2 decision processes (regardless of level, function, or context) is the commander's definition of "the situation," underlining the importance of psychology to the C^2-STA processes—psychology in the sense of perception, cognitive processing, and the interface between artificial intelligence and the study of the mind.

Given these characteristics of situation and threat assessment, it is clear that there would be no single way to perform these processes. Unlike Level 1 fusion processing, where correctness is easier to define and such a definition, in turn, provides a way to evaluate the quality of a given methodology (for position or identity estimation), correctness in assessing a situation or threat is a much less clear notion. Dockery [2] pursues this point in the context of defining fuzzy measures of effectiveness for C^2 systems (e.g., [3–5]). The formulation of standard measures of correctness for STA processes and C^3 (or weapon) systems is extremely important, in that such measures provide the basis for evaluating the "goodness" of fusion in the context of derived *force effectiveness*. If one could establish the successive links of correctness-effectiveness measures as in Figure 9.1, then the contribution of multilevel fusion processing and particularly STA analyses to force effectiveness could be assessed and the cost-effectiveness of particular approaches to STA and related system design trade-offs evaluated quantitatively.

Unfortunately, the discipline of data fusion has not reached this level of formality and rigor, especially for Level 2 or Level 3 (i.e., STA) analyses. Consequently, most STA methods have been developed to directly assist some commander in developing an estimate (i.e., a mind's eye view) of the battlefield picture. The results of SA analyses typically yield estimates of hostile force (and perhaps own force) composition, deployment, and behavior, whereas the results of TA analyses yield combat implications of the hostile force and provide information for decision making and action.

In this chapter, a few of the analytical approaches to STA are reviewed, as well as, from a case study review, some of the artificial intelligence approaches to STA used by a number of researchers. STA is a tough, multiperspective, multilevel problem; these factors, the notion of contextual analysis of Level 1 results (part of our definition of Level 2 analysis), and the somewhat fuzzy definition of what constitutes a "situation" or a "threat" makes the selection of particular solution approaches very difficult. Experience to date has shown that analytical approaches are dominated by what are called here *expectation templates* as well as figure-of-merit techniques, both of which are discussed in the chapter. Artificial intelligence approaches are generally dominated by expert system or knowledge-based strategies, and some prototypes are reviewed here to describe the range of approaches employed. In addition, some adaptive, AI-based solutions are reviewed. In general, the reader interested in STA should also review Chapter 12, which describes still other AI approaches, such as plan recognition techniques.

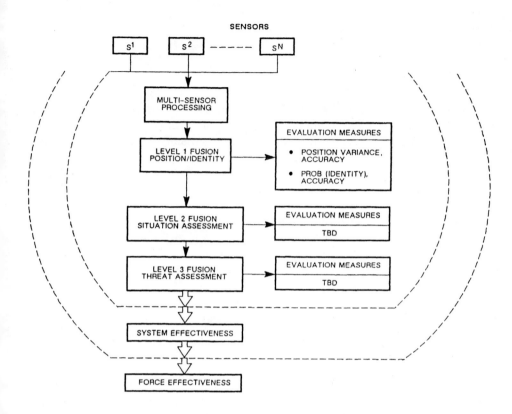

Figure 9.1 Connectivity of evaluation measures.

9.1 ISSUES AND PROBLEMS OF METHODOLOGICAL SELECTION

Before reviewing some approaches to STA used in the data fusion community, a number of the factors involved in selecting one or more methods to employ in STA analysis are discussed. Upon review of this section, the reader should conclude that the availability and employment of a wide variety of algorithms and techniques is no substitute for understanding the basic nature of the STA problem as described in Chapter 8. Effective solutions of STA problems result only from applying sound system engineering techniques over the requirements-to-test spectrum.

Lack of Theoretical Foundation

The fact that a "theory" of situation or threat assessment does not exist (although the ideas of Chapter 8 are available) makes it difficult to assess the suitability of various available processing methods for any particular assessment problem. This is analogous to assessing the suitability of decision aids of various types (see Hopple [1]), and in both cases we must prevent forcing an available solution onto a given problem. The key point is that, in considering solutions to assessment (or decision-aiding) problems, we should avoid the lure of the technology push (the familiar "solution in search of a problem" syndrome), and assure ourselves that our solution approach is based on a requirements-driven analysis. Not having a theory of these assessment processes increases our vulnerability to possibly forcing a solution onto the problem.

Requirements-Driven Methodology

In the analogous case of choosing a methodological basis for C^2 (or STA) processing, Hopple [1] describes the selection process as a hierarchical one, driven by requirements. In Figure 9.2, the choice of decision-aiding or STA methodology is driven by a matching process between the characteristics of the results provided by a collection or set of methods and those characteristics required by the situation or threat assessment "product" (i.e., processing product). Thus, if the required characteristics of a situation assessment product include both a statistical characterization of events and activities and a forecast of likely enemy courses of action, we will require methods from the taxonomy or inventory that provide such characteristics.

Need for Multiple Methods

This discussion points out another feature of situation or threat assessment processing: that such processing typically involves the use of *multiple* methods in what could be called a hybrid approach, to achieve methodological synergism. In the real world, with real engineering teams, this may be difficult to achieve, because methodology experts tend to operate from methodological "ghettos" (operations research experts avoiding artificial intelligence experts, *et cetera*).

Another way to appreciate the range of methods required to perform STA is to reexamine the elements and functions of STA described in Chapters 2 and 8. If those definitions are reasonable and applicable to any given STA problem, then the full range of analytical capability necessary to produce estimates of the constituent STA elements is required, and this would clearly comprise a large and mixed set of methods.

Problem Structuring

Yet another way to look at the problem of methodological selection is to consider that each method should aid in *structuring the problem* of situation or threat assess-

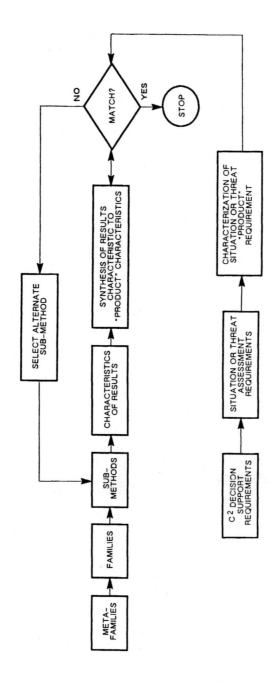

Figure 9.2 Requirements-driven methodology selection.

ment. If we consider that assessment generation is essentially an inference-generating process, then the method selection should also be based on a method's ability for deconflicting hypotheses, the degree and nature of assumptions required, and the extent of evidence that can be processed—the three building blocks of the inference process. This framework often leads to the choice of robust graphics processing methods to visually depict the problem structure, as well as the use of data management and data organization techniques. In this approach, the key is to facilitate the objective of revealing the interrelationships among the situation and threat elements of the problem.

Constituency-Dependency Relationships

The methodological structure appropriate for analyzing an assessment problem can also be derived on the basis of the examination of the various relationships between the situation or threat elements. Wohl *et al.* [6] suggest a set of relationships that depict either constituency or dependency relationships between the elements; this set is shown in Table 9.1.

Both *physical* and *functional* constituency representations generally make use of block diagrams that indicate, name, or otherwise identify for each block, either explicitly or implicitly (i.e., via connectivity), the following:

- Functions or processes executed,
- Inputs required,
- Outputs provided,
- Controls applied,
- Resource required.

Table 9.1
Categories of Representational-Decompositional Techniques (after [6])

Type of Relationship Represented	Question Answered	Examples
Physical Constituency	"... is composed of ..."	System block diagrams, specification trees
Functional Constituency	"... involves/requires/provides ..."	$IDEF_0$ decompositions, functional block diagrams, functional decomposition trees, Interpretive Structural Modeling
Process Constituency	"... performs the process of ..."	Mathematical functions, logical operations, rules, procedures
Sequential Dependency	"... occurs conditional upon ..."	PERT charts, Petri-nets, scripts, operational sequence diagrams
Temporal Dependency	"... occurs when ..."	Event sequences, time lines, scripts, operational sequence diagrams
All of the Above	All of the Above	Computer simulations, real-world systems

In addition, each individual block in a diagram can be further decomposed onto its consituent block diagram, and so on until the level of primitives is reached. To handle the added complexity associated with these different levels of description, decomposition trees are often introduced to represent the relationship between levels.

Operational constituency representations generally describe either mathematical or logical operations on inputs to produce outputs. They can also take the form of explicit rules ("if-then-else") and procedures.

Temporal and *sequential* dependency representations nearly always describe functions, processes, activities, and events on a timeline basis to indicate either their required sequence or timing or both.

Contextual Analysis

Another important aspect of STA processing techniques is that such methods attempt to perform a *contextual analysis* of the Level 1 fusion products. That is, the Level 1 processing results often depict the abstract (i.e., context-free) locations, tracks, and identities of battlefield entities of interest. In Level 2 and 3 processing such arrays or networks of located and identified entities are examined in the context of, for example,

- Battlefield topography and weather,
- Adjacent hostile and friendly force structures,
- Tactical or strategic operational doctrine,
- Tactical or stategic force deployments,
- Supporting intelligence data.

The implication of this contextual approach to STA is the significant database requirements. Just the terrain and weather database requirements can result in significant demands for computer storage and processing, and these problems can overwhelm the analytical processing perspective as it affects computer selection and architecture.

Database performance also can have an impact on the inference-generation process. Database operations must be balanced in terms of fast data insertions *versus* fast data retrievals, so that inference generation is derived from a temporally consistent data set; more is said on temporal effects in Section 9.2.4.

Similarities to Level 1 Processing

A final comment on STA processing is that it has certain similarities to Level 1 processing. For example, the individual processing steps of alignment, association, and correlation must still be carried out. In the STA case, we must still remove any positional or sensing geometry and timing effects from the data (alignment), develop some measure of closeness between entities or groups of entities (association), and use such measures for declaring hypotheses (correlation). These processes can be identified in most research efforts associated with STA analysis.

Current research and experimentation in situation and threat assessment processing has employed a subset of some of the methodologies described here but, in addition, some limited work has extended the concepts described to adaptive approaches that incorporate learning paradigms. Most of the work, as mentioned in Chapter 8 and here, has used the general concept of *expectation templates* and slot filling on the basis of cumulative observations from battlefield sensors. As pointed out in Chapter 7, such schemata facilitate the retrieval and understanding of information; often, knowledge structures are productive mechanisms for producing plans and decisions, but they also can impose excessive structure and lead to mental fixation—sometimes with deleterious or disastrous consequences (failing to foresee massive deception or a successful surprise attack, not anticipating enemy behavior on the battlefield, and the like). Adaptive techniques are gradually receiving emphasis to overcome these deficiencies.

In the sections that follow, past and current research in STA is reviewed. Section 9.2 examines the broad category of expectation template-based techniques, Section 9.3 discusses relatively recent work in cooperative and adaptive approaches, and finally Section 9.4 briefly reviews the important role that so-called performance models play in STA analysis.

9.2 EXPECTATION TEMPLATE-BASED TECHNIQUES

In a template-based approach, situation assessments are derived by developing *a priori* models of force structures, force deployments, event sequences, activities, or observations that collectively define some situational element of interest. These models, generally called *templates* here, can take various forms in specific implementations. On the one hand, they can be created in a tabular format, which may reflect the participants in a situation, the event sequences for each participant, and certain time and distance relationships between the events. Alternately, they can be created with knowledge representations from the field of artificial intelligence; for example, frames or scripts. Further, they can be represented using figures of merit and measures of correlation, which are generalizations of the concept of an association measure (as, for example, used in tracking and discussed in Chapter 6). In any of these constructs, fusion of the multisensor data is simply a result of the sensor data populating the template and thereby reflecting the various relationships of interest. That is, without data these structures form abstract frameworks reflecting relationships of interest. The fusion system designer, aware of the multisensor system available, designs the framework (templates) to exploit the availability of such data. However, there are not usually any algorithms that fuse the data in the sense of track or declaration fusion at Level 1. In a sense, one might say that templates and other constructs provide frameworks for fusing relationships among the data and parameters observed or calculated.

To varying degrees, all of these methods generally depend on the enemy behaving in a rational way and according to stated or practiced doctrine. That is, these models or templates are usually constructed from intelligence information derived from SIGINT, from field exercises, or from documented tactics and doctrine. This *a priori* knowledge also includes the so-called *order of battle* (OB) or *electronic order of battle* (EOB) that, respectively, depict the unit-level hierarchical associations of either hostile combatants or electronic equipment. A systematic procedure that employs templating concepts for situation and threat analyses and that is generally well known to the military is called *intelligence preparation of the battlefield* (IPB) [7].

9.2.1 Intelligence Preparation of the Battlefield

The IPB process is a military procedure designed to give battlefield commanders information about an enemy's past and present actions and future intentions. As shown in Figure 9.3, the process is composed of two major pieces, threat evaluation and threat integration. In threat evaluation, multisource intelligence data in a message-level report format is received and fused by an order of battle analyst with the order of battle holdings and enemy doctrinal information to provide an initial assessment of the situation. Erroneous reports, conflicts among reports, contradictions with known enemy doctrine and contradictions with the order of battle hold-

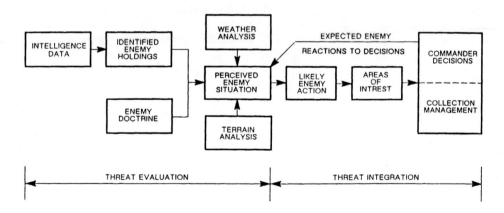

Figure 9.3 Intelligence preparation of the battlefield (from [7]).

ings are identified at this point and resolved if possible. Unresolved conflicts are maintained awaiting future intelligence reports that may resolve the conflict. This initial situation assessment is then fused with weather and terrain to provide the perceived enemy situation, which constitutes the end of the threat evaluation phase.

In threat integration, the perceived enemy situation is used as the basis for predicting further enemy intentions and identifying areas of interest that should be monitored for verification of those predictions. On the basis of this information the battlefield commander determines additional intelligence priorities known as *priority information requests* (PIRs). The PIRs and the current intelligence estimate are then used to determine the intelligence resource collection plan that forms the basis for assigning intelligence collection tasks.

Notice that, in the IPB process, the terms *situation* and *threat* are intermixed without specific clarification; additionally, these terms do not exactly comply with the definitions suggested here. For convenience to those familiar with IPB and to be consistent with the referenced literature, we will use the terms in the IPB context (only in this section).

The IPB process is continuous. It concentrates on building the IPB data base prior to hostilities and outlines its applicability in support of tactical operations. This results in an intelligence estimate and analysis that portrays probable enemy courses of action and intentions. Mission planning sets the IPB process in motion.

Graphics are basic to IPB analysis. Most intelligence information can be communicated with pictures. Annotated military maps, multilayered overlays, gridded photographs, microfilm, and large-scale map substitutes, all capable of computer-assisted display, are used in the IPB process. These graphics become the basis for intelligence and operational planning.

IPB provides a basis for collection management before the battle and guides the effective employment of collection resources during the battle. The graphic data bases developed and maintained through IPB, coupled with conventional data bases, provide a foundation for situation and target development. They provide a means for projecting significant battlefield events and enemy activities and for predicting enemy intentions. By comparing them with actual events and activities as they occur, the commander is provided timely, complete, and accurate intelligence.

Template construction begins in the first step of the IPB process. In the "classic" IPB process, a template, normally constructed to scale, is a graphic illustration of enemy force structure, deployment, or capability. IPB provides a means for seeing the battlefield and is the basis for command judgments and decisions affecting resource allocation. IPB is used as a comparative database to integrate what is known about the enemy with a specific weather and terrain scenario.

Templates enable the visualization of enemy capabilities, prediction of likely courses of action before the battle, and confirmation or refutation of them during combat. Templates also provide a means for continuous identification and assessment of enemy capability and vulnerability. Information graphically displayed on templates can be added to, changed, or deleted as the situation changes. Table 9.2 describes the four principal types of templates developed during the IPB process and explains how and when each is to be used [7].

Table 9.2
Intelligence Preparation of the Battlefield Templates (after [7])

Template	Description	Purpose	When Prepared
Doctrinal	Depicts enemy doctrinal deployment for various types of operations without constraints imposed by weather and terrain. Compositions, formations, frontages, depths, equipment numbers and ratios, and HVT are types of information displayed.	Provides the basis for integrating enemy doctrine with terrain and weather data.	Threat evaluation
Situation	Depicts how the enemy might deploy and operate within the constraints imposed by the weather and terrain.	Used to identify critical enemy activities and locations. Provides a basis for situation and target development and HVT analysis.	Threat integration
Event	Depicts locations where critical events and activities are expected to occur and where critical targets will appear.	Used to predict time-related events within critical areas. Provides a basis for collection operations, predicting enemy intentions, and locating and tracking HVT.	Threat integration
Decision Support	Depicts decision points keyed to significant events and activities. The intelligence estimate is in graphic form.	Used to provide a guide as to when tactical decisions are required to battlefield events.	Threat integration

Doctrinal templates convert enemy OB factors into graphic portrayals. They are models of how the enemy might deploy its force according to doctrine and training if not constrained by the weather and terrain. They portray various echelons and types of units for various capabilities and schemes of maneuver. They also graphically portray the composition and disposition, frontages and depths, and spacing and signatures of these echelons and units. Figure 9.4 shows an example (not real data) of a doctrinal deployment pattern.

Doctrinal templates may include portrayal of higher-echelon supporting elements or elements normally deployed with the unit templated. They may be further refined into doctrinal template subsets. These subsets might include battlefield functional systems or weapons and equipment deployment. Such templates, especially those depicting weapons and equipment deployment, are very useful in identifying types of enemy units and specific formations. Subsets may be equally useful in determining enemy intentions. Notice that such template processing assumes completion of Level 1 fusions processing (unit-level position and identity estimates). Fusion processing for these templates is that required to populate the template with additional units as defined from Level 1 processing. The IPB process does

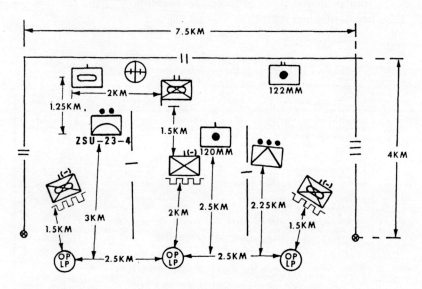

Figure 9.4 Representative doctrinal deployment of a motorized rifle battalion (from [7]).

not define this implementation but characterizes the overall process of intelligence analysis.

After establishing the correct doctrinal template, the next step of IPB is the evaluation of the area of influence. During this step, the scope of IPB is narrowed to a specific area of the battlefield. The evaluation considers all OB elements applicable to this specific area.

After the specific areas have been evaluated, the information and materials required to complete the IPB process are assembled. The next step is a detailed analysis of both terrain and weather (a *contextual* analysis of the doctrinal template data). Terrain analysis is focused on the military aspects of the terrain and the effects on friendly and enemy capabilities to move, shoot, and communicate. This includes the following five factors:

- Observation and fields of fire,
- Concealment and cover,
- Obstacles,
- Key terrain,
- Avenues of approach and mobility corridors.

The terrain analysis process emphasizes the use of graphics to portray the effects of trafficability and intervisibility on operations. A series of terrain matrices and overlays are prepared to develop a terrain graphic data base to facilitate threat integration.

In performing terrain analysis, we reemphasize that a significant number of what we call *performance models* are required; the term refers to those routines by which phenomenological and geometric-type relationships such as propagational effects on communications, optical *lines of sight* (LOS), lethal weapon envelopes, sensor coverage envelopes, *et cetera* are calculated. Thus, situation or threat assessment processes require fusion not only of symbolic-object type of data, but, to add the perspectives of intent or risk, also requires fusing these phenomenological-geometric views as well. These relationships (e.g., platform-emitter intervisibility) are depicted on the doctrinal template to complete its specificity.

The final step of the terrain analysis process selects the avenue of approach that supports the enemy's capability to move, shoot, and communicate. Weather has a significant impact on both friendly and enemy capability. Analyzing the weather in detail to determine how it affects friendly and enemy capability to move, shoot, and communicate is critical to this step of IPB. Because the weather has a tremendous effect on terrain, terrain and weather analysis are inseparable factors of intelligence.

A combined obstacle overlay combines all terrain and weather induced obstacles resulting from this analysis. It focuses on significant terrain areas. Next, avenues of approach and mobility corridors are identified, concentrating on those areas where the enemy can move. Once the most viable avenues of approach and mobility corridors have been selected, overlays are prepared depicting each. Through analysis, LOS for weapons, communications, target acquisition, intelligence collection, and *electronic countermeasure* (ECM) systems are developed for each option.

The last step in the IPB process integrates enemy force execution doctrine with weather and terrain data. The objective of the integration is to determine how the enemy will fight given weather and terrain conditions. Threat integration, a sequential process, is accomplished through the development of situation, event, and decision support templates. The "situation template," developed from the most current doctrinal templates and combined obstacle overlays, shows how threat forces might deviate from doctrinal dispositions, frontages, depths, and echelon spacing to account for the effects of the weather and terrain. It is the result of the fusion of unit deployment, terrain and weather data assessments.

As an enemy force moves along a *mobility corridor* (MC) it will have to do certain things at certain times and places as dictated by terrain, weather, and tactics. Based on this, the analyst selects *named areas of interest* (NAI) where certain activities or events are expected to have tactical significance. NAI, which may also be *target areas of interest* (TAI), are points or areas along a particular avenue of approach or MC where activity, or lack of it, will help confirm or deny a particular enemy course of action. TAI are areas or points on the ground, the successful interdiction of which will cause the enemy to either abandon a particular course of action or require the use of unusual support to continue along that particular route. The situation template and analysis of these potential events and activities provide the basis for "event template" development. These event templates are therefore templates of expected events or activities; multisensor data are fused to assess the occurrence of such events or activities, which are then used to "fuse" or populate the event template.

The event template provides the information needed to project what most likely will have to occur relative to enemy courses of action. This projection is based on an analysis of the relationship of NAIs to one another and to specific available courses of action. In Figure 9.5, which is an example of an event template,

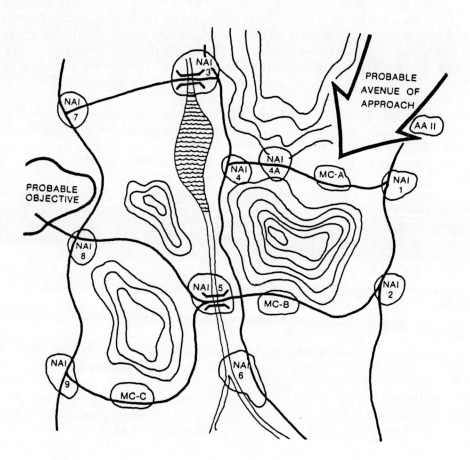

Figure 9.5 Event template (from [7]).

NAI 1 through 9 are areas where particular types of activity would provide indications of intent. For example, activity in NAI 1 would indicate whether mobility corridor A or B was being adopted as the route of advance. Other NAI in the example represent intermediate points for collection planning purposes or tracking for target development purposes. The event template and events analysis matrix allow for the initiation of precise sensor collection requirements, maximizing the use of limited collection assets against the vast array of potential targets on the future battlefield. By knowing in advance what the enemy can do and comparing it with what is happening, the analyst has the basis for predicting what the enemy intends to do next. Such information provides the basis for constructing "decision support templates."

Event and decision support templates, the most important products of the IPB process, represent a reduction of all the analysis and template construction tasks that have preceded them into analysis of the area of operations and an intelligence estimate. Figure 9.6 depicts an example of a type of decision support template.

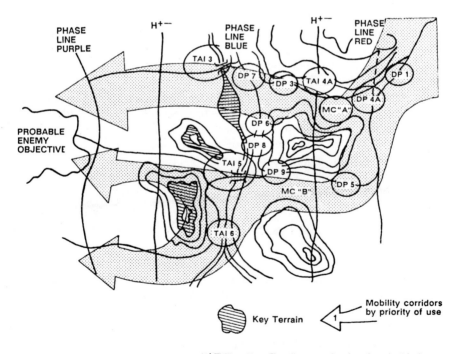

Figure 9.6 Decision support template (from [7]).

The decision support template relates events, activities, and targets of the event template to the commander's decision requirements. It is basically a combined graphical intelligence estimate and operations plan. Commanders and their staffs develop decision support templates by overlaying the event templates and illustrating enemy course of action with *decision points* (DPs) for friendly courses of action. The DPs are placed on the template at those points where a commander must decide which planned course of action to employ to have the desired effect on the enemy. DPs shown in the illustration represent areas chosen because of time-distance factors from target areas of interest. If a decision is not made by the commander before an enemy force reaches or passes a decision point, a set of options that had existed may be negated.

The original description of IPB in [7] was expanded on in [8], which serves as an excellent starting reference for the automation of the IPB template-oriented techniques.

9.2.2 Event-Activity Profiling

In the IPB process, events are templated by analyzing a specific area of interest and hypothesizing a set of possible "mobility corridors" and probable activities. This template then forms a framework for the decision template, as described in Subsection 9.2.1. IPB makes use of doctrinal information primarily for depicting probable force deployment patterns.

An alternate approach, usually called *event-activity profiling*, makes use of doctrinal or otherwise expected *patterns of action*. Such patterns would be derived from stated military doctrine on tactics and operations and from observations of hostile force exercises during peacetime or in crisis situations. This is yet another implementation of an "expectation template" philosophy. This type of an approach has been studied for some time (e.g., [9, 10]) and is related to work on decision templating (e.g., [11, 12]).

A SIGINT Application

This technique has been employed in studying possible methods for the fusion of SIGINT data as a basis for situation assessment [13]. In this approach, so-called event profiles were constructed; these were computer-based data structures that characterized an event (or activity) of interest in terms of its constitutent COMINT- and ELINT-related observations. The overall structure of the template system formed a hierarchy of such event profiles with appropriate indices and pointers. The situation domain was constrained to those events and activities related to air defense operations. As in most of these approaches, the situation assessment unfolded as a result of the fusion of cumulative multisensor based COMINT-ELINT observations "filling" the event hierarchy; that is, by such observations instantiating a set of event profiles. Figure 9.7 shows a notional event profile for this application.

The template structure shows the signal "events" (those expected to occur in conjunction with, and reflecting the occurrence of, tactical event XYZ) by numerical code and name, a weighting factor used in assessing confidence in the tactical event, and the expected logical occurrence of related signals (which could be used in a sensor collection strategy). As signal occurrences are observed by the multisensor COMINT-ELINT collection system, the signals are first checked against time and distance matrices (which are also part of the template specification); these matrices reflect the permissible relative time and distance relationships between signals (geolocation capability is assumed). Signals are also categorized as key or routine, reflecting a priority assessment found necessary to control template instantiation. Instantiation is the process of activating, in the computer system, a candidate template, suspecting it will "mature" (be filled with sufficient data) to reflect a confident estimation of the occurrence of a tactical event; more on this later.

The fusion processes involved in these methods result from analyses that model or typify the "primitive" or minor activities of interest in the domain. As noted earlier, such modeling relationships require quantification of time and dis-

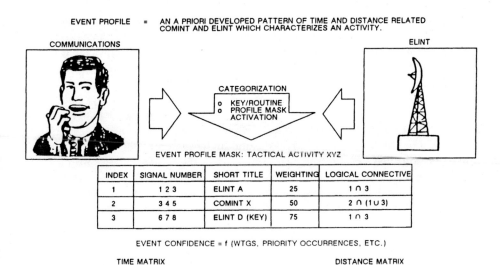

Figure 9.7 Event profile concept (from [13, 14]).

tance relationships (and possibly other relationships) among the entities constituting the activity. The issue of uncertainty and its management enters the problem because of the various modeling requirements. For example, there may be uncertainty in the very composition of the template (i.e., in the signal constituency) or in the time and distance relationships. Additionally, it is unlikely that templates, in real-world applications, will ever be fully populated with data. As a result, most such techniques employ confidence factors to reflect the uncertainty in the estimated event occurrence.

Instantiation, Pruning, and Granularity

A variety of system design issues must be dealt with in this type of approach. Two important issues are template instantiation and pruning control and template "granularity," which are related. Template instantiation control relates to the data management issue that typically arises because of a one-many mapping between observations and activities. That is, in many cases, an observation from the multisensor system can be feasibly associated with *multiple* possible events or activities (as in tracking problems, such multiple possible associations could also be the result of the variances of the observation and estimated target processes.) A single ELINT observation for instance may associate to a ground-controlled intercept activity or an air-controlled intercept. If this type of mapping exists in general, it is clear that a given set of observations will create numerous possible template instantiations, and thereby a possible data management problem. Additional data related to the true activity will eventually clarify the situation to a degree, but in combat situations there will always be uncertainty in the sensed data, and the data will be incomplete, so that templates may never be completely verified with certainty. Thus, even under the best circumstances, the system designer will have to develop an approach for template instantiation and pruning. In [14], the instantiation control scheme employed a template-specific priority scheme such that templates were instantiated only when so-called key emitters were observed in the SIGINT data. Instantiation and pruning were both done by monitoring the degree of match of the observations with the timing and distance relations specified in each template and deleting those templates that had "timed out," or for which infeasible distance relationships ultimately occurred.

A related design choice is the granularity or scope of the overall template structure. If a "fine-grained" approach is selected, in which the minor events and activities are made up of just a few elements, the system will be highly responsive in time. However, such an approach produces low-level inferences (i.e., not much fusion) and, at any given time, will have many low-level templates instantiated, a possible data management problem. In a "coarse-grained" approach, the primitive events are broader in character and the system produces higher-level inferences in general but at a slower rate. In either approach, the design choice must be consistent with the sensor system aggregate sampling rate. That is, the desired event-level responsiveness cannot exceed the composite sampling rate of the sensors. This is analogous to the observability condition for estimation techniques that establishes criteria for relating available measurements to estimated system state.

Another feature of template-based approaches is that they provide a framework for the prediction of battlefield events and, thereby, a framework for sensor collection management. If it is felt that a given activity is taking place, the appropriate template can be used to predict expected future activities and provide inputs to sensor scheduling algorithms.

A Naval Application

Noble [15], takes a similar approach to situation assessment for naval operations. He characterizes the underlying data fusion algorithm as a *multiple hypothesis fuzzy logic system*. This label simply means that, at any given time, multiple candidate templates (hypothesis) are instantiated and that uncertainties in the ability to specify various template parameters (e.g., time relationships) leads to the use of fuzzy arithmetic techniques to operate on the parameters. The fuzzy mathematics technique is just another way to handle the uncertainties described earlier. The templates are modeled after the "fuzzy schemata" examined in [16, 17]. Figure 9.8 represents a summary view of an "operation" template used for situation assessment in [15].

The summary representation of this template shows the sequence of hostile actions most typical of a hostile operation. In this representation, the name of the operation is printed at the top, each of the operation participants is associated with a row, and time is depicted along the horizontal axis. Significant hostile activities, the tactical events, are drawn as rectangles on the row of the participant performing the activity. The left and right edges of the rectangles indicate typical beginning and ending times for the tactical events.

The data in this summary representation provide most of the information needed about hostile objectives and courses of action. The hostile objective is printed at the top of the template. The identities of the participants are printed on the rows, and the roles of the participants are given by the actions performed by each of these participants. The purposes of signals and reports received can be inferred by associating them (fusing them) with the events specified on the template.

Elastic Constraints

The data in a summary template, such as the one shown in Figure 9.8, describe only one way an operation may occur. As previously discussed, a data fusion sys-

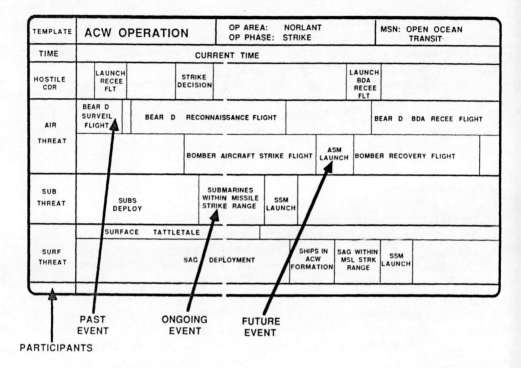

Figure 9.8 A summary template for anticarrier warfare. This template shows participants and events typical of anticarrier warfare (from [15]).

tem that relies only on this summary data is likely to be unable to recognize this operation unless it occurs almost exactly the way it is depicted in the figure, which is unlikely to happen very often. The template's *elastic constraints* enable the system to recognize many variations of this type of operation.

The elastic constraint data specify constraints among the template "slots," which are the template's events and rows. These constraints specify the typical and required characteristics of hostile assets that may be used as a row in the operation, specify the typical and required characteristics of events in a template, and specify the typical and required relationships between assets, between assets and events, and between events. (The time and distance matrices used in [14] represent the elastic constraints there; clearly this idea can be extended to accommodate variances in expected events and activities as well.) Noble also uses time relationships and variances in his approach [15].

Thus, there are two important levels of uncertainty in template structure: within-template uncertainty, accommodated by template-specific elastic constraints; and between-template uncertainty, accommodated by the very design structure of the template hierarchy or network and by instantiation and pruning techniques.

Template Confidence Factors

Any data fusion algorithm applied in these situations would compute a "degree of fit" or association between the set of received reports and each template in the system library. This degree of fit represents the extent to which each template is consistent with and able to account for the received reports. The degree of fit is related to "degree of belief" or the "possibility" that the template represents the ongoing situation. Accordingly, various types of uncertainty measures, including fuzzy measures, can be formulated to estimate intra- and intertemplate associations. These confidence factors, as described previously, are usually ad hoc constructs that attempt to quantify the influence of particular pieces of evidence (i.e., sensor data) on the estimation of an event or activity. As noted on Figure 9.7, for example, the confidence factor can be a function of the weights assigned to particular data, the number of priority data occurrences, perhaps the reliability assigned to the sensors from which the data came, *et cetera*—there are no set formulas for this. It is also important to keep in mind that such factors will probably need to be communicated to a human operator who must be able to interpret them. Note, too, that it may be desirable or required to assign intertemplate confidence factors as another means to control instantiation and pruning and the flow of system logic; again, these factors are usually developed on an ad hoc basis from simulation-based analyses.

Typical Information Processing

Figure 9.9 shows the representative steps in processing an input report. When the report arrives, the algorithm checks to see if the report is of a known type by comparing it with templates in the system library. If the report is of a known type, then an instance of the template is created and its elastic constraints are initialized (step A). Constraint initialization is accomplished by copying the default elastic constraints from the system library template to the template instance, and then restricting these constraints using data in the report.

The newly created template instance then attempts to find existing indicator instances able to fill its slots (step B). It searches the set of existing instances for those that best fit the slots. The algorithm strives to fill all slots with events or assets typical for that slot, but if necessary will fill the slot with events or assets that barely meet the slot constraints (low degree of association) in the fashion of a "default reasoning" approach. After each slot is filled, the algorithm adjusts the elastic con-

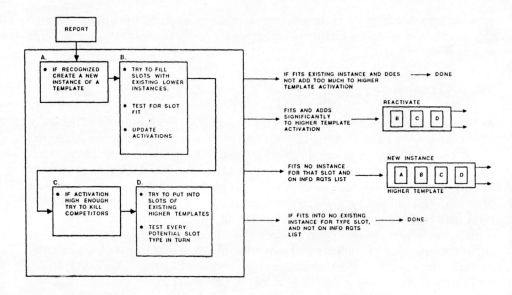

Figure 9.9 Steps for processing a situation report (after [15]).

straints for related instance slots. After all slots are filled, the algorithm recomputes the aggregate degree of fit (called *activation* in Figure 9.9) between the template and received reports.

If the template instance can account for the received reports sufficiently well, then the algorithm will delete any competing instances that account for the data poorly; that is, prune the template set (step C). Two templates are competitors if any of their slots are filled by the same lower instance (a result of the one-many mapping previously discussed). Competitor deletion (pruning) removes from the instance data base poorly supported tentative situation hypotheses (based on the confidence factors previously described).

After attempting to fill the template instance slots with existing instances lower in the template hierarchy, the algorithm attempts to fit the template instance into slots in template instances higher in the template hierarchy (step D). The algorithm examines all operation templates in the system template library, identifying each slot in each template that can be filled by the type of template being processed. It then searches for existing template instances with an empty slot of that type, and attempts to fill the slot with the new template instance. If the marker instance fits well enough, the algorithm fills the slot and adjusts the elastic constraints of the newly filled higher template instance. If the template instance does not fit well enough or if no slot of this type can be found in an existing template instance, the algorithm may create a new template instance with a fillable slot of this type (i.e., nominate a related template).

After it creates a new template instance, the algorithm processes this template instance using the previously described steps (small rectangle to right in Figure 9.9). It fills the slot of the newly created instance with existing lower instances, adjusts the instance constraints, calculates the degree of fit for the new instance, attempts to prune poorly supported competing template instances, fills slots of higher templates, and creates new instances of higher templates. The processing flow just described (from [15]) is motivated by a few simple information processing principles. These principles guide the algorithm in its search for the best global fit between templates and observed data.

The first principle concerns the method of problem decomposition. Decomposition is effected by treating each template as an independent entity trying to "prove" that it can account for the observed data better than other templates. In trying to "prove itself," each template makes decisions considering only two factors: which instances of templates at the next lower level in the hierarchy can best fill its slots, and which slots in template instances at the next higher level can it fill. This process decomposes the overall data fusion algorithm because the algorithm does not explicity consider how filling each template slot affects how well slots in all other templates can be filled. It makes "slow filling" decisions without considering how these actions affect the global optimization for matching reports to templates.

The second information processing principle concerns global optimization issues and methods for ensuring global internal consistency. Global optimization and internal consistency is supported by three algorithm features: restricting each template slot to be filled by only one lower instance at a time, permitting each lower template instance to fill multiple higher templates, and permitting well-supported template instances to delete poorly supported competitors. The first two features cause template instances to proliferate, for they permit a single lower instance to generate many higher template instances. Each of these instances will account for the observed reports to varying degrees. The last feature, the pruning of competing templates, reduces the template proliferation, leaving only those template instances that can account for the observed data reasonably well.

The third information processing principle used in [15] employs the same template processing algorithm for templates at any level. This principle permits the algorithm to be recursive and simplifies the data fusion logic and software implementation.

A last comment: it can be easily seen that templates have a strong analogy to "plans," each event or activity set (template) representing a plan fragment in effect. Plan recognition techniques for situation assessment, described in Chapter 12, thus represent methods closely related to templating strategies and are not discussed separately in this chapter. Table 9.3 summarizes the advantages and disadvantages of templating approaches of the type discussed here.

9.2.3 Figure-of-Merit Techniques

Figure-of-merit techniques embody flexible sets of algorithms through which the strength of entity or event relationships can be measured. By employing a hierarchy of figures of merit, the existence of an entity or event hierarchy can be represented in a quantitative way. In essence, an FOM can be considered an association metric or a classification metric. As for many situation-estimating approaches, the general logic is to associate a set of multisensor (or single sensor) reports with an entity of a class, and then to associate single entities to one another, that is, establish set memberships, and continue this logic through a hierarchical network of interest, typically formulating a platform or event hierarchy.

The basis for the FOM approach is the desire to formulate association or classification metrics that employ as many observables as possible. Rather than using just location measurements, FOM approaches attempt to formulate a relationship among several variables for an improved association or classification strategy. The approach bears similarity to weighted decision formulas commonly described in the decision theory literature.

Whatever the relationship postulated, the general method of comparison of observations also must be considered. Wright [17] suggests there are basically three approaches: exact match, tolerance check, and FOM. The exact match technique

Table 9.3
Features of Template-Based Approaches

ADVANTAGES

- EASILY UNDERSTOOD; RELATIVELY EASY TO DEVELOP TEMPLATE STRUCTURES
- STRAIGHTFORWARD REPRESENTATION, INTEGRATION OF LARGE AMOUNTS OF DATA
- "GRANULARITY" (SCOPE, COMPLEXITY) OF TEMPLATES IS SELECTABLE

DISADVANTAGES

- STATIC CONSTRUCT, RELATIVELY INFLEXIBLE
- COMPLEX DATA BASE MANAGEMENT PROBLEMS
- COMPLEXITIES CAN ARISE FROM MIS-MATCH OF TEMPLATE GRANULARITY AND DATA OR DATA RATES (HARD TO ACHIEVE SCENARIO ROBUSTNESS)
- AD HOC WEIGHTING FACTOR STRATEGIES CAN BE DIFFICULT TO INTERPRET

is self-explanatory: two pieces of data must match in an exact comparison or else no association is declared. In the tolerance check technique, the data comparison or difference must be within a prespecified range to declare association. The FOM technique calculates a value (usually a weighted value) of a parameter that is itself a function of parameters specific to a type of observation (e.g., frequency or PRI for radar) that, based on *a priori* tests, calibrations, doctrinal information, and simulations, represents the "likelihood" of an association between reports or entities (or events, activities).

To implement this logic with FOMs, sets of relationships are established among all of the observables anticipated; for a given entity, say, a radar, the FOM frequently is made up of the weighted sum of several FOMs, each of which depicts a relationship between observations or parameters associated with the radar. The total (summed) value is then used as an association metric to support a decision strategy that determines if the multisensor data, on which the FOM is based, reflect the existence of a unit, activity, or event. These FOM values are similar in concept to the elastic constraints used for template processing.

At the lowest level in a hierarchy, an FOM associates reports or observations to entities in the same manner as an association metric for tracking associates observations to tracks. At the next level, FOMs are developed that associate units or events to one another. This is done by again describing allowed variances in parameters through which unit associations can be made; for example, interunit distance or number of units admissible to a cluster. The process continues up the hierarchy, subsequently associating clusters to one another, groups to one another, and so on.

This approach seems to have been fostered during the development effort for the *battlefield exploitation and target acquisition* (BETA) system [18, 19], a system that formed the developmental foundation for both the limited operational capability—Europe (LOCE) system, and the all-source analysis system-enemy situation correlation element (ASAS-ENSCE) system. In BETA and these other systems, FOMs are used both to fuse reports (report to report association), usually called the *self-correlation process* in the related literature, and to associate sensor data with entities and entities to groups to establish a battlefield hierarchy—this is usually called *cross correlation* in the literature.

The entity hierarchy assumed in the BETA approach defines

- *Simple elements*—Single ground equipment such as guns, radios, radars;
- *Compound elements*—Composite of multiple simple elements such as a tank = (vehicle + gun + radio);
- *Complex elements*—Composites of simple and compound elements such as artillery batteries, battalion command posts, *et cetera*.

Figure 9.10 shows the conceptual nature of the BETA entity hierarchy.

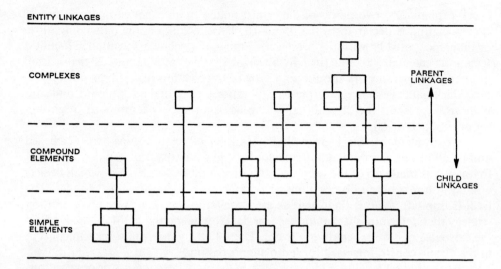

Figure 9.10 BETA entity hierarchy.

General FOM Approach

Wright [18] defines some general guidelines appropriate to FOM calculations:

- All figure-of-merit calculations should have the same form.
- All figure-of-merit calculations should have values such that $0 \leq FOM \leq 1$.
- All figure-of-merit calculations should be developed independently.
- Each figure-of-merit calculation should be dependent on the type and nature of the particular data involved.

These guidelines allow the development of a system with numerous FOMs, the set of which would be determined by the number of observables, entities, and entity groups in the hierarchy. A representative form of the FOM, shown in Figure 9.11, has the following properties (T could be either an observable or the difference, or other function, between observables measured or sensed in two reports, units, events, *et cetera*, being compared):

1. A value T_1 exists at which the figure of merit equals 1.
2. A value T_2 exists at which the figure of merit equals or approaches 0.
3. For any value of T between T_1 and T_2 the figure of merit is a positive decreasing function that has a value between 0 and 1.

The key to all this of course (as in all association or classification problems) is in the selection of the functions and thresholds associated with decision making (correlation). As mentioned earlier, these are usually developed from physical mod-

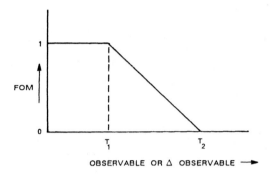

Figure 9.11 Representative form of an FOM.

els, simulations, and doctrinal or other reference information, and must be properly tested and validated.

After all the various figure-of-merit calculations have been made, the total result of the comparison process must be calculated. This value, variously called the *comparison score* (CS), *measure of correlation* (MOC), or *association measure* (AM), represents the total "likelihood" of the two entities being the same. Each figure of merit is weighed in some manner so that the CS also has a value between 0 and 1. There are many possible ways of combining the individual figure-of-merit calculations but the usual approach is a weighted sum:

$$CS = W_1 FOM_1 + W_2 FOM_2 + \cdots + W_n FOM_n$$

where

FOM_1 = figure-of-merit calculation for observable (parameter, unit, event) number 1;
W_1 = the associated weight for the figure-of-merit calculation for observable number 1, *et cetera*.

This allows addition or subtraction of figure-of-merit calculations by just renormalizing the weights. The weights must be chosen carefully and refined by real-world experiments.

The comparison score would then be compared with preset values of acceptance and rejection thresholds. If the CS value is higher than the acceptance threshold then the two reports are declared as associated with the same entity. If the CS value is lower than the rejection threshold, then the two reports do not represent the same entity. If the CS value lies between the acceptance and rejection threshold, the result is ambiguous and further processing must be done.

Self-Correlation

Self-correlation is the process of combining new sensor data with existing data that describes the same entity to obtain a single record that contains the "best" information about the entity. In BETA, self-correlation is performed for the five entity types: radars, radios, shooters, compounds, and complexes. The sixth entity type, movers, is automatically added to the data base without correlation processing.

Self-correlation processing is activated by the receipt of each input entity data record (EDR). The process is divisible into three main subprocesses: (1) the determination of a list of candidate EDRs from the correlated data base with which the input record may be combined; (2) the comparison of these candidates and the possible determination that one or more of them represents the same entity as the input EDR; and (3) the update of the correlated data base reflecting this determination. Only the FOM operations in step (2) will be discussed here.

Comparison processing is performed to determine if any of the candidates from the previously developed list represent the same entity as the input EDR. The comparison of a candidate with the input EDR is based on a calculated *measure of correlation* (MOC) that, as described earlier, is a weighted sum of various figures of merit.

In the case of radars, we can express the MOC as

$$\text{MOC} = \sum_{i=1}^{5} \delta_i W_i \text{FOM}(i) / \sum_{i=1}^{5} \delta_i W_i$$

where FOM(1) is the figure of merit obtained from the comparison of input and candidate location; FOM(2) is time; FOM(3) is frequency; FOM(4) is pulse duration; and FOM(5) is pulse repetition interval. The W_i are weights previously assigned to each FOM that reflect its importance to the overall measure of correlation. The δ_i are flags set to 0 or 1. The δ_i is set to 0 when FOM(i) is not available or invalid. For example, if a radar report does not include pulse duration, we set $\delta_4 = 0$. This flag enables the comparison processing routine to utilize incomplete data records.

For radios, we begin by determining the modulation type of both input and candidate EDRs if they are available. If both are available and incompatible modulation types, the candidate is rejected. Otherwise, we determine the measure of correlation using three FOMs: FOM(1) for location, FOM(2) for time, and FOM(3) for frequency.

For shooters, compounds, and complexes, only the location and time FOMs are used in comparison processing to simplify the computational processing by avoiding the combinatoric aspects of such comparisons.

Cross-Correlation

Cross-correlation is the process of associating self-correlation results (unit declarations) with existing higher-level entities to obtain a battlefield hierarchy. For cross-correlation, both the candidate parent and child entities are already defined; only the proper subordination (of all those possible) is being determined in this step. An input EDR is received by the cross-correlation module after the self-correlation module has determined that the EDR represents a newly observed entity or an updated EDR with "significant" data changes. As for self-correlation, the process of cross-correlation also involves three subprocesses: (1) the determination of a list of possible parent candidates for the input EDR; (2) the selection of the "best" parent candidate; and (3) the resulting data base update. Figure 9.12 illustrates the cross-correlation process.

Note in Figure 9.12 that a "template" is used. This template, similar to those discussed in Subsections 9.2.1 and 9.2.2, is the list of components, observables, and their relationships (e.g., spatial relationships) necessary to define a "compound" or "complex" with acceptable probability. When the input EDR is received, the template directory is searched and those templates retrieved that contain a component of the input EDR subject type. For each template that contains a unit of the type estimated by the self-correlation process, calculations are made of

- The uncertainty in the reported unit location as determined from available *(a priori)* observation process errors,
- The uncertainty in the interunit spatial relationships in the doctrinal templates.

Together, these calculations yield "search bounds." If the combined uncertainties result in nonoverlapping search bounds (uncertainty bounds) then no cross-correlation is attempted.

For those entity-parent pairs that pass this test, a weighted FOM, called in BETA the *association measure,* is calculated to guide the cross-correlation process. The AM for complexes (e.g., command posts) is a weighted sum of three figures of merit: an affiliation score, a time decay score, and a location score. The weights are given in the template. The association measure score for a compound has no affiliation component. For example, a tank is compounded of a tracked vehicle, a radio, and a gun.

The *affiliation score* reflects the probability that the parent entity indeed has another element of the input type that has not yet been found. For example, if an artillery battery is "known" by doctrine and specified in the template to have at least two radars, yet none has been found, then the probability it has some still to be found is very high. Conversely, if it is "known" to have at most four radars, and at least four have been found, then, assuming the four radars already found have

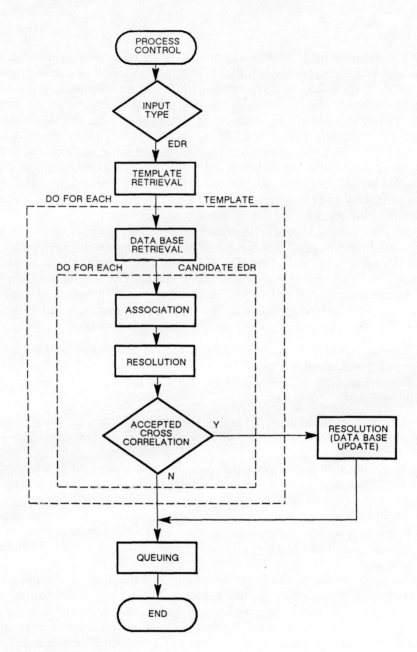

Figure 9.12 The BETA cross-correlation process.

been associated correctly with that parent, the probability that there still is another radar to find is very low. Representative relationships between an affiliation score and the unit template are illustrated in Figure 9.13.

To calculate the affiliation score, let max be the maximum number of entities of the input type expected and min be the minimum number, both values taken from the template. Assume the input EDR does actually belong to the candidate parent, and let n be the resulting number of children of the input type. Then

Affiliation Score = $(\max - n)/(1 + \max - \min)$

where, when n falls between max and min, AS is equal to 0 if n is greater than or equal to max, and equal to 1 if it is less than min. Thus, in the example (Figure 9.13), suppose a new radar is found and a candidate parent already has two radars. From the template, min = 2 and max = 4, so

Affiliation Score = $[4 - (2 + 1)]/(1 + 4 - 2) = 0.33$

For VHF radios, because min = 2 and max = 3, a candidate parent that has zero or one VHF radio, with a new sighting, has less than or equal to two VHF radios so the affiliation score is 1; one that already has two or more, with a new sighting, has greater than or equal to three VHF radios and the affiliation score is 0. As these examples indicate, the inclusion of the +1 in the denominator and the *or equal to* noted earlier makes the affiliation score "jump" to 0 when n is still within the template specified limits. This is illustrated in Figure 9.14. The resulting nonlinearity in the score means that the AS function does not describe entities as coming from a uniform distribution; that is, it is not considered equally likely that any number of entities in the template-defined range have been observed.

The *time decay score* is based on the closed intervals representing the elapsed time over which the entity has been observed for both candidate and input EDR. As illustrated in Figure 9.15 if these intervals for the input and potential parent EDR overlap, the time decay score is 1. If not, and the elapsed time between the two intervals, t, is less than a time constant C given in the template,

Time Decay = $1 - t/C$

If t is greater than C, this score is 0. Figure 9.16 shows the influence of this constant on the time decay score.

The *location score* is calculated by one of two methods, again governed by the template. The first, used if a nonzero maximum deployment radius for that parent relative to that subordinate type is given, is illustrated in Figure 9.17. For this method, if the distance between the estimated locations of the input and candidate parent EDRs, d, lies between that maximum radius and the template-sup-

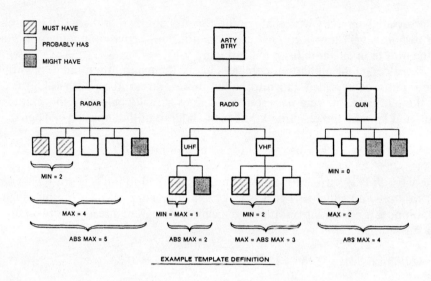
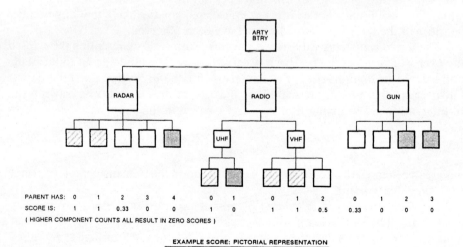

Figure 9.13 Affiliation score: a = template definition; b = pictorial representation of score (from [19]).

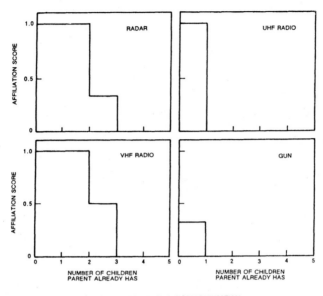

Figure 9.14 Affiliation score: graphic representation (from [19]).

plied minimum deployment radius, the location score is 1. If d is less than the minimum radius (min, r), then

$$\text{Location Score} = \exp\left[-(r-d)/2\right]$$

and if it is greater than the maximum radius, R, then

$$\text{Location} = \exp\left[-(d-R)/2\right]$$

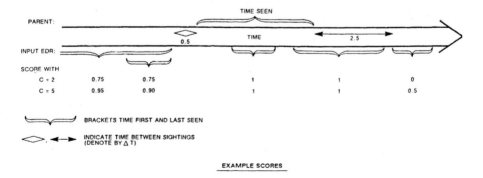

Figure 9.15 Example time decay scores (from [19]).

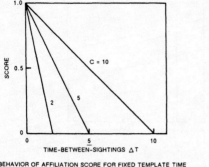
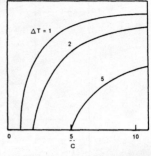

Figure 9.16 Dependence of time decay scores on constant C (from [19]).

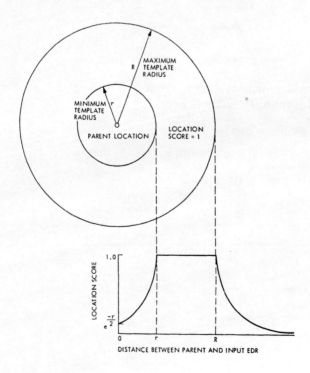

Figure 9.17 Location score (from [19]).

Note that this score employs no knowledge of the uncertainties in the observation processes, even if known.

If no maximum deployment radius is given, then the second method is used, setting the location score equal to

$$\text{Location Score} = \exp(-X/2)$$

where X is the quadratic from $d^T c d$. In this expression, d is the difference between the postulated parent location and the reported location, and c is the covariance of that difference. Thus, when no maximum doctrinal deployment distance can be estimated in a parent template, the location score depends on the observation or sensor error characteristics.

Following these equations, an association measure (a weighted sum of the preceding scores) is calculated for each such potential parent. If this association measure exceeds the acceptance threshold given in the template, the potential parent EDR is a candidate for the "acceptable" list. If measure falls below the rejection threshold, processing continues, looking at the next potential parent. And if the measure falls between the two thresholds, the potential parent EDR is a candidate for the "ambiguous" list. If the acceptance threshold is set equal to the rejection threshold, there will be no ambiguous candidates. The acceptable candidate with the highest association measure is usually linked with the input EDR, becoming its parent.

The BETA examples have been included to give the gist of an FOM approach at a moderate level of detail. Figure-of-merit techniques in general have good flexibility because the FOM algorithms and MOC algorithms can be changed in many ways. Like event profiling or IPB processes, they also are dependent on doctrinal and other *a priori* data. Their flexibility can result in MOCs whose behavior is not easy to predict, and extensive simulation may be required to train the algorithm sufficiently to achieve good performance over a range of conditions. Table 9.4 summarizes the advantages and disadvantages of FOM-based approaches, all of which have been mentioned in the preceding discussion.

9.2.4 Knowledge-Based and Expert-System Techniques

The resurgence and proliferation of certain elements of artificial intelligence technologies during the 1980s has had an impact on the overall C^3I community, including those in the data fusion domain. Many laboratory prototypes of decision aids, intelligence support systems, and mission planning systems have been built over the last decade, although relatively few have found their way into operational settings. This section will review a few of the systems that have been experimented

Table 9.4
Features of Figure-of-Merit Approaches

ADVANTAGES

- RELATIVELY SIMPLE, EASILY UNDERSTOOD
- FLEXIBLE VIA FORMULATION OF MOC, ADJUSTMENT OF WEIGHTS
- RESULT IN A QUANTITIVE ASSESSMENT OF A SITUATION/THREAT
- STRAIGHTFORWARD COMPUTATION; THRESHOLDING PERMITS FAST PROCESSING

DISADVANTAGES

- FREQUENTLY NOT FOUNDED IN FORMAL PROCESS MODELS (AD HOC)
- DEVELOPMENT, VALIDATION CUMBERSOME (EXTENSIVE SIMULATION)
---

with for situation or threat assessment. These systems are classified in the template category because they rely on the expectation template concept, as do the other methods discussed in Section 9.2, but they implement the concept using the *expert system* (ES) or *knowledge-based system* (KBS) paradigm (see Chapter 12 for a broader discussion of artificial intelligence methods).

In many, perhaps most, of these systems, the goal is to produce an estimate of the hostile *force structure* or order of battle. Although this knowledge is an important part of the situation or threat assessment process and product, we suggest that many other components are missing, in accordance with the discussion of Chapter 8.

ESIAS and DECADE

Of the work we have seen, we believe that the work on the *Expert System for Intelligence Analysis Support* (ESIAS) project [20–22] is among the most comprehensive, at least in its conceptual treatment of the situation assessment problem. Only a few other efforts have been as broad in scope and explore as many of the research issues involved; among those is the ANALYST program and programs related to ANALYST [23–25], the PENDRAGON project [26–29], and some work on adaptive systems [30–32]. Whereas the other efforts range over a broader set of issues in the application of KBS technologies to C^2 system design, the ESIAS work focuses more on the specifics of situation assessment, and so we will focus on it and its treatment of the problem. Much of what follows is from [20–22].

The work in [20–22] actually spans three efforts: the development of ESIAS and the situation assessment KBS [20]; the explanation feature for ESIAS [21]; and the development of a system called *decision aid development and evaluation* (DECADE), which is a development environment for DBS-based aids for situation assessment [22]. Although DECADE was the last item reported on, we feel it is most instructive to discuss it first, as it embodies the highest-level concepts in the overall approach.

The goal in this work was consistent with our suggested definition of situation assessment—to develop an environment to efficiently build aids that
- Evaluate threat capabilities and actions,
- Infer threat intentions,
- Predict undetected and future situations.

To design DECADE, a technique was developed called *conceptual knowledge modeling*, which is a modeling methodology for representing domain knowledge at an implementation-independent level of abstraction [33]. Using this method, a C^3I conceptual knowledge model was developed, with two components:
- The C^3I conceptual structure model,
- a C^3I situation assessment behavior model.

We will focus here on the latter.

The C^3I situation assessment process is represented in the C^3I situation assessment behavior model. This model presents both functional and procedural views of the situation assessment process, as practiced in C^3I domains. The situation assessment process depicted in the model was derived studying the manner in which operations personnel assess situations. The functional view of the C^3I situation assessment behavior model is intended to serve as a requirements baseline, whereas the procedural view provides a high-level, implementation-independent model of the procedural knowledge to be represented in the knowledge base. The functional and procedural views of the behavioral model are extended to provide a detailed view of the situation assessment process in specific C^3I domains, to support application development in those domains.

The top-level *functional* view of the C^3I situation assessment behavior model is shown in Figure 9.18. In this model, rectangles represent functions, parallelograms represent external data objects (such as knowledge bases, displays, and reports), ovals represent internal data objects, small circles represent events, and rectangles under the small circles represent triggers in terms of logical operations. Representation of the C^3I situation assessment process in this manner provides a high-level statement of the basic functionality that a C^3I situation assessment expert system must possess. Functions are further decomposed to reveal additional details of the process.

Figure 9.19 shows a portion of the *procedural* view of the C^3I situation assessment behavior model that describes the procedural knowledge involved in the assessment of threat behavior. Much of the symbology is the same as that of the functional view. The procedural view, however, specifies a function in more detail: sequences are represented in layered rectangles that specify preconditions in the top layer, postconditions in the middle layer, and actions in the bottom layer. Through

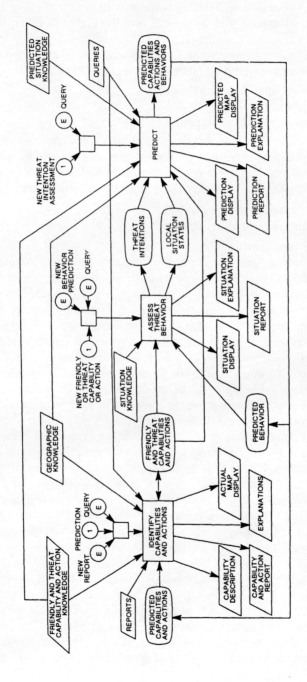

Figure 9.18 C³ situation assessment behavior model: top-level functional view (after [22])

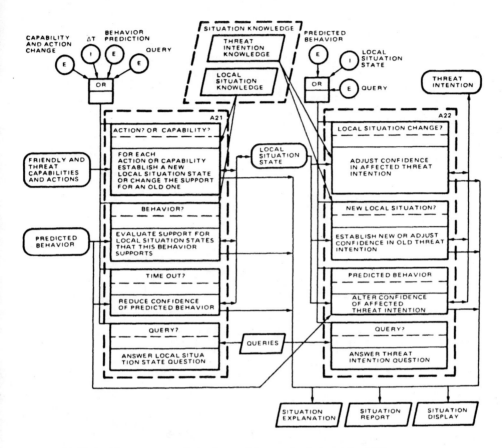

Figure 9.19 C³I situation assessment behavior model: procedural view (after [22]).

this representation, such concepts as causality, time orientation, and event occurrence may be fully described.

Based on the C³I conceptual knowledge model, an application-independent approach to developing situation assessment expert systems was developed to support a wide range of C³I domains. Using this approach, the C³I conceptual structure model provides a conceptual view of the declarative knowledge (i.e., objects) that a C³I situation assessment expert system is to possess, whereas the C³I situation assessment behavior model describes the procedural knowledge (i.e., rules) in the system.

Given the knowledge represented in a domain-specific model, a C³I situation assessment blackboard model, shown in Figure 9.20, is used to structure the domain knowledge and reasoning steps of the expert system as it will be imple-

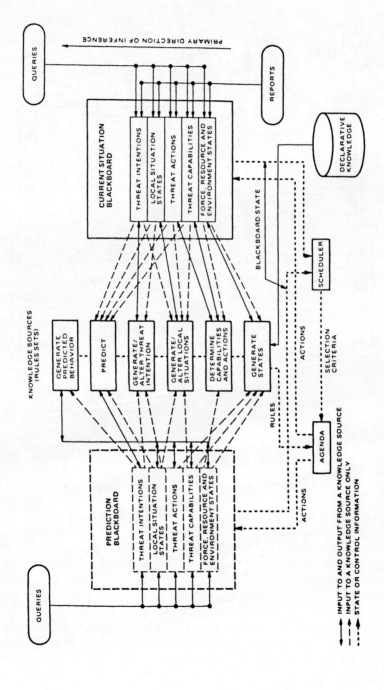

Figure 9.20 Situation assessment blackboard (after [22]).

mented. A blackboard is a global, dynamic data structure maintaining the current state of the problem domain. Logically, independent knowledge sources (i.e., rule sets) respond to changes in the blackboard in an opportunistic and self-activating manner to incrementally develop a solution to a problem. The order in which knowledge sources contribute to the solution and post intermediate solutions on the blackboard may be controlled through agenda and scheduler functions shown on the figure.

The situation assessment blackboard model that has been developed represents the reasoning process used to infer current and expected intentions of a threat. Each blackboard is divided into several levels of abstraction to represent the major steps involved in assessing threat behavior. At the lowest level, force and resource states are maintained; that is, the current state of threat forces, own forces, terrain, and environmental conditions (for example, weather). From those states, threat capabilities, such as destruction potential and strength are inferred.

Based on the inferred force and resource states and threat capabilities, threat actions, such as formations and movements are detected (analogous to the IPB process described in Subsection 10.2.1). From these actions, local situation states are inferred (for example, enemy superiority indicators). Finally, based on prior levels of inference, intentions of the threat are inferred, such as moving into defended areas and preparing for attack. At each level of reasoning, uncertainty reasoning is supported using the Dempster-Shafer theory of evidence (see [34, 35] and Chapter 7 for explanations of the Dempster-Shafer approach).

The quantitative evaluation of capability of a threat has been the subject of numerous research activities; a frequently used method is *multiple-attribute utility theory* (MAUT), which is employed to assign threat values to battlefield entities (see [36]). The MAUT approach basically involves constructing hierarchical models of how measurable parameters relate to objective functions, specifically called *utility functions,* of interest. In the case of threat quantification, the objective function would be, for example, a quantitative representation of threat intent and lethality. In turn, intent may be modeled by threat platform proximity and heading, among other factors, developed as a weighted combination of proximity and heading "attributes," to assess the magnitude of the "utility" function; that is, the threat value. Simulation and sensitivity analyses are usually performed to both better understand and optimize the models employed.

This approach therefore produces results which span over our definition of situation and threat assessment. By incorporating multiple levels of abstraction, various situation elements are estimated and force capability and intent are also estimated.

In STA problems and in many solutions, we assume implicitly or explicitly that the problem state (i.e., the situation or threat state) is at least "piecewise static." That is, we assume that over a fixed time interval "dynamic" observations accrued over that time can be used to update a (locally static) situational or threat state.

Otherwise, we have totally dynamic observation sets and states, both evolving asynchronously, an extremely difficult problem type to deal with. In KBSs, this has to do with so-called temporal reasoning problems for which the problem states are dynamic. In a C^3I system, intelligence reports are continually being received, modifying the state of tactical units; that is, the situational state. Therefore, the state of a tactical unit and an inference that references it may not coincide at a given time epoch. This presents a potential problem when the analyst selects to view the data of a tactical unit from within an inference explanation. The state of the tactical unit data displayed must coincide with the state of the inference explanation from which the data query originated. Even in the piecewise static approach, data-state correlations must be time tagged to control the flow of inferences; ESIAS accounts for such factors through the use of state-dependent slots and an updated numbering scheme to assure consistency between data and inferences.

The overall ESIAS and DECADE efforts have thus been focused entirely on the situation assessment problem and have dealt with problems in a top-down fashion, ranging from abstract architectural issues to the specifics of the explanation capability.

AMUID and PENDRAGON

Two other research prototypes form interesting case studies: the *automated multi-unit identification* (AMUID) project and the PENDRAGON project [26–29]. AMUID is a KBS approach to force structure identification whereas PENDRAGON detects nodes and networks in the battlefield and monitors all activities related to those entities. This section will just highlight some of the implementation and processing concepts for these systems.

AMUID is a knowledge-based system for land battlefield analysis by integrating COMINT, ELINT, imagery and MTI radar data. This system attempts to classify the target of interest and organize those targets into higher-level units (e.g., battalions, regiments, fleets). In each case, data arrives over time, and the system must accomplish real-time analysis and situation updating.

AMUID utilizes a "blackboard" system for control. In this system, data and hypotheses are maintained in a centralized location and the decision-making process is accomplished by a set of "knowledge sources." Each knowledge source is an "expert" at accomplishing one task in the overall system.

The AMUID program architecture is shown in Figure 9.21. Again, the major components are (1) a set of knowledge sources, (2) a controller (or set of rules) to decide which module to access next, (3) a data base or knowledge base of information and hypotheses, and (4) a database manager. For AMUID, the system's hypotheses are its model or explanation for the behavior observed in the sensor data. A hypothesis organization that has worked very efficiently in several sensor-fusion problem domains is a hierarchical structure.

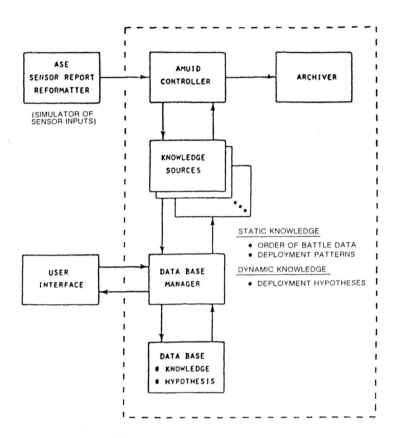

Figure 9.21 AMUID program architecture (after [26]).

The AMUID system is structured in this manner. Figure 9.22 shows the hierarchical levels based on the force structure of military units. The "bottom-up" analysis within this system involves associating sensor data with components, organizing the components into companies, companies into battalions, and so on.

AMUID has been installed in the U.S. Air Forces's *Advanced Sensor Exploitation* (ASE) testbed at the Rome Air Development Center [37] and serves as a prototype information fusion system as well as a testbed for other fusion methods.

PENDRAGON [27–29] is a much more sophisticated system, designed to support a division-level G-2 analyst; it is intended to detect and identify entities (as other systems such as AMUID do; see also [38]), but also to monitor all activities related to those entities. PENDRAGON has four main systems that operate in parallel: a data system that provides data; a "structure" system that detects nodes and networks and builds hierarchies; a "behavior" system that analyzes activities being

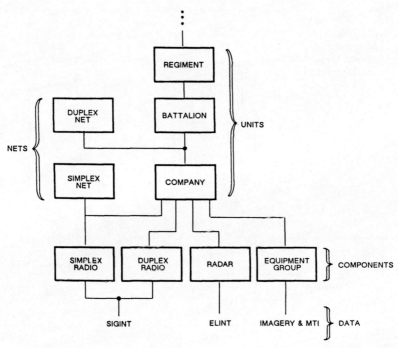

Figure 9.22 AMUID hierarchical hypothesis structure (from [26]).

carried out by detected structures; and an interface system that links other systems to the user for explanation and interaction. These systems provide evidence and guidance to one another. Much of the description that follows is derived from [27].

The basic approach used in PENDRAGON views fusion reasoning as the creation and management of "arguments" about entities in the battlefield. An "argument" is a self-contained environment that deduces and then displays conclusions tied to some aspect of the enemy entity (e.g., node, network). An argument is defined as both

1. The reasoned structure that represents the conclusion of an item of interest as well as the entire evidential basis that supports that conclusion.
2. The reasoning process whereby the conclusion is explained, defended, and revised as required.

The reasoned structures, also called *argument structures,* are the product of the reasoning process; they are a structural record of the reasoning process used for later explanation, problem solving, and revision. For example, PENDRAGON tries to form a strong "echelon argument" for every new entity it detects. This argument collects and studies data and concludes the echelon for the entity to which it is assigned. The underlying notion is that fusion reasoning employs a host of semi-independent techniques that differ dramatically, based on the task at hand. The argument formation approach is suggested as a way to manage the diversity of reasoning methods across problems that are substantially independent while supporting the necessary interactions in reasoning for problems that interact.

Very subtle design decisions are made in this framework. It is necessary to decide whether a capability, such as echelon determination, is locally "owned" (i.e., as a procedure or algorithm) by an argument or whether the capability is endowed with data acquisition, derivation control, memory, and communication functions and thus rendered an "autonomous" argument.

Argument formation in PENDRAGON is distributed, tightly controlled evidential reasoning. It is distributed in two senses: numerous capabilities are realized as autonomous arguments, and collections of arguments operate as unities with respect to other, similar collections. It is tightly controlled in two senses: each argument manages its derivations (as well as any communications mandated by the problem solving and the conclusions), and metaarguments can suppress or augment the arguments beneath them. Finally, it is evidential in a much broader sense than is common for most fusion systems. Arguments are constructed in such a way that they know what consitutes evidence for them, so that they can acquire this knowledge as required. Conclusions generally have both content and a degree (according to some measure) of likelihood; in turn they provide evidence for later arguments in the train of reasoning. Arguments, as reasoned structures, have a set of conclusions and pointers back to the evidence that the conclusions rest on.

Although PENDRAGON employs a robust approach to situation assessment, its novel methodology raises some significant technology issues:

- Assurance that arguments are coordinated and consistent; this is analogous to the knowledge base consistency problem for conventional KBSs.
- Correct control of reasoning; this robust approach to argument formulation can lead to a large control problem—also, the associated issue of reason maintenance must be dealt with.
- Uncertainty management; again robustness in methods for uncertainty representation yields complexity of management.

PENDRAGON introduces a number of creative ideas and techniques for attacking the situation assessment problem, although the effort as reported to date is still in the early stages of development and experimentation. Table 9.5 summarizes the features of expert or knowledge-based approaches.

9.3 COOPERATIVE AND ADAPTIVE APPROACHES

We said at the outset of this chapter that situation and threat assessment processes are generally multilevel activities. This factor results in requirements for systems

Table 9.5
Features of Expert and Knowledge-Based Approaches

ADVANTAGES

- MODELS EXPERT ANALYST BEHAVIOR
- EMPLOYS SYMBOLIC REPRESENTATION, SYMBOLIC INFERENCE, HEURISTIC SEARCH – AVAILABLE IN SOFTWARE
- EMPLOYS EXPLANATION FEATURES
- ARCHIVES THE EXPERTISE
- INDIRECT TRAINING
- DEPLOYABLE

DISADVANTAGES

- DIFFICULT TO CONSTRUCT (ESP. KNOWL. ENGRG.)
- OFTEN REQUIRES SPECIAL COMPUTING EQUIPMENT
- REQUIRES SPECIAL DEVELOPMENT STAFF
- NEAR REAL-TIME PERFORMANCE MAY BE DIFFICULT TO ACHIEVE
- UNCERTAINTY MANAGEMENT TECHNIQUES UNPROVEN

---

that must, at a minimum, communicate with one another or, ideally, cooperate with one another. The limitations of expectation-based approaches, even with some adaptive characteristics inserted (via multiple hypothesis approaches, confidence factors, elastic constraints, FOMs, *et cetera*) generate the need for adaptive or learning-based approaches. Most of the work in this area is still of a research nature and found primarily in laboratories or exploratory research settings. There is, however, a fairly large body of work in the general area of adaptive systems and machine learning, the review of which is beyond the scope of this book. However, some preliminary comments will be made to set the stage for the review of a couple of studies that focus on C^3I-type problems.

When we speak of cooperative approaches to STA today, we usually mean cooperating expert systems or KBSs because such systems are designed to extend or augment the cooperative processes among people in a C^3I system. An important distinction in such systems is whether the system provides a communication framework or backbone or whether cooperation occurs at the knowledge level.

In communicating systems, the simplest type of communication comes in the form of rote data transfer; one expert system directly passes needed data to another expert system. The goals, functionalities, and processes of the respective systems are completely independent. An ideal and yet realistic form of "communication" among expert systems can also be obtained by providing a cooperating software environment. In such an environment one expert system is permitted to share data, information, functions, reasoning processes, *et cetera* with other expert systems. Control of the flow of information and processing is maintained by an arbitrator.

Other simple methods for communication include various message-passing techniques and associated protocols.

Much research has of course occurred in the use of blackboard architectures previously mentioned. Much of the current research involves systems descended from the HEARSAY experiments [39] and blackboard methods studied therein. Of the available models, the blackboard-based system demonstrates great potential as a general communications paradigm among expert systems.

Blackboards feature the use of two major types of information. The first, referred to here as *data,* includes both facts from an external source and inferences (or hypotheses) drawn by the system itself. The second is called *control information.* In a conventional system this would be the program itself. In a blackboard-based cooperating expert system structure it would include both rules for use in the inference engine of each expert and goals passed down from users or experts in the system. In a blackboard-based system, this information would be placed on an agenda and processed in an order dependent on various factors.

In cooperating expert system environments there is a third major type of information. This, referred to here as *flow control,* determines how the other two types of information move between the various experts in the system. Not only does flow control regulate what information is available to each expert, its state will affect the efficiency with which the communications system is used and other issues related to processing and data redundancy.

Another aspect of cooperating expert systems is the concept of focus of attention. This describes a set of metarules that control which of the set of possible actions the individual systems will undertake next. The use of such a system makes each expert essentially autonomous, allowing local interruption. Controlling the focus of attention among component expert systems permits faster resolution to difficult tactical STA problems.

One last technique should be mentioned at this point: a script-based approach (see [40]). In such an approach, the complex time and order relationships of military situations can be efficiently represented using script templates. The work in [40] represents some preliminary research using this format for exhibiting knowledge for situation assessment.

Finally, we will make some preliminary remarks concerning learning systems. Because some military theorists assert that many data bases developed from peacetime surveillance may be worthless or at least of significantly lesser value in actual combat, it would be prudent to provide some degree of adaptability or learning in an STA analysis system. Whether such approaches are warranted or feasible can be debated but there is no doubt they are more challenging and more complex. Studies of adaptive systems, self-organizing systems, associative systems (e.g., neural networks), and learning systems presently tread right on the boundary between basic and applied research; we judge that practical systems with such capabilities will not be seen very soon.

9.3.1 ALLIES, STARS/PRM, and TAES: Cooperating and Adaptive Approaches

ALLIES

In the period 1981–1985, advanced research models (generally AI-based) of a force structure estimating system ANALYST—see [23, 24], and of a blue force planning system (OPLANNER—see [41]) were developed. Both these systems interacted with and were driven by an object-oriented simulation called the *battlefield environment model* (BEM). It was natural therefore to explore the integration of these systems, as they had both a conceptual and practical affinity. This exploration was conducted under a research program called the air land loosely integrated expert systems (ALLIES) [25].

The basic objective of the experiment was to determine the extent of the changes to each expert system required to establish communications conventions and capture the knowledge of the other systems so the required real-time information could be passed. The goal was to make the links among the systems as loose as possible and still retain the desired information transfer; hence, we have the "loosely integrated" phrase in the project name.

The intended functional interaction and cooperation was such that OPLANNER, given an operational goal, would develop a plan based on (1) an enemy situation elicited from ANALYST, and (2) on an assumed initial state of the friendly units obtained from the BEM data bases. The plan would be disseminated to the actors in the BEM simulation, who would carry out the diverse planned actions against an enemy force and report the status of those actions to OPLANNER. At the same time ANALYST, continually receiving intelligence reports from the friendly controlled sensors in the BEM, would continue to send intelligence to OPLANNER related to critical plan items that had preconditions dependent on the enemy situation. OPLANNER would notice any deviations and notify the user of these deviations. Figure 9.23 shows the top-level functionality of ALLIES.

At the end of the reported (first phase) work, ALLIES achieved a communications capability but not a knowledge-level cooperating capability; it could be described as a cooperative inquiry or question-and-answer system. It established application-specific message-passing protocols based on the fact that it was a simpler approach (than universal protocols) and messages had to be legible by people as well as machines (i.e., constructed according to formats of the intended receiving organization). Because these systems operated asynchronously and were data-driven as regards their KBSs, natural delays in messages were acceptable.

ALLIES is a prototype that electrically links STA analysis to military planning and battlefield sensor systems. While communications-related architectural

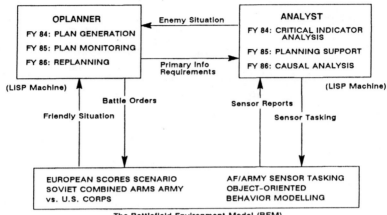

Figure 9.23 ALLIES: Air land loosely integrated expert systems (after [41]).

issues were solved, further experiments will be required to study possible synergistic effects or performance improvements.

STARS/PRM and TAES

Researchers working on these projects are among the few who seem to be studying the applicability of learning paradigms to data fusion and STA analysis. Their work is summarized in [30–32], which discuss their efforts to develop two research prototypes, one called the *situation threat assessment response strategy/perceptual reasoning machine* (STARS/PRM) [32], and one called the *tactical assessment expert system* (TAES) [30, 31]. TAES is an extension of the basic ideas put forward in STARS/PRM and is oriented to a real-time, airborne, pilot/crew-aiding application. STARS/PRM is a robust exploration of the applicability of associative memory for learning and recall, iterative evidential reasoning, and other ES and KBS techniques for STA analysis.

The *perceptual reasoning machine* (PRM) is a learning system framework that provides mechanisms for the system to learn dynamically over time [32]. This structure utilizes inference mechanisms that incorporate representation of total ignorance as opposed to assuming equally likely prior distributions about the evidences. Learning is accomplished by reinforcing existing knowledge via information gathering and reasoning with uncertainty about the observations in a non-Bayesian framework, based on Dempster-Shafer theory.

A functional architecture for a learning system concept based on the perceptual reasoning machine was developed by researchers. The expert system framework decision aid learning scheme is depicted in Figure 9.24. Current information derived from knowledge sources is collected and processed in a gather/assess module, whose evidence functions and algorithms generate beliefs and hypotheses about the observations, using both noniterative and iterative forms of Dempster-Shafer theory.

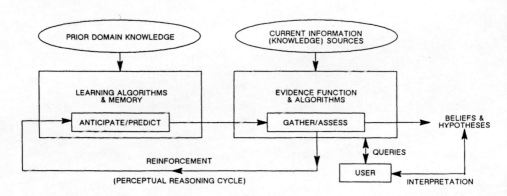

Figure 9.24 Perceptual reasoning machine (from [32]).

The output of the gather/assess module is split between evidence interpretation and feedback to an anticipate/predict module, which contains learning algorithms and knowledge bases in memory acquired from prior domain knowledge and learning updates from the gather/assess module. Part of the input data is also used to recall prior knowledge from the anticipate module. The output of the anticipate/predict module in turn "drives" the gather/assess module closing the feedback loop. It is noted in Figure 9.24 that the PRM actually consists of interactive imbedded feedback loops: the outer control feedback loop, whose function was depicted earlier, and an inner loop within the gather-assess module that controls evidential reasoning.

The information derived from the anticipate/predict module consists of most likely associations and the next best plans. The next best plans aid the selection of the knowledge sources (i.e., sensor data of various levels of processing and abstraction), whereas the most likely associations become additional evidence to be combined with the incoming evidence (to check the convergence of domain hypotheses) and, at the same time, aid the selection of the support in evidential reasoning within the frame of discernment of the observations. This completes the "perceptual reasoning cycle," which is a form of learning by reinforcement (a combination of learning by being told and learning from examples).

The anticipate/predict module of the PRM is implemented as an expert system. The anticipation knowledge base is constructed from *a priori* learning from simulations and represented as a parametric model. As current information enters the inference engine of this ES, the parameters of the model are updated, based on this new evidence. This operation is equivalent to an associative mapping followed by memorization of the new information [42]. As a result, project researchers have implemented this module using a linear associative memory paradigm; an alternative would be to use a neural network approach.

PRM is designed to be the central control mechanism for the convergence of hypotheses in the STARS system. Figure 9.25 depicts a functional block diagram of the situation assessment and decision aid expert system component of STARS. As shown in Figure 9.25 eight major interacting systems interface with the sensors, monitor and control, and crew vehicle interface subsystems. The detailed operation, organization, software constructs and rapid prototype status of this system is depicted in Figure 9.25. The bold lines highlighting subsystems and their interconnects provide a guide to trace a function of the PRM. In this case the PRM supports the classification/identification component of the system.

Some preliminary results of applying this construct to the problem of ship classification were reported in [31]. Rapid convergence of the hypothesis (ship class declaration) was shown to occur with or without any contribution by the anticipation/predict module. The study explored the feasibility and design issues involved in the application of several techniques, including learning, to the STA problem. Although the initial results are encouraging, more work is required on implementation complexity (specifically for avionics architectures), software implications, computing stategies (parallel and distributed processing), and technology alternatives (e.g., neural networks, optical computing).

9.4 THE ROLE OF PERFORMANCE MODELS IN STA

To complete the description of the tools and techniques necessary to build systems that aid in STA analysis, this section will very briefly recount the important role of performance models. It is important to mention these because they can represent a nontrivial aspect of developing real-world STA-aiding systems.

Recall that these models aid in the contextual analysis of the data and in interpreting and assessing force capability or intent. Each model can be developed at varying levels of complexity and fidelity, and thereby development cost. Many of them require a capability for relatively involved processing of topographic (e.g., Defense Mapping Agency data, which is available at varying levels of resolution),

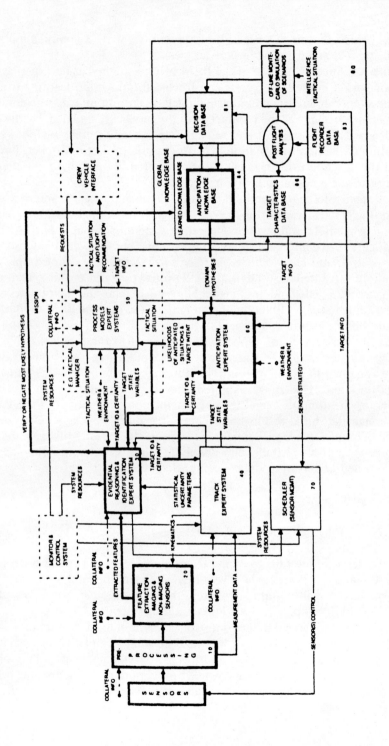

Figure 9.25 Situation assessment and decision aid expert systems (after [32]).

mapping data (e.g., CIA World Data Bases I, II), or weather data, *before* the functional processing of the model is even begun.

Some of the models typically employed in STA analysis include the following:

- *Sensor or weapon line of sight*—Models that can calculate and display point-to-point LOS or a swept LOS capability from a given point, all as a function of elevation.
- *Concealment and cover analysis*—Models that, for example, show areas of forest canopy cover in a given area of interest.
- *Obstacle effects and trafficability*—Models that can compute feasible regions or paths of trafficability as a function of vehicle type and the given terrain and weather.
- *Field of fire analysis*—Models that calculate feasible fields of fire, considering actual interunit spacings and fratricide factors.
- *Intervisibility*—Analogous to LOS modeling, these models compute intervisibility between two (usually friendly or hostile) sensors or weapons.
- *Communication or other propagation*—Models that compute electromagnetic wave propagation characteristics for given emitters.
- *Lethal weapon envelopes*—Models that compute and display the "kill spaces" of weapons, probability of kill envelopes, *et cetera.*
- *ECM models*—Models that estimate the ECM-Jamming-Deception effects of emitter-on-emitter or platform-on-platform engagements.

Other models could be added to this list but the preceding are considered representative. Some of these models will usually be necessary to support the multilevel inference processes of STA analysis discussed in Sections 9.2 to 9.4. As noted, their development or integration can represent a nontrivial aspect of STA system implementation.

9.5 COMMENTS

The fusion of information associated with performing situation and threat analyses is a complex process, requiring "deep knowledge" of military operations and doctrine, military equipment characteristics, the effects of terrain and weather on operations and equipment, and a host of other factors including even such intangibles as a sense of the enemy's will to fight. Current techniques in automated data fusion for STA are providing helpful but only partial solutions to what are clearly very complex problems. Implementation of such component solutions draw on a variety of methodological categories; for example, artificial intelligence methods and methods from decision theory and cognitive psychology. Integrated employment of such methods, in necessarily methodologically hybrid architectures, is just now being researched and experimented with and should provide for broader based automated STA support.

REFERENCES

1. Hopple, G.W., "An Assessment of the State-of-the-Art of Advanced Analytical Methodologies for C^2 Decision Support," in *Principles of Command and Control*, J.L. Boyes and S.J. Andriole (eds.), AFCEA International Press, Washington, DC, 1987.
2. Dockery, J.T., "Why Not Fuzzy Measures of Effectiveness?" *Signal Magazine*, May 1986.
3. Bouthonnier, V., and A. Levis, "Effectiveness Analysis of C^3 Systems." *IEEE Trans. Systems, Man, and Cybernetics*, Vol. SMC-14, No. 1, January-February 1984.
4. Dockery, J.T., "Theory of Effectiveness Measures," SHAPE Technical Center Tech. Memo. STC-TM-729, January 1985.
5. Lammers, G.H., "C^3 Effectiveness Studies," in *Advances in Command, Control, and Communication Systems*, C.J. Harris and I. White (eds.), Peter Peregrinus Ltd., London, 1987.
6. Wohl, J.G., *et al.*, "Human Decision Processes in Military Command and Control," in *Advances in Man-Machine Systems Research*, Vol. 1, JAI Press, Greenwich, CT, 1984.
7. *IPB-Intelligence Preparation of the Battlefield*, Sup R 6600-A, U.S. Government Printing Office, Washington, DC, June 1983.
8. "Intelligence Preparation of the Battlefield," FM 34-130, Headquarters, Dept. of the Army, May 1989.
9. Reed, J.D., *et al.*, "Tactical Operations System Applications and Software Experimentation—Tactical Templating," TRADOC Test Activity, TCATA-FM-271-4, July 1976.
10. Griesel, M.A., *et al.*, "Cross-Correlation: Statistics, Templating, and Doctrine," Jet Propulsion Laboratory, JPL-D-184, February 1984.
11. Kelly, C.W., *et al.*, "The Decision Template Concept," *Decisions and Designs*, VA, PR-80-17-99, September 1980.
12. Weiss, J.W., and C.W. Kelly, "RSCREEN and OPGEN: Two Problem Structuring Aids Which Employ Decision Templates," *Decisions and Designs*, VA, TR-80-4-97, October 1980.
13. Romano, Gen. F.V., "Fusion Centers," *Signal Magazine*, May 1984, pp. 131-133.
14. Personal communications with Dr. Richard Harriss, HRB-Systems, State College, PA.
15. Noble, D.F., "Template-Based Data Fusion for Situation Assessment," paper presented at the First Tri-Service symp. on Data Fusion, Johns Hopkins University, Baltimore, June 1987.
16. Noble, D.F., and J. Truelove, "Schema-Based Theory of Information Presentation for Distributed Decision Making," Tech. Rep. NTIS # AD A163150, Engineering Research Associates, Vienna, VA, 1985.
17. Noble, D.F., D. Boehm-Davis, and C. Grosz, "A Schema-Based Model of Information Processing for Situation Assessment," Tech. Rep. NTIS # AD A175156, Engineering Research Associates, Vienna, VA, 1986.
18. Wright, F.L., "The Fusion of Multisensor Data," *Signal Magazine*, October 1980.
19. Griesel, M.A., *et al.*, "Cross-Correlation: Statistics, Templating, and Doctrine," U.S. Army Intelligence Center and School (NTIS AD-A155-624), February 29, 1984.
20. King, W.H., *et al.*, "A Prototype Expert Assistant for Tactical Intelligence Battlefield Situation Assessment," MIT/ONR Workshop on C^3I, 1986.
21. King, W.H., *et al.*, "Implementation of a Time-Dependent Explanation Capability for Tactical Situation Assessment," WESTEX-87 Conf. on Expert Systems, 1987.
22. Ruoff, K.L., *et al.*, "Situation Assessment Expert System for C^3I: Models, Methodologies, and Tools," in *Science of Command and Control*, S.E. Johnson and A.H. Levis (eds.), AFCEA International Press, Washington, DC, 1988.
23. Bonasso, R.P., Jr., "ANALYST: An Expert System for Processing Sensor Returns," MITRE Rep. MTP-83 W00002, February 1984.
24. Antonisse, H.J., *et al.*, "ANALYST II: A Knowledge-Based Intelligence Support System," MITRE Rep. MTR-84 W00220, April 1985.

25. Benoit, J.R., et al., "ALLIES: An Experiment in Cooperating Expert Systems for Command and Control," in *Proc. IEEE Computer Society Expert Systems in Government Symp.*, October 1986.
26. Spain, D.S., "Application of Artifical Intelligence to Tactical Situation Assessment," EASCON Conf., Technology Shaping the Future, Washington, DC, September 1983.
27. Courand, G., "PENDRAGON: A Highly Automated Fusion System,"*Proc. 1st Tri-Service Symp. on Data Fusion,* Johns Hopkins University, Baltimore, May 1987.
28. Levitt, T.S., G. Courand, et al., "Intelligence Data Analysis," Final Rep. TR-1056-6, Advanced Decision Systems, October 1984.
29. Courand, G., "A Framework for Automated Consensual Problem-Solving," Diss., Stanford University, Dept. of Engineering-Economic Systems.
30. Kadar, I., and G. Eichmann, "Non-Bayesian Optical Inference Machines," *Proc. Int. Optical Computing Conf.,* SPI Vol. 700, Jerusalem, July 1986, pp. 275–276.
31. Kadar, I., and E. Baron-Vartian, "Process Modeling: A Situation Assessment Expert System," AIAA Computers in Aerospace Conf., Wakefield, MA, October 1987.
32. Kadar, I., "Data Fusion by Perceptual Reasoning and Prediction," *Proc. 1st Tri-Service Symp. on Data Fusion,* Johns Hopkins University, Baltimore, May 1987.
33. Yasdi, R., "A Conceptual Design Aid Environment for Expert Database Systems," in *Data and Knowledge Engineering,* Vol. 1, North-Holland Press, New York, 1985.
34. Dempster, A.P., "A Generalization of Bayesian Inference," *J. Royal Statistical Society,* Series B, Vol. 30, 1968.
35. Shafer, G., "A Mathematical Theory of Evidence," Princeton University Press, Princeton, NJ, 1976.
36. Hiestrand, D., et al., "An Automated Threat Value Model," *Proc. 50th MORS,* March 1983.
37. "Detailed Design Description of the ASE Testbed," Final Tech. Rep. Vol. I, RADC, October 1982.
38. Levitt, T.S., et al., "A Model-Based System for Force Structure Analysis," *Proc. SPIE Conf. Applications of Artificial Intelligence,* April 1985.
39. Erman, L.D., et al., "THE HEARSAY-II Speech Understanding System: Integrating Knowledge to Resolve Uncertainty," *Computing Surveys,* Vol. 12, No. 2, June 1980.
40. Laskowski, S.J., and E.J. Hoffman, "Script-Based Reasoning for Situation Monitoring," in *Proc. 6th Nat. Conf. Artificial Intelligence,* August 1987.
41. Benoit, J.W., "Artificial Intelligence Planning Systems in Command and Control," MITRE Corp. Rep. MP85-W00010, September 1985.
42. Kohonen, T., *Self-Organization and Associative Memory,* Springer-Verlag, New York, 1984.

Chapter 10
DATA FUSION SYSTEM ARCHITECTURE DESIGN

Translation of the *functional* model to physical architecture requires assigning numerous requirements (i.e., specific functions and quantitative values) to functions and subfunctions within the data fusion system. In this chapter, we introduce the data fusion systems engineering process and discuss several of the system designer's major computational, database and architectural *implementation* considerations.

The implementation process applies the normal sequence of system engineering activities:

- Definition of system-level (mission or operational) requirements in the form of a specification.
- Definition of system functions and parametric performance of each function to meet system-level requirements. Allocation of system performance to functions (the functional allocation process) and then assignment of functions to physical system components such as sensors, processors, software, and communication links (the "allocated performance baseline system"). This must occur for each of the three levels of data fusion introduced in Chapter 2. Sensor functions, allocated performance: Level 1 functions, allocated performance; Level 2 functions, allocated performance; Level 3 functions, allocated performance.
- Development of candidate system solutions that implement the functional requirements.
- Analysis of the candidates to evaluate relative achievement of the performance requirements against some quantitative measures of merit. (Refer to Chapter 11 for a discussion of this process.)

The very nature of data fusion systems poses a number of special design difficulties to the system engineer, suggesting the need for this disciplined design process rather than ad hoc solutions. In particular, real-time systems are characterized by different processing timelines for each stage of fusion processing, often requiring concurrent access to common data bases. At the first stage (Level 1), sensor processing is generally performed at fraction-of-a-second intervals to convert raw sampled measurements into preprocessed detection reports. At the next stage (still within Level 1), detected target data is processed to perform data association or state estimation and identity fusion for display, at sensor scan rates usually on the order of seconds. At higher stages (Levels 2 and 3) inferences are performed over larger time scales (seconds, minutes, and even hours in some cases) to provide synoptic assessments on human time scales for aid to command decision making. These processing timelines often require the balanced allocation of fixed processing resources to each processing timeline, ensuring that the design does not, for example, yield to the temptation to allocate too many resources to keep up with the demands of sensors at the expense of supporting the human decision process. These and other issues related to the most effective allocation of resources require the specification of a functional, quantifiable model prior to its implementation in a physical system.

10.1 THE DATA FUSION SYSTEMS ENGINEERING PROCESS

The systems engineering process applied to the design of data fusion systems follows a general sequence of engineering activities that must relate mission-level requirements to functional requirements. Physical designs are then synthesized to meet these functional requirements. The process involves tightly coupled trades between the selection of sensors and application of appropriate processing algorithms to most efficiently associate and combine sensed data. For most systems designs, the process follows the structured steps of design delineated in guidance documents, such as military standards, MIL-STD-490 and MIL-STD-499A [1, 2], which impose formal systems engineering discipline on the design process.

A general structured design process is depicted in Figure 10.1, showing the *flowdown* of requirements from the mission (operational) level to the functional requirement level, which can then be implemented in hardware and software. The steps of the process include five distinct stages: definition of mission requirement, definition of functional requirements, analysis of sensor requirements, subsystem design, and design synthesis.

10.1.1 Definition of Mission Requirements

Mission requirements are defined by the military user in terms of operational effectiveness criteria based on a formal *statement of operational need* (SON). The requirements are defined in the highest level system specification, usually referred to as the *A-Level*. This specification is at the top of a hierarchy of specifications (Figure 10.2) that allocate these requirements to the components making up the

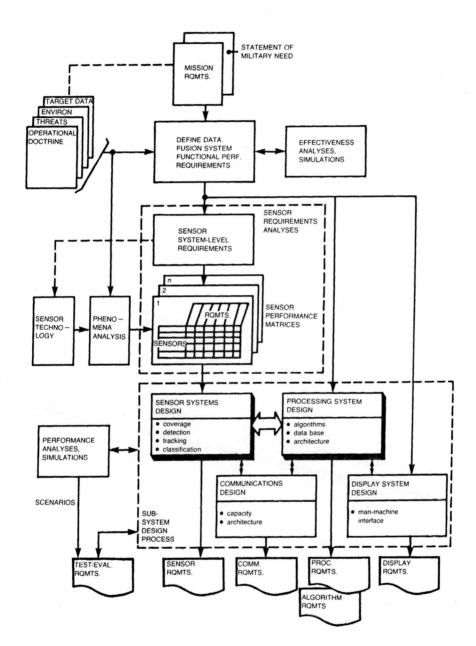

Figure 10.1 Flowdown process of data fusion system design requirements.

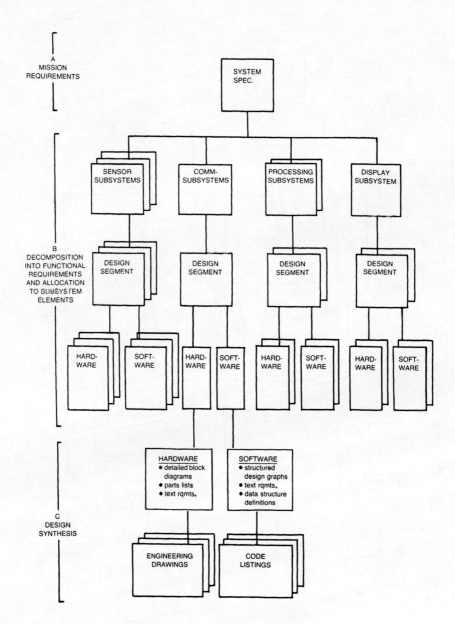

Figure 10.2 Hierarchy of system specifications for a data fusion system.

system. These A-level requirements describe effectiveness in the broadest terms, such as,

- Surveillance volume and target densities,
- Target nomination rate,
- Targeting accuracy, leakage, and timeliness,
- Countermeasures immunity,
- Target identification accuracy.

In support of the mission requirements, the operational doctrine of users and supporting forces are defined as well as the environment in which the system must operate. The environmental definition includes physical (terrain, weather, *et cetera*), electromagnetic (natural, electronic orders of battle of friendly and hostile forces), and composition of forces. Environmental data may be expressed parametrically or in the form of specific scenarios in which the system must operate. A definition of the attributes of targets and threats is also provided to support the mission requirements.

10.1.2 Definition of Functional Requirements

The mission requirements are converted to functional performance requirements at the system level, often requiring *decomposition* of the system into functional subsystems (e.g., spacecraft sensor, airborne sensor, communications, fusion center). This decomposition process requires the development of quantitative *performance* parameters that, when combined, equate to the *effectiveness* requirements established in the mission-level specification. For example, the system surveillance volume (ground to 350,000 feet) requirement may be partitioned into airborne coverage (ground to 75,000 feet) and spaceborne coverage (75,000 to 350,000 feet) requirements for each subsystem. This process will spawn many specific functional requirements and often must be supported by system simulations to verify that the candidate functions and performance requirements will achieve the mission-level requirements.

10.1.3 Sensor Requirements Analysis

The target characteristics, operating environment, and remote sensing requirements (e.g., range, accuracy, spatial resolution, update rate) usually constrain the selection of sensors from among available technologies. Target characteristics (e.g., radar cross section, IR or RF emissions, signatures, effective radiated power) and associated sensing phenomena are analyzed to develop detectability performance data. This is usually expressed in the form of receiver operating curves, which relate the trade-off between sensor detection and false alarm performance. Sensor spatial coverage capabilities are critical factors for consideration at this stage, often defining specific roles such as search, cueing, location, and identification to individual sensors. The requirements *versus* sensor capabilities analysis is depicted on a set of sensor performance matrices. Each matrix, as depicted in Figure 10.1, compares achievable performance *versus* requirements for a combination of operating conditions: in bad weather, heavy jamming, densely packed target scenario, *et cetera*. A study of these matrices can generally eliminate unsuitable sensors and permit an initial selection of candidate sensor suites for further analysis.

10.1.4 Subsystem Design Process

Following a preliminary selection of candidate sensor suites, the primary task of functional design synthesis is to specify the sensor, processing, communication, and display functional requirements that must be met by the physical design of the system (by *physical,* we mean hardware and software realizations that implement the functions).

This process requires a strong interaction (i.e., trade-off) between sensor and processing requirements. This trade-off will result in a selection of the key characteristics of the system, including

- Selection of hard- or soft-decision sensors and the method of reporting data from sensors to processors.
- Definition of spatial and temporal performance of individual sensors (coverage, detection, tracking, and classification).
- Selection of central, distributed, or hybrid architecture for sensor-level and fusion-level processing.
- Selection of appropriate position location, tracking, and attribute combination functional algorithms to achieve fused performance requirements defined at the A-level. This includes decisions such as single- or multiple-hypothesis operation; heuristic, statistical, or syntactic classification; combined or sequential tracking-attribute combination; report-report or report-track association, *et cetera.*
- Definition of required data bases and processing system architecture to achieve throughput capacity requirements.

In addition to the sensor-processing trade-off, the analysis must consider communication and display requirements, which are closely related to sensor-processing issues. The trade-offs are usually supported by performance simulations that verify the sensor-processing concepts will meet system performance requirements. These simulations permit the comparison of alternative sensor suites and levels of sensor performance as well as the performance of various processing algorithms. The results of the system design process are functional (or B-level prime-item and critical-item) specifications that have flowed down the mission requirements to functional requirements that, once physically realized, will meet the operational need.

This flow-down process develops parametric requirements that can be related more directly to physical system design parameters. During this phase, quantitative analyses and simulations support the flowdown process by developing lower-level parametric requirements and verifying that candidate designs meet those requirements. The process of decomposing higher-level requirements to lower-level requirements is illustrated in the following common examples.

Target Capacity

The total number of real targets within the surveillance volume of the system directly influences the size of the target track data base. Target densities, spatial distributions and total quantities must be determined from the operational scenario data supplied in the mission requirements. The data base must accommodate not only all confirmed targets, but must also store tracks in formation, multiple hypotheses (if provided) and tracks being continued following loss of data. Track performance requirements, size of each track record, and update rates therefore must be considered, with capacity, to define the structure, size, and speed of the database management system required to meet target capacity requirements.

Target Update Rate

The overall rate at which the data on any individual target must be updated directly influences the rate at which sensors must revisit targets and report data, as well as the rate at which all related fusion processing functions must be performed to update the target estimates (e.g., identity, position). This requirement directly imposed constraints on sensor operation, processing throughput, and database access parameters.

Figure 10.2 illustrates the common hierarchy of specifications required to flow mission requirements to physical systems. The subsystem design process results in five categories of requirements that permit the design synthesis and test of a physical system. The following paragraphs summarize the typical key parameters contained in these B-level specifications.

Sensor Requirements

The primary characteristics that specify the operation and performance of individual sensors include the following items:

1. Spatial coverage
   a. Volumes of coverage (FOR, FOV)
   b. Instantaneous field of view (IFOV)
2. Search operation
   a. Spatial search pattern
   b. Temporal update (revisit) rate
3. Active or passive operation
4. Cooperative or noncooperative operation
5. Spatial (position) measurement characteristics
   a. Dimensionality
   b. Spatial accuracy, resolution
   c. Boresight, alignment
6. Attribute measurement characteristics
   a. Sensed parameters (attributes or features)
   b. Measurement accuracy, resolution
7. Range of operation
   a. At detection
   b. At continuous tracking
   c. At classification or identification

These are the primary characteristics and parameters that must be specified for any sensor. In addition, many sensor-unique parameters may be required as well as specific target and environmental conditions. In most cases, sensor performance parameters are specified as a function of target, environmental, and countermeasure conditions.

Communication Requirements

The sensor-processor communication structure, whether for collocated sensors (a bus architecture) or distributed sensors (a communication network) must be specified in terms of the following characteristics:

1. Message data description
   a. Message paths between all nodes
   b. Message structures, formats, data contents
2. Communications performance
   a. Data rates, channel capacities
   b. Statistical channel characteristics (bit and burst error rates, *et cetera*)

    c. Spatial performance (range, LOS)
    d. Temporal performance
3. Message timing
    a. Message protocol at all levels
    b. Timing relative to sensor-processing events

Processing Requirements

The functional processing requirements describe the following items that will drive the design of signal and data processing hardware and software:

1. Processing architecture
    a. Structure and functional sequence
    b. Operating modes
2. Database architecture and contents
    a. Definition of all data types
    b. Storage capacity for all data types
    c. Storage-retrieval rates
3. Processing algorithms
    a. Signal preprocessing algorithms
    b. Association-tracking functional algorithms
    c. Attribute combination functional algorithms
    d. Sensor management algorithms
    e. Situation-threat assessment algorithms
    f. Decision aids, processes, and criteria

Display Requirements

Human-machine interface requirements are specified, based directly on mission and processing requirements. These are specified in terms of

1. Operator input characteristics
    a. Input controls, messages
        (1) Manual switch and keyboard inputs
        (2) Cursor control inputs
    b. Natural language inputs
    c. Rates, contents, and accuracy of inputs

2. Operator output characteristics
   a. Visual display characteristics
   b. Aural and synthetic voice characteristics

Test and Evaluation Requirements

The objectives and methodology for verifying performance and effectiveness of the data fusion system is finally specified. These include a test plan that summarizes the series of laboratory and operational field tests to verify that the system meets every detailed functional specification. The test plan is supported by detailed documentation of test requirements and procedures that define the specific test and analysis methods for each performance requirement. Because the mission effectiveness requirements are so broad, for real-world evaluation, a limited number of selected scenarios are often used to test specific values under the most stressful conditions.

10.1.5 Design Synthesis

The final stage is implementation or design synthesis, in which the requirements are allocated to real hardware and software components. This *segmentation* process further partitions functional requirements between hardware and software, establishing new interface requirements between the two. The sensor communication, processing, and display hardware or software designs are synthesized. Design reviews are held to ensure that all functional requirements are being met by candidate designs prior to full commitment to build and test the resulting data fusion system. The design is documented in the final, lowest level of fabrication specifications (the C-level specifications), which include engineering drawings (hardware) and code listings (software) as well as supporting documentation for each.

10.2 DATABASE MANAGEMENT FOR DATA FUSION

10.2.1 Level 1: Association and Attribute Refinement

Target State Data

Target state information has traditionally been maintained in a single-sensor *track file* that contains records for each entity (target) under track. The state data, described in Chapter 6, is generally dynamic, being recursively updated as sensors make sequential measurements on targets. With the introduction of multiple sensors, associated data is generally appended to the target records to expand the role of the track file to that of a target data base containing, for each target, all associated sensor, identification, priority, and sensor management data.

Target Attribute Data

Attributes of targets may be expressed directly as features; statistical (parametric), nonparametric, or semantic descriptions of features; or feature-target production rules. This information is stored in the form of decision rule data used to classify the measured attributes into target classes as described in Chapters 4 and 7. The set of these attribute-target relationships for all target categories must be stored for use by sensor classifiers. The size of this data base is a function of the number of target classes, features measured (classifier dimensionality), and the complexity of feature descriptions.

10.2.2 Levels 2 and 3: Situation Assessment and Threat Refinement

Situation and Threat Assessment Data

Higher-level representations of the data fusion surveillance scene are created by Level 2 and 3 processing, which may be efficiently organized as knowledge bases in knowledge-based systems, as described in Chapter 9. These knowledge bases generally maintain higher-level entities (e.g., aggregated target sets, target relations, threat intervals, command nodes) than those maintained at Level 1 (e.g., individual targets and events). The knowledge bases can contain knowledge in a variety of forms and representations, including

1. *Entities, events, attributes and relationships*—Rules, semantic networks, frames, or other hierarchical methods used to describe entities or events to be identified on the basis of attributes.
2. *Decision rules*—Rules for making hard-decisions for presentation of situation, threat data to the user.
3. *Temporal behavior descriptions (scripts)*—Time sequence patterns (see Section 9.2) of events, activities that uniquely identify behavior.
4. *Spatial relationship descriptions*—Spatial patterns of entities and events (see section 9.2) that identify behavior.
5. *Hypotheses*—Storage of intermediate *scene-level* hypotheses that describe the situation or threat.

Antony [3] has described a distributed, hierarchical blackboard architecture KBS for battlefield data fusion. In this organization he suggests a classification of knowledge representations and data bases into three classes:

1. *Short-term knowledge*—Incoming sensor and source data that is preprocessed (this could include, for example, sensor-level tracking) and buffered for Level 1 processing is classified as short term.
2. *Medium-term knowledge*—The organization of short-term knowledge into a perception of the current situation formulates this category. Such knowledge is maintained and dynamically changed on the blackboard as the situation assessment process estimates the state of targets, events, and activities. The blackboard provides a shared data base with which multiple assessment processes may interact and mutually evaluate alternative (and conflicting) interpretations of the data.
3. *Long-term knowledge*—Static factual and procedural knowledge that supports control and reasoning processes is the final class.

Tables 10.1 and 10.2 enumerate the major categories of specific data sets that must be maintained for Level 1 and Level 2 or 3 processing, respectively. These sets (bases) of data may be maintained in a variety of forms, appropriate to the application: static look-up tables, dynamic files, and complete *database management systems* (DBMS).

10.2.3 DBMS Considerations for Data Fusion Systems

The normal database management system functions of query, data sharing (concurrent, multiple function access), and report management are required in all but the simplest of data bases. Introduction to DBMS processes and the methods to implement such systems is beyond the scope of this text. The reader is referred to DBMS texts [4-6].

10.3 DATA PROCESSING FOR DATA FUSION

The data fusion processes described in previous chapters have included both numerical and symbolic computational requirements. Numerical processing was described for sensor signal processing, sensor-target association, data association–target tracking, and attribute combination in Chapters 4 through 7. Symbolic processing was described for overall data fusion control, sensor management, situation and threat assessment in Chapters 2, 5, and 9, respectively. For all but the smallest of multiple sensor systems, these processes can impose heavy computational demands on the system architecture. In this section, we describe the categories of processing architecture that efficiently perform both numerical and symbolic data fusion processing functions.

Table 10.1
Categories of Data Bases Required by Level 1 Data Fusion

Data Base	Typical Contents of Data Base	Uses of Data
Target Attribute Data Base	Target-feature relationships; classification data (parametric or nonparametric), rules, networks, frames, or templates	Sensor preprocessing for target classification by attributes
Target *a priori* Data	*A priori* quantities of targets by class, predicted target locations and trajectories, orders of battle	Prior data used in Bayesian classifiers Prior target locations used for sensor management
Target (Track) Data Base	Typical data for each target: Target (track) index Current state estimate (e.g., $x, \dot{x}, y, \dot{y}$,) Statistics of state estimates (e.g., covariance matrix of errors) Track state (initiate, confirm, loss) Time of initiation, last update Level 1 ID (friend-foe-neutral) Level 2 ID (target type) Level 3 ID (target class) Confidence data, each ID (e.g., measure of uncertainty in ID) Sources of identity (sensors, contributions and reports) Target priority for sensor management	Data association, tracking, and attribute combination processes maintain this data base for sequential processing Multiple hypothesis association and identification algorithms may also maintain candidate hypotheses in this data base, or in a separate (relational) data base This data base forms the input to levels 2 and 3 fusion processing Related fire control, weapon guidance, and sensor-management data may also be included
Track history	Chronological sequence of sensor reports and association, classification decisions (same data as above)	Used for batch association and attribute processing in which stored time periods are processed at one time
Sensor Model Data	Range, LOS, detection-ID performance data against various target types, in varying environments	Used by sensor manager to predict the performance of sensors against targets for assignment

10.3.1 Processing System Architecture

The functional and physical allocation stages of the data fusion systems engineering process described in Section 10.1 partition the fusion functions into physical items connected by intercommunication interfaces. Because of the computationally intensive nature of data fusion, this partitioning process requires a thorough understanding of the trade-offs of computer architecture design and the implications of

Table 10.2
Categories of Data Bases Required by Level 2 and 3 Data Fusion

Data Base	Typical Contents of Data Base	Uses of Data
Behavioral Data Base	Target and event behaviors and characteristics; temporal-spatial templates, tactics, combat doctrine, *et cetera*	Used for target and aggregated target set identification, situation assessment. Also used for sensor management to predict sensor view opportunities
Terrain Features	Topology, hydrology, road-rail networks, cultural features, obstacles, locations of cover and concealment, *et cetera*	Used for identification of individual targets by behavior and for aggregation of targets. Sensor management uses for LOS prediction
Airway Doctrine	Corridors, restricted zones, routes, and schedules (flight plans)	Used to identify targets by planned behavior
Own-Forces Status	Disposition of forces, strength and condition, reserves, logistics, command and control, planned course of action	Both situation assessment and threat refinement functions use this data to determine course of action hypotheses and the effects on sensor management, and data fusion performance
Intelligence on Enemy Forces	Disposition of forces, strength and condition, reserves, logistics, command and control, probable intent, targets of interest and course of action	
Threat Capabilities	Sensor and weapons performance capabilities: Performance envelopes Detection, warning, track, ID capabilities Countermeasure vulnerability	Assess threat capability against own forces, predict threat behavior; sensor manager predicts enemy detection of own emissions to reduce exploitation
Situation Data Base	Location of all entities, relationships and predicted courses of action; identification of potential threats and opportunities for all forces	Maintains the current hypotheses of threat situation assessment; dynamic data base is sequentially updated as assessments are updated by completion of Level 2 assessment
Threat Data Base	Assessment of threatening entities, events, and activities; estimate of own-force vulnerability (sensors and weapons) based on enemy opportunities, means, motives from Level 2	

partitioning functions to physical processing elements. Considerations in functional allocation and system architecture include

- *System throughput*—Latency, concurrency, sustained operation rates
- *Centralized versus distributed detection and estimation*
- *Centralized versus distributed database storage and access*
- *Message traffic contraints*—Routing, capacity, delays

Figure 10.3 illustrates several possible top-level data fusion architectures that partition the functions of sensor processing, association-tracking, attribute combination, sensor management, and situation assessment to physical processors in different ways. The most basic system allocates all functions to a single central processor. Additional complexity is introduced as the functions are allocated to distinct processing elements to handle increased processing loads, introduce functional redundancy (increased reliability), and permit physical separation of the functions (e.g., sensor networks). In Sections 10.4 and 10.5 we illustrate the centralized and integrated, distributed architectures, respectively.

10.3.2 Methods of Applying Parallelism to Data Fusion

The very nature of mutiple sensor processing algorithms makes a number of data fusion functions candidates for parallel processing. Functional parallelism occurs in several areas of the data fusion process: (1) multiple sensors receiving reports and preprocessing data in parallel, (2) association of multiple reports in parallel, (3) computation and maintenance of multiple association and identification hypotheses, and (4) parallel searches for possible interpretations of data in knowledge-based reasoning for situation assessment. At least two levels of functional parallelism may be considered within data fusion systems for the introduction of parallel processing: system-level functions and algorithm-level functions.

System-Level Functions

The top-level structure of most data fusion systems are inherently parallel due to the use of multiple sensors, with concurrent sensor preprocessing prior to association and combination of data. The U.S. Air Force's Pave Pillar avionics architecture [7] illustrates system-level parallelism, where parallel and closely interconnected sensor preprocessors and mission processors provide functional redundancy and resource sharing as well as fault tolerance. In this structure, sensor-level parallelism includes full interconnection among all sensors and all preprocessors as well as the capability to exchange processed data among any preprocessors or the mission avionics data multiplex bus (Figure 10.4). Multiple mission devices (e.g., displays, weapons, navigation equipment) may require different forms of fused

CENTRAL FUSION PROCESSING
- Sensors (S) provide report-level data to central processor (P)
- All levels of fusion are performed at the single processor (P) which outputs data to display (D)

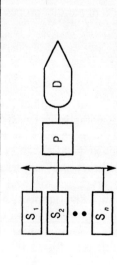

DISTRIBUTED (FEDERATED) LEVEL 1 PROCESSING
- Sensor-level tracking and preprocessing is performed in sensor proc. (SP)
- All levels of fusion are performed in single (or multiple) processors (P)

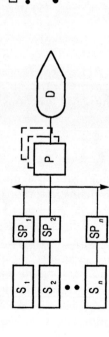

PARTITIONED LEVEL 1 PROCESSING
- Imaging sensors (S) perform segmentation and feature extraction at the sensor
- Imagery and feature data are passed to fusion processing on separate paths
- Image processor (IP) registers imagery and fusion processor (P) classifies feature data

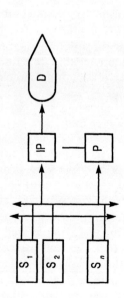

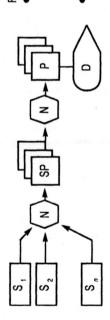

Figure 10.3 Representative allocations of Levels 1, 2, and 3 fusion functions to processors.

FULLY INTEGRATED
- Switch networks (N) permit sensor processors (SP) to be switched between all sensors (S) to load share, provide redundancy
- Multiple processors (P) perform all levels of fusion processing; levels may be partitioned among processors

DISTRIBUTED SENSOR/FUSION
- Sensor nodes ($S_1 .. S_n$) are physically distributed, with each node having a local, central processor (NP) which performs local data fusion (at level 1)
- Each node processor (NP) data links information to a global processor(s) (P) which performs level 1 association and level 2, 3 processing

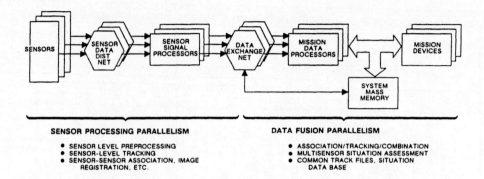

Figure 10.4 Sensor processing and data fusion processing parallelism in the Pave Pillar architecture.

data, and it is possible to assign distinct, concurrent data fusion processes to parallel mission data processors. This architecture is described in additional detail in Section 10.5.

Algorithm-Level Functions

The decomposition of specific data fusion functions into parallel, independent algorithms is required to permit efficient operation of physically parallel processors. The large number of alternative parallel architectures possible provide a wide range of decomposition methods for fusion functions. Table 10.3 illustrates the primary characteristics of parallel architectures [8], the alternative design approaches, and the properties of each that are most applicable to the decomposition or implementation of fusion functions in parallel algorithms. Several specific implications of this chart include

- Vertically organized structures are most appropriate only for synchronized, similar source systems where common sensor processing functions are performed in time synchrony for all sensors or for systems where multiple target operations are performed in parallel.
- The choice of an interprocessor communication structure depends most on the degree to which a common data base must be used by multiple processes. Buses between independent units are efficient when the processing sequence flows data from unit to unit with a minimum of access to common data (e.g., the sensor-preprocessor, preprocessor–sensor tracker, sensor tracker–central processor path).
- The topology of interconnection between sensor processors and data processors influences the functional redundancy of the system: the degree of flexibility to reroute data from sensors-processors or processors-processors to

Table 10.3
Parallel Processing Approaches and Implications for Data Fusion

CHARACTERISTIC	ALTERNATE APPROACHES	APPROACH DESCRIPTION	IMPLICATIONS FOR DATA FUSION PROCESSING
LOGICAL STRUCTURE	VERTICAL (MASTER-SLAVE)	A SINGLE UNIT ACTS AS A MASTER AT ANY ONE TIME COORDINATING COMMUNICATION AND OPERATIONS OF OTHER UNITS (SLAVES).	APPLICABLE WHERE SENSOR DATA COLLECTION IS SYNCHRONIZED OR FUSION IS GOAL DRIVEN, WITH COMMON CONTROL OVER ALL PROCESSES.
	HORIZONTAL	ALL UNITS ARE LOGICALLY EQUAL AND CAN INITIATE COMMUNICATION WITH ANY OTHER UNIT.	APPLICABLE WHERE SENSOR DRIVEN PROCESSING BY MULTIPLE OPERATIONS REQUIRES CONCURRENT UNSYNCHRONIZED INPUT, ASSOCIATION, COMBINATION OPERATIONS.
INTERPROCESSOR COMMUNICATION	COMMON MEMORY	CENTRAL MEMORY IS FOCAL POINT OF ALL MESSAGE AND DATA EXCHANGE.	COMMON DATA BASES (SITUATION, TARGET ATTRIBUTE DATA, TERRAIN, ETC.) ARE REQUIRED BY NUMEROUS FUNCTIONS. THE SIZE, UPDATE RATE AND ACCESSIBILITY OF THESE BASES ARE INFLUENCED BY INTERPROCESSOR COMMUNICATIONS.
	BUS STRUCTURE	DISTRIBUTED UNITS COMMUNICATE VIA MESSAGE LINKS WITH VARYING TOPOLOGIES.	
INTERCONNECTION TOPOLOGY	COMMON BUS	SINGLE, OPEN PATH TO ALL UNITS.	INTERCONNECTION BETWEEN PROCESSORS, AS WELL AS SENSORS INFLUENCES: • FUNCTIONAL REDUNDANCY FOR RELIABILITY AND DEGRADATION PROPERTIES • EFFICIENCY OF DATA EXCHANGE BETWEEN PARTITIONED OPERATIONS
	STAR	ALL UNITS CONNECT TO ONE CENTRAL UNIT.	
	HYPERCUBE	ALL UNITS CONNECTED IN N-DIMENSIONAL CUBE STRUCTURE.	
	FULLY CONNECTED	NXN MATRIX CONNECTS ALL UNITS.	
MODE OF INTERACTION BETWEEN UNITS	TIGHTLY COUPLED (MULTIPROCESSORS)	SYNCHRONIZED UNITS SHARE RESOURCES AND COMMON MEMORY UNDER A COMMON OPERATING SYSTEM WHICH COORDINATES ALL ACTIVITIES.	MOST APPLICABLE TO A LARGE NUMBER OF SIMILAR, RELATIVELY INDEPENDENT PROCESSES WHICH REQUIRE MINOR DATA EXCHANGE.
	DISTRIBUTED	AUTONOMOUS UNITS WITH DISTRIBUTED TASKS, UNEQUAL UNIT CAPABILITIES, AND DATA-LEVEL COMMUNICATION.	GENERAL APPLICABILITY TO PARTITIONING OF FUNCTIONS (CORRELATION, SIT ASSESS, COMBINATION) TO PROCESSORS SHARING A COMMON DATA BASE.
	LOOSELY COUPLED (NETWORKS)	INDEPENDENT, AUTONOMOUS UNITS ARE COUPLED ONLY BY MESSAGE LEVEL COMMUNICATION.	MOST APPLICABLE TO DISTRIBUTED SENSOR SYSTEMS (NETWORKS) WITH DISTRIBUTED DECISION MAKING PROCESSING, LIMITED COMMUNICATIONS.
MODE OF PROCESSING IN EACH UNIT	SISD	SINGLE INSTRUCTION SINGLE DATA STREAM	IMPLEMENTATION OF CONTROL AND DATA FLOW WITHIN ARCHITECTURE DRIVES THE METHOD BY WHICH ALGORITHMS CAN BE EFFICIENTLY DECOMPOSED INTO PARALLEL OPERATIONS. MISD = PIPELINED SIGNAL PROCESSING SIMD = HIGHLY REPETITIVE (ARRAY, ASSOCIATIVE) OPERATIONS MIMD = MOST COMPLEX, INDEPENDENT FUNCTIONS.
	MISD	MULTIPLE INSTRUCTION SINGLE DATA STREAM	
	SIMD	SINGLE INSTRUCTION MULTIPLE DATA STREAM	
	MIMD	MULTIPLE INSTRUCTION MULTIPLE DATA STREAM	

maintain operations despite processor or path failures. It also influences the efficiency of the transfer of data between units, which is dictated by the needs of the parallel algorithms.

A number of data fusion functions have been proposed for parallel decomposition into separate algorithms for concurrent operation on parallel processors, including (1) multiple hypothesis attribute combination, (2) multiple hypothesis association-tracking, and (3) logical inferencing for rule-based reasoning.

Multiple Hypothesis Attribute Combination

Sensor reporting methods that provide soft-decision data, as previously described in Chapter 4, permit multiple hypotheses of target type to be computed for each sensor measurement. For large numbers of distinct target types, the processing and reporting requirements can be excessive. Consider, for example, a modest three-sensor system, with each sensor reporting a measure of confidence (a probability) for 32 possible target types. For a report rate of 40 observations per second, the system must compute and evaluate 3840 hypotheses per second. In a Bayesian inference system, each hypothesis requires over 250 operations per hypothesis, demanding 100 kops just to perform combination. Probability interval approaches such as Dempster-Shafer, although not requiring computation of all hypotheses for each observation, demand even more operations per hypothesis.

Multiple Hypothesis Association-Tracking

Dynamic target tracker-correlators that retain multiple association hypotheses effectively spawn a geometric growth in track hypotheses that must be updated and tested for association with new data until deleted. Hypotheses grow as a function of the number of observations per scan, the number of existing tracks, and the rate at which hypotheses are pruned. The maintenance of large and rapidly growing numbers of hypotheses can tax current processors for even moderate target loads. For example, three sequential data sets of two observations each can spawn over 500 hypotheses to be maintained for the fourth data set [9].

Logical Inferencing for Rule-Based Reasoning

The computational requirements for symbolic (AI or knowledge-based) inferencing operations are also directly related to hypotheses, report rates, and numbers of sensors. Linas and Hall have estimated that symbolic association for 3 sensors viewing 20 tracks can require 20 Mops if 100 inferences are assumed per association [10]. Specific approaches to this problem were previously discussed in Chapter 2.
   The recent capability to produce cost-effective parallel processing machines gives reason to consider the methods by which these and other data fusion functions may be most efficiently implemented in parallel architectures. In the following paragraphs, we suggest several means by which parallel systems are well-suited to several categories of data fusion problems.

10.3.3   Parallel Computing Architectures Applied to Data Fusion

A number of predominant parallel architectures have been developed from the many structures conceivable, and these may be expected to become more available to data fusion applications as very large scale integration achievements and complementary parallel software development tools [11] make their use cost effective. The organization of *single* (S) or *multiple* (M) *instruction* (I) and *data* (D) streams provides four categories of physical architectures characterized as SISD, MISD, SIMD, and MIMD processors. The most common parallel architectures and their most effective applications to logical, algorithmic, and symbolic data fusion functions are described in the following paragraphs.

SISD and MISD Pipelined Processors

SISD machines may be structured to overlap instruction execution in functional elements of the processor (pipelining) to achieve *temporal* parallelism. Highly repetitive, short signal processing functions (e.g., FFT butterfly, local image filters) required in sensor preprocessors have been effectively performed by pipelined processors. Similarly, the sequential and repetitive computations for data association and attribute combination may be speeded by pipelined processors.

SIMD Processors

Physical arrays of processors operating under the control of a single data stream achieve *spatial parallelism*. Array processors are well-suited to the multiple hypothesis attribute combination operations, where identical operations are performed on all elements of large vectors (e.g., vectors of probability values). A special class of SIMD array machines use content addressable memory to provide associative processing capabilities, providing rapid access to data on the basis of *content* rather than memory address. Associative SIMD processors are applicable to database search operations, such as those in template matching or track association. In one such application, an SIMD machine has been considered to efficiently search for track association candidates in a large data base, prior to calculation of "fine" association on a SISD machine [12]. The most recent and highly interconnected machines ("fine-grained SIMD") include a wide class characterized by large numbers (16,000 or more) of tightly coupled processors that permit concurrent, independent operations and direct data exchange. Hypercube machines [13] of dimension n provide the potential for 2^n concurrent operations (nodes) with $(2^n)^2$ communication paths for exchange of data among operations. Multiple hypothesis association-tracking has been suggested as a possible application [12], by partitioning the problem in a hypothesis-per-node fashion.

MIMD Tightly Coupled Multiprocessors

MIMD machines with a high degree of interaction among processors are referred to as *tightly coupled*. Data flow [14] and systolic array processors are such machines that may be most effectively used in algorithmic applications, as previously identified for fine-grained SIMD machines, as well as for symbolic processing for knowledge-based reasoning.

MIMD Loosely Coupled Multiprocessors

Loosely coupled MIMD machines have a low degree of processor interaction and can be considered a set of interconnected, relatively independent SISD processors that share access to common memory spaces. This is the most common form of multiprocessor system in which easily separable functions are allocated to independent processors. The Pave Pillar architecture previously described, for example, can be considered to be such an architecture. That architecture, however, may loosely couple *within its own structure* a variety of processor types (e.g., SISD processors for sensor management, SIMD processors for attribute combination, and MIMD processors for situation assessment).

10.3.4 Connectionist Architectures Applied to Data Fusion

The so-called connectionist architectures [15] implement neural processing networks with the following properties that distinguish them from conventional algorithmic processes operating sequentially or in parallel:

- The topology of the interconnections between nodes (processors) and the node characteristics fully define the performance, rather than algorithms decomposed and allocated to the nodes.
- Each node operates independent of all others, except for receipt of inputs and forwarding of outputs to those specific nodes to which it is uniquely connected.
- The processing in all nodes is identical and simple: adaptive weighting, summation, and thresholding of input values to produce multiple outputs.

The neural network concepts presented in Chapter 7 for identity fusion are expanded here to describe the implementation of such architectures. Figure 10.5 illustrates a general neural network structure made up of four *layers of nodes* or processing elements. In this example, the layers are fully interconnected in a feed forward direction (every node in each suceeding layer accepts inputs from all nodes in the preceeding layer). The figure illustrates the processing function performed by each node:

$$z_j = F_j\left(\sum_i y_j w_{ji} - \theta_j\right) \qquad (10.1)$$

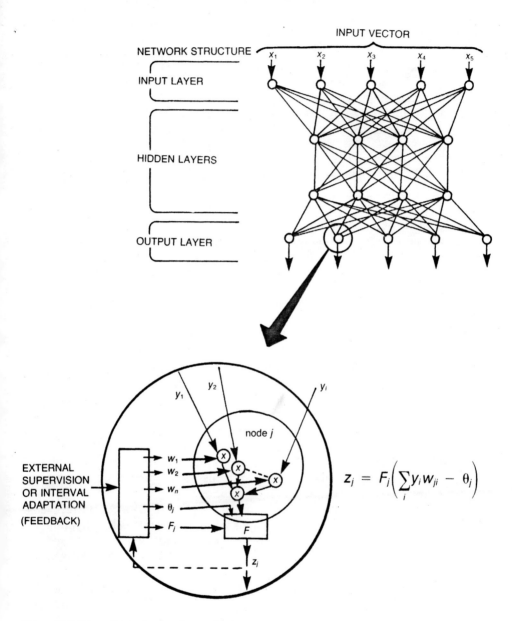

Figure 10.5 General neural network organization.

where

z_j = scalar total output of node j;
F = output function that conditions results for presentation for next layer;
y_i = scalar output from node i, the node state;
w_{ji} = weight applied to input from node i to j;
θ_j = threshold value, below which node j provides no output to suceeding nodes.

The nodes can perform binary or continuous-valued arithmetic and can include linear or nonlinear output functions. The model shows how the control of the weight vector, threshold, and output function can be used to adjust (or "fix") the output. This permits the node to "learn," as the output function can be controlled to achieve a specified output for a known input (supervised learning). Successive adaptation to a wide variety (i.e., a statistical set) of inputs permits the net to dynamically arrive at weights, thresholds, and output functions that provide the desired output pattern for given input patterns. If the control of the nodes are networked to be internal feedback connections, the network can become self-organizing, performing unsupervised learning.

The short-term knowledge of the network at any time is in the current states of all nodes in the net. Long-term knowledge acquired by learning is stored in the current weights, output functions, and thresholds of all of the nodes in the net. The network *structure* (when fully connected) can be entirely modified by changing weights, because a weight of zero effectively disconnects two nodes. In effect, the network node functions construct a complex representation of the input patterns presented. Although conceptually simple, the development of network structures and node functions is very complex and the subject of extensive research. The focus of research is the development of efficient learning processes (net functions and structures) that permit networks to rapidly establish accurate and robust representations of input patterns for subsequent recognition.

Connectionist architectures can be readily applied to data fusion problems requiring the virtually simultaneous analysis of multiple, competing hypotheses. The strength of these machines lies in their adaptive and robust performance of both nonparametric supervised classification and nonsupervised classification. Supervised classification has been demonstrated with Hopfield nets [16] and results have been reported for the hull-emitter classification applications for ship identification [17]. Unsupervised nets are useful for automatic clustering of large data sets into distinct categories based on defined similarities. One reported application is the frame-frame correlation of point-source targets in sequential imagery [18]. The immediate application of these architectures appears to be in sensor classifiers (perhaps with a capability to provide soft-decision measures of confidence for Bayesian or Dempster-Shafer combination functions). Ultimate applications could even include sensor-sensor combinations when nets are capable of processing extremely large feature spaces (dimensionalities of the order of 20–100).

Bowman [19] had described a representative hybrid data fusion architecture (Figure 10.6) that combines numeric and symbolic processing as well as neural networks. The system applies neural networks in two tiers: (1) target classification for each individual imaging sensor, and (2) combined classification across all imaging sensors. Nonimaging sensors (electronic support measures) and tracking data are processed by symbolic means and are also combined with target class data by a symbolic processor.

10.4 CENTRALIZED SYSTEM ARCHITECTURES

The data fusion requirements for naval shipboard combatants provide a tactical application that accepts both onboard and offboard sensor data for fusion into air, surface, and subsurface tracks over a vast surveillance volume. Feldman [20] has described an *advanced combat direction system* (ACDS) being developed to provide real-time track management and decision support for surveillance beyond four times the stand-off range required by the surface battle group. Sensor data is supplied by battle group sensors (stand-off range), battle group surveillance aircraft (two-time stand-off range) and other outside sources, such as ocean surveillance spacecraft out to four times the stand-off range. The external sensor data is linked to the ship via tactical data links 4, 11, and 16 (TADIL J).

Figure 10.7 shows the functional ACDS architecture that partitions the data fusion functions into the following four elements: (1) sensors and sources; (2) track management processor; (3) ACDS data base (five categories of data are maintained in the central database: (a) target tracks for air, surface, and subsurface targets with associated identification and threat status, (b) tactical status results from situation or threat assessments, (c) intelligence libraries containing knowledge of threat activities, plans, and orders of battle, (d) a doctrine rule base consisting of established criteria for identifying tracks on the basis of behavior, ranking threats, and aiding engagement decisions, and (e) maps of the world, possibly including ocean floor terrain, land-surface coasts, political boundaries, and commercial air routes; (4) decision support processor.

This architecture is characterized by a number of features that are common to many current generation data fusion systems. First, the system maintains an interface with *existing* sensors that are asynchronous and have independent interfaces. The sensors and data links have different reporting protocols (e.g., track coordinate systems, uncertainty and identification representations), requiring the track management processor to normalize data in addition to aligning time and space

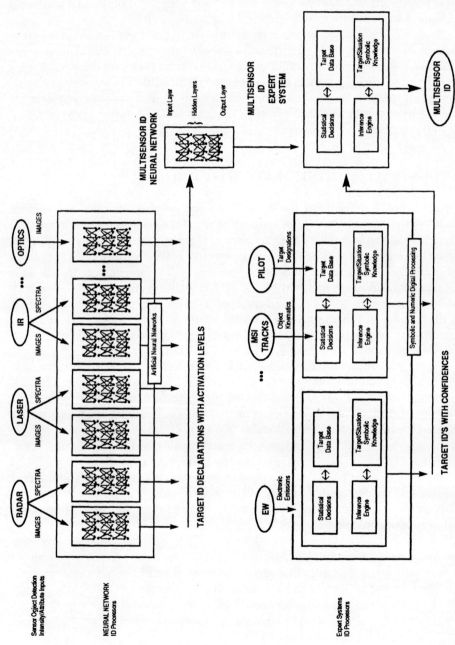

Figure 10.6 Integrated data fusion architecture with numeric, symbolic, and neural processing (from [19]).

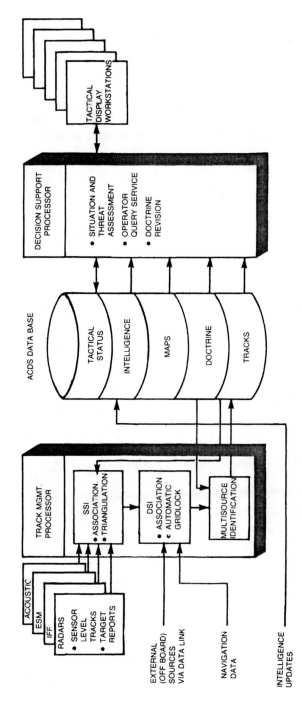

Figure 10.7 Advanced combat direction system architecture (after [20]).

coordinates. Second, the major processing partition is between the Level 1 and Level 2 and 3 processes, effectively separating the numeric and symbolic functions into two parallel processors. In the case of ACDS, identical processors are used for each function, assuming a relatively balanced allocation of resources. (Note that, in some cases, the numerical and symbolic processors may be of different types, each selected for its performance in the appropriate type of processing.) Third, the partitioning of functions provides a common data base requiring concurrent access to the track data by both the Level 1 and Level 2 and 3 processors. In this application, the track management processor can be expected to have access to the track data at a rate significantly higher than the decision support processor, because each SSI and DSI track-track association or identification resulting from the arrival of new observations (from multiple sensors) requires retrieval of this data. The decision support processor, on the other hand, retrieves the data at much lower (periodic) rates to assess the tactical situation and update displays.

10.5 INTEGRATED SYSTEM ARCHITECTURES

The multiple sensor and data fusion processing requirements for fighter aircraft provide a representative tactical application that demands an efficient architecture to detect, track, and identify multiple targets in the most demanding dynamic environment. The avionics required to perform the sensor and data fusion functions have been studied extensively by the U.S. Air Force under its Advanced System Integration Demonstration Program, also designated "Pave Pillar." In this technical development program [21], several system objectives directly related to data fusion were established for system studies:

- The development of fully integrated multimode architectures to improve operational effectiveness, availability, and survivability of multiple sensor systems.
- The incorporation of integrated apertures and sensors such as
  1. *Integrated communications navigation and identification* (CNI) functions.
  2. Integrated electronic warfare functions.
  3. Integrated, common-aperture imaging sensors.
- The application of wideband bus, high-speed, very large scale integrated circuitry, fault-tolerant processing, and high-order language technologies to achieve a high degree of functional redundancy.

Because the Pave Pillar application is representative of most autonomous multiple sensor systems, the architecture is described in some detail to illustrate the integration approach and partitioning of functions to processors.

10.5.1 Integrated Sensors

Before describing the system architecture, we briefly summarize the characteristics of the primary integrated sensors subsystems.

Integrated CNI

The integration of CNI functions [22–24] offers sensor level functional redundancy when common RF (receivers, exciters, transmitters, *et cetera*) and signal processing elements can be designed to perform multiple waveform functions in an architecture that permits switching and sharing of elements. The CNI functions in the 30 to 1500 MHz bands are excellent candidates for such common processing and element sharing because of the waveform and processing similarities of these functions: HF, VHF, and UHF voice communications; tactical data link communications; IFF; area navigation data links (TACAN, distance measuring equipment); and satellite navigation data links (e.g., global positioning satellite). Proposed integrated architectures are characterized by common multiband receivers, signal preprocessors, *communications security* (COMSEC) elements, and data processors that can be programmed to share resources and adapt to mission loading for these CNI functions. Some data fusion functions that may be allocated to an integrated CNI sensor to reduce sensor input-output to a mission processor include

- Use of data link information to perform sensor management (cueing) of IFF to acquire and validate friendly targets.
- Maintenance of sensor-level tracks for friendly IFF and data link targets. This can include IFF–data link association to maintain a common track file.
- Fusion of navigation data from satellite, area navigation, and data link sources to refine navigation state variable estimates.

Integrated Electronic Warfare System

In a manner similar to that described for CNI functions, the higher band electronic warfare functions (extending from 3 GHz to millimeter-wave frequencies and even including optical threat bands) can also be integrated [25]. Electronic support measures preprocessing functions (described earlier in Chapter 4) as well as some data fusion functions may be allocated to such an integrated sensor:

- Maintenance of sensor-level emitter tracks (both bearings-only and passively ranged tracks or static target locations).
- Sensor-level situation assessment, based on sensor-level track-file data for ESM sensor management (e.g., ESM sensor-sensor cueing).

The allocation of some Level 1 data fusion functions to integrated sensors (tracking, sensor management) may reduce the communication traffic between sensors and the central data fusion processor, but does not preclude control and override by the central processor.

10.5.2 Integrated Avionics Architecture

The studies resulted in a generic avionics architecture (Figure 10.8), which is characterized by the following features of integration:

1. Resource sharing is permissible between sensor and mission processing functions to provide functional redundancy at the highest level of the system. The failure of preprocessing on any one sensor, for example, can be replaced by

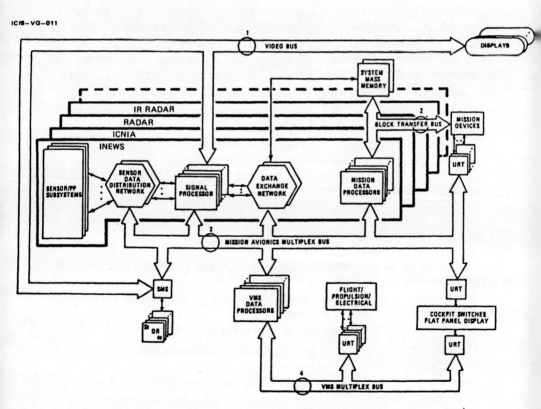

Figure 10.8 Pave Pillar avionics architecture.

redundant signal processors servicing other sensors of lesser importance or not being used at the current time in the mission.
2. Modularity is achieved using common hardware modules, which are used across all system applications to minimize the number of unique modules to be designed and spared. This also provides module-level redundancy (e.g., receivers, computing elements) through using switch networks or busses between modules.
3. Degradation in performance can be controlled and graceful as redundant resources are adapted throughout the flight as a function of changing mission requirements.
4. All major functions are connected by a common wideband bus and share access to a single mass memory. A systemwide operating system governs all systems, effectively operating as autonomous (loosely coupled), distributed processors in data-level communication.
5. A variety of interconnection methods are used to achieve the high degree of resource accessibility necessary for resource sharing. The methods include switched (matrix) data networks, data busses, shared memory and video switches. Klos [26] and Hoyt [27] have summarized the various approaches to implement interconnection networks in avionic systems that generally are more applicable to most data fusion systems applications.

The principal elements of the architecture are now described with specific emphasis on the methods by which data fusion processing can be performed in such a structure. Although Pave Pillar has a specific assignment of functions as well as implementation constraints, this discussion will describe the more general methods by which such a highly networked architecture can implement data fusion.

Sensor Subsystems

The elements of all sensors, prior to signal processing, are included in the sensor subsystem. These elements include optics and detectors, antennas and RF components (receivers, transmitters), and other physical apertures and detection devices that result in electrical signals related to the detected phenomena. In many sensors, the preprocessing functions of low-level signal conditioning are also included within the sensor because the physical distance from the aperture to the sensor data distribution network may prohibit the transmission of low-level signals. Conditioned signals are digitized and time tagged if necessary for transmission to the sensor data distribution network. Within integrated sensors, the common elements may include muliplex paths to permit interchanging functions among the common elements for resource sharing (e.g., arrays of multimode receivers with input-output switching to permit reassignment of reception tasks).

Sensor Data Distribution Network

This bidirectional network permits digital data from any sensor to be transferred to any *signal processor* (SP), as well as permitting any processor to route control messages to the sensors. Although sensor management is coordinated by the mission data processors, the control messages must be routed through the SPs to the appropriate sensors. Redundancy in the distribution network may be provided to safeguard against catastrophic single-point network failures that would lead to a failure of all sensors. The sensor-signal processor assignments can be changed to accommodate several categories of reconfiguration:

1. Failure of a sensor, in which case an SP becomes available as a spare to back up other sensors or to support modes of other sensors that partition processing tasks across multiple SPs.
2. Reconfiguration for mission modes during which sensor-SP assignments may be changed to allow some sensors to use the resources of several SPs while other sensors may be not operating.
3. Failure of an SP, which causes reassignment on the basis of sensor priority access to SP resources.

Signal Processors

The sensor SPs perform the functions described in Chapter 4, which convert raw sensor data into attribute and location measurements suitable for transfer to the mission processors (for data fusion) and video displays (imagery). The use of common digital processors [28] provides a small set of common modules that provide the desired reconfigurability and reduction in unique modules across all sensors. These common modules include several basic signal processing functions, interfaces, and memory elements:

- Standard fixed-point computer (i.e., MIL STD 1750A instruction set processor) for general purpose data processing, scalar arithmetic, and logical operations.
- Floating point processor for digital signal processing.
- Global memory.
- Bus interface and sensor network interface modules.
- Internal data network module to interconnect modules.
- Supervisory control modules to control all modules.

The SPs may also perform security functions (e.g., communications security, transmission security, requiring separation of encrypted (red or cyphertext) data and unencrypted (black or plain text) data paths and functions in accordance with security design guidelines. These functions are required for secure communication links.

Video Bus to Displays

The SPs processing imagery can provide processed sensor video output data to two locations:
1. Pilot displays for visual representation of sensed data. This may include overlaid video from multiple sensors where the imagery has been registered.
2. Stores management system for cueing weapons that require input imagery cueing weapons to designate targeting.

Stores that provide image data (e.g., EO-IR seekers on missiles) may also provide data that may be routed to the pilot displays, although the resolution will generally be poorer than that of prime sensors.

Data Exchange Network

The exchange of data between SPs, mass memory, and the mission avionics multiplex bus is performed by this network. The network also permits SPs to exchange processed sensor data to perform sensor-sensor fusion functions, such as image registration, track association, or classification, if beneficial. The access to mass memory permits downloading SP programs for task assignment upon initialization and for reassignment upon reconfiguration.

Mission Avionics Muliplex Bus

This bus is common to the complete fusion data path and permits the mission data processors to control the sensors, the allocation of SP resources to sensor processing, and the interconnection of all SPs. The bus provides a common path for processed sensor data to the mission processors as well as control data to sensors, SPs, and switch networks.

Mission Data Processors

These processors permit multiple, independent fusion processes to occur if beneficial. Navigation fusion, air-air target data fusion, air-ground target data fusion, and terrain-following sensor data blending (fusion) are among the candidate functions that could be performed independently by parallel mission data processors.

Vehicle Management System (VMS)

The VMS monitors and controls the flight, propulsion, and electrical-hydraulic systems as well as other utilities aboard the aircraft. It provides status display data to the pilot and accepts and processes manual inputs from the pilot.

Stores Management Subsystem (SMS)

The SMS controls all stores (weapons) including the transfer of targeting information and weapon release.

10.6 DATABASE AND PROCESSING PARAMETRIC REQUIREMENTS

The previous sections have described design implementation issues and alternatives, but also important is to define, for any given application, the parametric requirements imposed on database and processing systems. The methods of defining and quantifying those requirements are summarized in this section and with parametric curves for representative applications.

Figure 10.9 summarizes the principal factors that contribute to the database and processing requirements in typical fusion systems. Application parameters generally are directly stated as mission-level requirements whereas algorithm factors are derived from mission-level requirements, such as tracking or classification performance. These factors can be directly related to the processing performance parameters, once the system has been functionally defined. The performance parameters can then be related to computational parameters (e.g., operations per

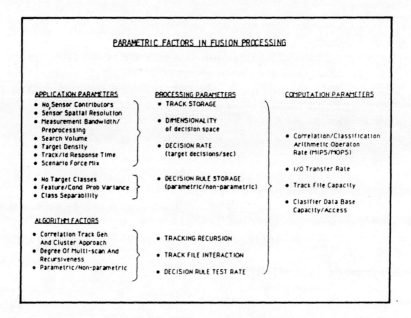

Figure 10.9 Major operational parameters that flow down to database, processing parameters.

second, storage capacities in words, transfer rates) once the physical system architecture has been defined by partitioning the functions into hardware and software components.

Figure 10.10 illustrates a general throughput model structure to estimate computational requirements (normalized number of operations per second, in terms of instructions per second, floating point operations per second, logical inferences per second, *et cetera*) for a system in which each function's contribution to processing is independently modeled. This model assumes a single processor architecture (no parallelism) with common access to the data base by all functions. Latency values are included in the appropriate functional models.

The data association-tracking model estimates *operations per second* (OPS) as a function of the number of sensors, targets, and the rate at which sensor reports are received, R. This report rate is a function of the revisit rate of each sensor (intentional looks by active sensors or intercepts by passive sensors of intermittent target emissions). The association-tracking model determines OPS required for all hypotheses associated, updated, and estimated for each sensor report received. This is a function of the target density model (generally, a statistical distribution), which directly influences the number of spatially "close" targets that may require many association tests and the maintenance of multiple hypotheses.

The track file database management model determines the number of OPS required to sustain the processing of the report rate, which is a function of the required access to report and target data by both the association-tracking and combination functions. The data combination model performs an estimate of target class for each sensor report and the required OPS is primarily a function of the number of class hypotheses computed for each report. Sensor management actions are modeled as a function of the number of sensors to be managed and the number of targets to be serviced.

The symbolic processing operations, assumed in this model to be backward or goal-directed searches through the data base to perform situation assessment and infer possible threats, are a function of the number of targets and the number of searches estimated per target. A conversion factor, J, between *logical inferences per second* (LIPS) and OPS is applied to normalize the computational requirements for common machine for all processing.

Such a model permits input variables to be adjusted to develop parametric curves relating system-level performance requirements to computational loads. Such computational models are useful in the functional allocation processes described in section 10.1, where alternative design approaches may be modeled and analyzed to support physical design trade-off studies. The results of computational models can guide the selection of processing architectures, data structures, and physical components (e.g., processors, memory, busses, and switches). Figure 10.11 illustrates a comparison of Bayesian and Dempster-Shafer processing requirements as a function of sensor report rate (R), number of sensors (N), and number of

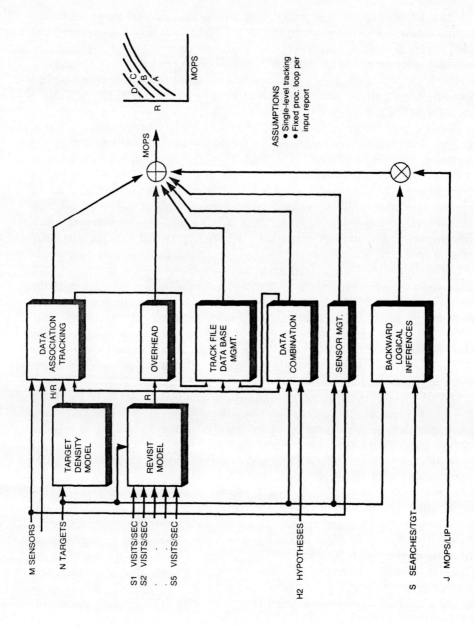

Figure 10.10 Basic data fusion throughput model.

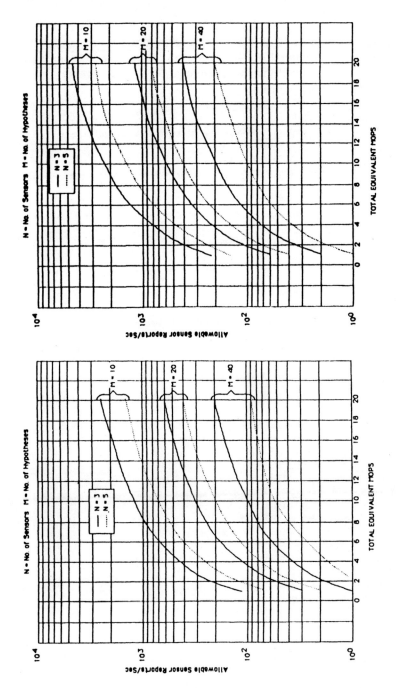

Figure 10.11 Comparison of Bayesian and Dempster-Shafer computational requirements.

hypotheses computed (M). In this case the number of iterations per report is a function of the product of M and N, whereas the iteration rate is then a function of R. The operations per iteration are based on memory access-storage operations as well as numerical computations required to convert sensor measurement vectors into combined hypothesis vectors.

REFERENCES

1. MIL-STD-490, *Specification Practices,* June 4, 1985.
2. MIL-STD-499A (USAF), *Engineering Management,* May 1, 1974.
3. Antony, R., "A Framework for Automated Tactical Data Fusion," *Proc. 1987 Tri-Service Data Fusion Symp.,* Vol. 1, pp. 380-396.
4. Ullman, J., *Principles of Database and Knowledge-Base Systems,* Vol. 1, Computer Science Press, Rockville, MD, 1988.
5. Ceri, S., and G. Pelagatti, *Distributed Databases, Principles and Systems,* McGraw-Hill, New York, 1984.
6. Parsaye, K., M. Chignell, S. Khoshafian, and H. Wong, *Intelligent Databases,* John Wiley and Sons, New York, 1988.
7. Miller, D., and L. Klos, "Pave Pillar Avionics—An Architecture for the Future," *IEEE Proc. NAECON 1984,* pp. 156-162.
8. Fathi, E.T., and M. Krieger, "Multiple Microprocessor Systems: What, Why and When?" *IEEE Computer,* March 1983, pp. 21-32.
9. Blackman, S.S., *Multiple Target Tracking with Radar Applications,* Artech House, Norwood, MA, 1986, p. 290.
10. Hall, D., and J. Llinas, "Data Fusion and Multisensor Correlation," course notes, Technology Training Corp., 1987, p. 327.
11. Denning, P.J., "Parallel Computing and Its Evolution," *IEEE Comm. of ACM,* Vol. 24, No. 12, December 1986, pp. 1163-1167.
12. Rosen, J.A., and P.S. Shoenfeld, "Parallel Processing Applications to Multi-Source Correlation and Tracking," *Proc. Data Fusion Symp. 1987.* Vol. 1, pp. 206-213.
13. Gurd, J.R., *et al.,* "The Manchester Prototype Data Flow Computer," *IEEE Comm. of ACM,* Vol. 28, No. 1, pp. 34-52.
14. Frenkel, K.A., "Evaluating Two Massively Parallel Machines," *IEEE Comm. of ACM,* Vol. 29, No. 8, pp. 752-758.
15. Fahlman, S. E., and G.E. Hinton, "Connectionist Architectures for Artificial Intelligence," *IEEE Computer,* January 1987, pp. 100-108.
16. Lippman, R.P., "An Introduction to Computing with Neural Nets," *IEEE ASSP Magazine,* April 1987, pp. 4-22.
17. Marchette, D., and C. Priebe, "An Application of Neural Networks to a Data Fusion Problem," *Proc. Data Fusion Symp. 1987,* Vol. 1, pp. 230-235.
18. McCurry, M.E., "Neural Network Implementation of a Scan-to-Scan Correlation Algorithm," *Proc. SPIE High Speed Computing,* Vol. 880, 1988, pp. 85-87.
19. Bowman, C., "Artificial Neural Network Approaches to Target Recognition," IEEE Digital Avionic Systems Conf., 1988, pp. 847-857.
20. Feldman, S., "Real-Time Automated Track Management," *Proc. 1988 Tri-Service Data Fusion Symp.,* May 1988, pp. 298-305.
21. Ostgaard, J.C., and R. Szkody, "Pave Pillar Avionics Designed for Dependability," *Defense Electronics,* May 1986, pp. 79-89.

22. Harris, R.L., "Technology Considerations of Integrated Communication Navigation Identification Avionics (ICNIA)," IEEE, *Proc. 1984 NAECON,* pp. 1138–1148.
23. Ropelewski, R.R., "USAF, Army Plan ICNIA Flight Testing by 1987," *AW&ST,* October 4, 1985, pp. 125-131.
24. Harris, R.L., and D.L. Howell, "Modeling Mission Reliability of Advanced Communication, Navigation and Identification Avionics Systems," AIAA/IEEE, *Proc. 8th Digital Avionics Syst. Conf.,* 1988, pp. 815–820.
25. Schultz, J.B., "INEWS Ignites Technology Tradeoffs and Competition," *Defense Electronics,* September 1985, pp. 136–144.
26. Kloss, L., "The Switched Network Approach to High Speed Communications," IEEE Digital Avionics Systems Conf., 1984, pp. 114–118.
27. Hoyt, R.R., and W.J. Kenny, "A Network Architecture," IEEE Digital Avionics Systems Conf., 1984, pp. 135–141.
28. Lee, W.H., "The Common Signal Processor: First All-VLSI Signal Processor for DoD," *Technical Directions,* IBM Federal Systems Div., September 1985, pp. 1–9.

Chapter 11
SYSTEM MODELING AND PERFORMANCE EVALUATION

The functions of any element of a complex system must be understood, modeled, and quantitatively evaluated to determine the contribution that it provides to the effectiveness of the overall system. This process of quantitative assessment is often applied to the data fusion elements of command, control, and communication systems to determine the contribution that fusion provides to the military effectiveness of those systems.

These assessments are required to answer important system-level issues that are frequently raised, such as,

- What is the best combination of sensors and sources to meet a given set of detection probability, target discrimination, and target location requirements?
- What level of detection, discrimination, and location performance can be achieved by fusing a given set of existing sensors? What improvements are accrued by adding sensors or improving the performance of individual sensors?
- What is the relative contribution to military effectiveness of various candidate data fusion system configurations?
- What trade-off must be made between improvements in information performance (due to data fusion) and weapon performance? (For example, should X dollars be spent on data fusion to extend a weapon system's detection range by 40 miles or should those same dollars be spent to increase the weapon's kill probability by .15 through warhead improvement?)

In this chapter we will describe the methods to model data fusion systems, define measures of merit to quantify performance and effectiveness, and conduct tests and analyses to explain the results. This chapter introduces the *methodology* of systems analysis, specifically as applied to data fusion systems, but is not a substitute for the numerous references that are necessary to develop the finer points of analysis. The reader is cautioned that the relationship beween the data fusion function and its contribution to improved military value is complex. Therefore, serious analyses must demonstrate an appropriate selection of the modeling approach, a thorough understanding of all contributing factors and functional relationships, attention to detail, an ability to relate results to real-world data to validate accuracy of model elements, and sound judgement in evaluating the meaning and extent of results.

11.1 THE BASIC THEORY OF C^3 SYSTEMS

Fundamental models of warfare and the contribution of command, control, and communication (including data fusion) have been developed to characterize the behavior of various elements of military operations. Indeed, warfare is one of the most complex of all human endeavors to model because of the presence of the unexpected, the deliberate use of countermeasures and deception, and behavior of humans operating under emotional duress. In spite of these difficulties, mathematical models have successfully represented warfare and operational research studies [1, 2] and have validated the general ability of these models to describe the functional behavior of actual combat.

11.1.1 Lanchester Models of Combat

The classic representation of the attrition of opposing forces is provided by Lanchester's [3] differential equations. These equations relate the size of each force in time as a function of attrition rate coefficients that describe the effectiveness of each force's ability to wear down the opposing force:

$$dx/dt = -ay, \quad y(t = 0) = \text{initial size of force } y \quad (11.1)$$
$$dy/dt = -bx, \quad x(t = 0) = \text{initial size of force } x \quad (11.2)$$

where

$a = $ attrition rate of x produced by y;
$b = $ attrition rate of y produced by x.

The attrition rate coefficients a and b describe the total effectiveness of each force relative to its size and include the effects of firepower as well as command and control. In this basic model there are numerous assumptions such as homogeneity of forces and their relative effectiveness.

This basic model permits the attrition of opposing forces to be studied and compared to actual combat data to assess the factors that contribute to higher attrition coefficients [4]. Quantitative analyses of combat behavior usually plot force sizes as a function of time, showing the attrition of each side as combat progresses. These graphs are referred to as *draw-down curves*.

Because the two equations are coupled, a simultaneous solution is required to directly relate the two force sizes, x and y, at any point in time. The simultaneous solution results in the relationship known as Lanchester's *square law*:

$$b(x_0 - x^2) = a(y_0 - y^2) \quad (11.3)$$

where

x_0 = size of x force at $t = 0$;
y_0 = size of y force at $t = 0$.

This equality shows that the strength of a combat force is proportional to the square of the force size entering the engagement. Additionally, the required initial force difference required to win an engagement can be derived as a function of the two attrition coefficients. This simple model uses the single measure of force effectiveness, attrition, to lump together the many individual factors that contribute to a force's ability to wear down ("attrite") the opposing force. We now consider ways to decompose the single coefficient into these more detailed contributing factors.

Taylor [4] has summarized the general methods to refine the model further to consider heterogeneous forces, aggregated force models and the approaches to include additional factors such as command and control in the attrition rate coefficients. Following Everett's approach [5], we can show the first-order effects of command and control on combat by expanding Lanchester's equations to include coefficients that introduce three important contributions of C^2. These three factors include the accuracy, resolution, and speed with which each side is capable of determining the location of opposing forces. The equations can be expanded to represent these effects by the addition of modifying coefficients:

$$dx/dt = -F_y \left[\frac{\prod_k R_{ky}^2}{\prod_l (K_{3y} R_{ly})^2} \right] Y \cdot K_{1y} \left(\frac{1}{K_{2y}} \right) \quad (11.4)$$

$$dy/dt = -F_x \left[\frac{\prod_k R_{kx}}{\prod_l (K_{3x} R_{lx})^2} \right] X \cdot K_{1x} \left(\frac{1}{K_{2x}} \right) \quad (11.5)$$

where

F_y = rate of fire of y;
R_{kx} = radius of lethality of weapon k of force x;
R_{lx} = radius of sensor measurement area;
K_{1x} = attrition rate improvement for force x due to improved targeting by target acquisition sensors;
K_2 and K_3 = attrition rate degradations due to delay in sensor measurements, for each force.

In this equation, the attrition coefficients have been replaced by the product of the rate of fire (F, a measure of force capability with time) and the single-shot kill probability (the ratio of the weapon's lethal area to measurement error area). In addition, the coefficients K_1 and K_2 have been added to adjust the effectiveness of the attrition rate. Everett has shown, in this simple case, three quantitative contributions of targeting data on force effectiveness:

- *Measurement accuracy* improves force effectiveness by the square of the reduction in measurement error. In the context of Lanchester's square law, the strength of the combat force is related to the square of it's sensor's targeting accuracies as well as the square of its force size.
- *Resolution* of targeting information effectively improves the rate of fire by allowing firepower to be placed on higher concentrations of targets, when targets are not homogeneous. The coefficient, K_1 multiplies the fire rate by the effective reduction in shots required to kill a single target within the sensor's measurement data (for multiply resolved targets that are concentrated).
- *Time delay* in detecting and reporting targets degrade the effectiveness of firepower in two ways: (1) delays in kill assessments cause rounds to be wasted on destroyed targets expressed by K_2 and, (2) delays in reporting the locations of moving targets has the same effect as increasing measurement error, as expressed by the K_3 multiplier.

These modifications to Lanchester's basic model show the important contribution of target information provided by data fusion systems to force effectiveness. The basic model even allows the quantitative analysis of the influence of fusion performance factors on the effectiveness of military forces. Figure 11.1 provides an example draw-down graph for two opposing forces of equal size, but with differing levels of C^3 performance. The figure shows the vivid improvement in combat performance from superior C^3.

The application of advanced numerical computing techniques to greatly expanded forms of Lanchester's equations are reported by Dockery and Santaro [6], in which four aspects of C^2 are recommended for model fidelity improvements:

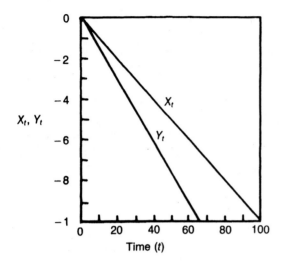

Figure 11.1 Effects of C³ on combat attrition.

1. Use of stochastic ordinary differential equations to model noise and randomness.
2. Use of fuzzy set variables to model uncertainties.
3. Introduciton of time and space dependence to loss rates to model the effects of movement.
4. Replacement of the homogeneous with heterogeneous forces capable of complex competitive-cooperative interactions.

11.1.2 Command, Control, and Communication Models

More complex functional models of the entire combat process have been proposed for explicitly representing the effects of command, control, and communications on the combat process. Whereas the Lanchester models tend to lump C³ and firepower into a common coefficient, these models attempt to describe individually the interaction of the numerous C³ functions, including:

- *Sensing*—The methods used to search, acquire, identify, and track targets including both one's own and opposing forces.
- *Communications*—The links between one's own forces to communicate the location and status of each other's forces as well as the transmission of sensor and intelligence data ("crosstell").

- *Processing*—The fusion of sensor and source data to create an accurate assessment of the combat environment.
- *Command*—The assessment of the possible meanings (military implications) of the situation: hypothesizing and assessing alternative courses of action and deciding on the most favorable alternatives.
- *Control*—The development and dissemination of tasking orders to forces under control.

Lawson [7] and Rona [8] have summarized several approaches to modeling the C^3 process and the issues involved in accurately representing the interactions between the functional elements of C^3, including data fusion. Wohl's model [9], presented earlier in Chapter 8, is a basic model of the stimulus-response relationships in C^3 systems. Three functional models are now described to show candidate approaches to represent and quantify the individual processes in more detail than Wohl's model.

Thermodynamic Model

Lawson proposed a model that divides the C^3 process into sense, process, compare, decide, and action functions (Figure 11.2), which sequentially process information in a closed loop that proceeds from sensing the combat environment to the actions placed upon it by own forces. The sense function (S) includes the collection of information about own and hostile forces via sensors as well as receipt of communications from one's own forces reporting their actual status, location, and plans. The process function (P) includes fusion of the sensed data and assessment of the situation to develop hypotheses of the actual conditions of the combat environment. In this function, additional external data may be supplied, including *a priori* order of battle information, attributes of forces, fighting doctrine, *et cetera*. A comparison (C) of the current situation is made with the desired situation to provide alternative responses that will correct the differences between the current state and the desired state. The command function then decides (D) on the course of action to take and issues the orders to its own forces to act (A). The figure shows how multiple C^3 loops may be networked: (1) to allow a higher-level commander to issue desired state (command) information to subordinate commanders, and (2) to permit coordination of multiple forces to meet a common objective by coordinating the decision process (dotted line).

A general representation of state is suggested by Lawson, which follows the form of thermodynamic gas law equations of state for pressure applied to a volume. Because the military pressure (P) is applied to the volume of the environment (V), the product of PV is proportional to the number of forces (N), multiplied by their effectiveness (K) and tempo of operations (T):

$$PV = (NK)T \qquad (11.6)$$

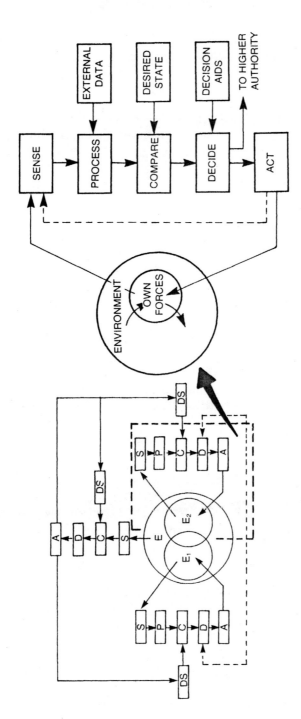

Figure 11.2 Thermodynamic model of C^3 by Lawson [7].

Lawson equates the pressure applied by each side to the attrition rate of the opponent to derive the basic Lanchester equation:

$$dX/dt \; V = yK_xT_x \tag{11.7}$$

In this form, Lawson relates the effects of C^3 by showing that the tempo influences the attrition rate linearly whereas time delays have a quadratic influence.

Canonical Model

In this model developed by Rona [8], a multidimensional transform operator converts an input set of stimuli into a resulting military effect. The transform operator (Figure 11.3) includes the provision for feedback from the results of actions, allowing for learning within the operator. The transformation is initialized by the mission objective and output stimulus selection (sensor management) and control. The stimulus and effect links are the sensor-fusion and C^2-weapon systems, respectively, which link the transformation to the stimulus and effect state vectors.

State Vector Model

Another representation offered by Zracket [10] follows the state space methods of control theory (Figure 11.4). The combat process is modeled by a dynamic state equation (which is similar to Lanchester's equation):

$$\dot{\mathbf{X}} = F(\mathbf{x}, \mathbf{u}_1, \mathbf{u}_2, \mathbf{w}, \mathbf{t})$$

where

$\mathbf{x}$ = state of both forces' location, size, and mission;
$\mathbf{u}_1, \mathbf{u}_2$ = control states of each force;
$\mathbf{w}$ = random influencing factors (weather, communication, sensing noise, *et cetera*);
$\mathbf{t}$ = time.

The true state of the environment ($\mathbf{X}$) is measured by each side's sensors, which communicate their measurements to the command and control functions. The C^2 function transforms sensed data into control states ($\mathbf{U}_n$). These are then communicated to forces and the difference is entered into the combat process model. The C^2 function also generates sensor management commands to its own sensors (included in $\mathbf{U}_n$) and issues electronic countermeasures ($\mathbf{V}$) that influence the opponent's sensors.

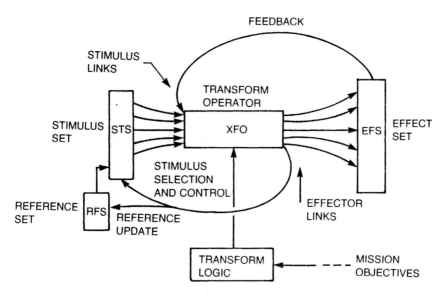

Figure 11.3 Canonical model of C^3 by Rona [8].

The purpose of these models is to develop a means of functionally and quantitatively describing the influences of command, control, and communication (including data fusion) on the combat process. These descriptions are essential to the development of meaningful tools that can relate the value of accurate information to the effectiveness of combat forces against hostile forces.

11.2 FORMAL MODELS OF THE DATA FUSION PROCESS

The work of Goodman [11] and Goodman and Nguyen [12] has attempted to model the data fusion process within C^3 systems as a combination of evidence problem in which information (the evidence), represented with quantitative uncertainty, is combined by a mathematical operator using classical, multivalued, or general logic. The emphasis of these efforts is to specify the set of primitive mathematical operations necessary to combine evidence and develop formal algebraic logics to implement fusion.

The mathematical functions proposed by Goodman to describe the functions of data fusion are

1. Mental imaging and cognition processes accept the inputs of sensor displays (visual) and human communications (aural), recognizing the meaning of inputs and translating them into operator actions (or inaction).

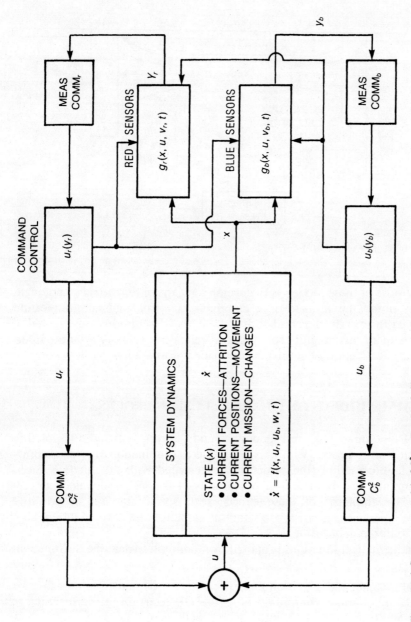

Figure 11.4 State space model of C^3 by Zracket [10].

2. Natural-to-machine language transformation converts human perceptions to machine inputs. This process includes the machine processing of human narratives (e.g., intelligence and activity reports) to parse and categorize data.
3. Symbolic formulation assembles the incoming data (both directly from automatic sensors and from natural-to-machine language conversion) into strings that meet the criteria of machine processing syntax. The primitive strings include data assembled with the typical formal connectors: conjunction (and), disjunction (or), negation (not), and implication (if-then).
4. Formal language formulation then applies the rules of syntax to the accepted data to prepare data for evaluation. This process brings the data to a common format.
5. Semantic evaluation is finally performed to compute hypotheses with quantitative values for a subsequent decision process.

The final two stages of this sequence are the focus of Goodman's work, which has been used for mathematically evaluating candidate algebraic logic descriptions to represent (language formulation) and combine (semantic evaluation) the evidence obtained directly from sensors and indirectly from human inputs. These candidates have included [11] classical Boolean algebra, fuzzy logic, probability logic, and a conditional probability logic.

11.3 ANALYSIS OF DATA FUSION SYSTEM PERFORMANCE

The classical systems analysis approach may be applied to the analysis of data fusion systems to compare alternative approaches (sensors, algorithms, architectures) or estimate the relative merits of candidate systems. Applying this analysis methodology includes five essential steps.

11.3.1 Definition of Objective

The objective of the analysis must be clearly stated in terms that specify the quantitative and qualitative information to be obtained, decisions to be made as a result of analysis, and the criteria for decision making, scope of the analysis, independent and dependent variables, and basic assumptions. An objective of a typical data fusion analysis example follows:

Example: Compare and rank the target detection-location capabilities of candidate data fusion systems that may use a four-sensor suite, employing sensor types S1, S2, S3, S4, S5, or S6, and fusion algorithms A1 or A2.

Determine the relative performance for both jamming and benign environments with jamming effects to equally degrade sensors S1, and S5 in detection capability. Consider four representative scenarios, which range from large homogeneous target distributions to small concentrated target clusters.

Assume that all sensors are collocated and that there is no delay between measurement time and report time for sensors S1, S2, S4, S5, and S6. Assume a 5 second delay between S3 measurement and report time.

11.3.2 Construct Alternatives

Using the available system resources, candidate data fusion system architectures are constructed. The resource variables (usually sensors, sensor performance levels, sensor locations, and processing techniques) usually provide a large number of possible combinations; and a subset of candidates is intelligently selected to represent the primary categories of architectural alternatives.

Example: For the preceding example, a set of system candidates may be defined by the following combinations of sensors, data links, and data fusion algorithms:

System Candidate A: S1, S2, S3, S4, A1
System Candidate B: S1, S3, S5, A2
System Candidate C: S1, S5, S6, A2

11.3.3 Establish Evaluation Criteria

Criteria for evaluating the various alternatives must be defined quantitatively in the form of measures of merit that can be determined for each candidate. The measures must allow discrimination between alternatives and be appropriate to answer the questions stated in the analysis objective.

Example: "Target detection-location performance" in the example objective could be quantified by the following standard measures:

Detection probability: $Pr(D)$

False alarm rate: $Pr(FA)$

Detection range as a function of $Pr(D)$: $R1 = F[Pr(D)]$

Target location accuracy as a function of detection range: $r = F(R)$

11.3.4 Develop Modeling Approach

The candidate systems must be modeled by defining inputs, process models and outputs (the measures.) The typical data fusion model architecture is depicted in Figure 11.5, showing the input variables that control the behavior of targets (both

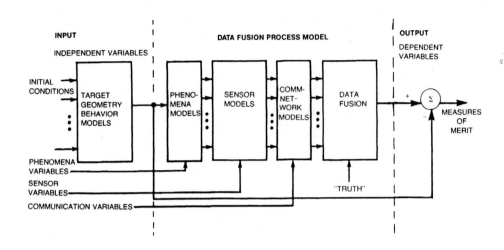

Figure 11.5 Data fusion model architecture.

kinematics and signals in space) and vary parameters within the process models. Target behavior models transform true behavior (*truth* is the actual location and activities of all participants) into variables that change with time. The phenomenology model transforms "truth data" into phenomena variables that are entered into the sensor models. Sensor models, in turn, transform phenomena variables (e.g., signal and noise energies) into sensor outputs (e.g., target detections), which are passed through communication network models to a data fusion process model. This model displays estimates (e.g., target locations, identifications) that are then compared with truth data to compute the measures of merit.

An important aspect of the modeling process is the development of scenarios representative of the real world. The scenario variables must include all critical entities, behaviors and events that influence the outcome of the analysis. Because these variables are usually numerous (target geometrics, electronic environments, target quantities and mixes, physical environments effecting sensor and communication performance, *et cetera*, the selection of a manageable set of truly representative scenarios is critical to the meaningfulness of the modeling process.

The combinations of candidates to be evaluated, scenarios to be applied and other independent variables define the test cases to be analyzed and are often expressed in a test matrix (Figure 11.6). For each case ("bin") in the matrix, the set of independent variables ("factors") is defined and the model determines the mea-

SCENARIO	1		2		3		4	
B/J \ CANDIDATE	B	J	B	J	B	J	B	J
A	1	2	3	4	5	6	7	8
B	9	10	11	12	13	14	15	16
C	17	18	19	20	21	22	23	24

B/J = BENIGN/JAMMED ELECTRONIC ENVIRONMENT

TEST CONDITION 15 = SCENARIO 4 (BENIGN) CANDIDATE B

Figure 11.6 Sample test matrix.

sures of merit unique to that case. In the example shown in the figure, eight sets of measures are computed for each candidate.

11.3.5 Analysis and Results

The analysis is performed by running the model for each case (full factorial experiment) or for carefully selected cases (partial factorial experiments) in the test matrix. The numerical results are then analyzed to determine the statistical significance of the results, often using the method of *analysis of variance* (ANOVA). In this method [13], a null hypothesis (which states that no difference exists between the populations of results for different treatments) is tested to determine the probability of rejection of the hypothesis as a function of the variances in the results of the experiments (e.g., analyses or simulation runs).

The analysis of data also includes the evaluation of results to

- Define any uncertainties in the outcome.
- Specify the significance of marginal differences between performance in candidates as a function of input variables.
- Explain relations between output and input variables to understand the effects of complex processes within the model.
- Specify the sensitivity of measures of merit to input variables (scenarios, sensor suites, environment, algorithms, *et cetera*).
- Recommend a candidate and provide the rationale for selection.

11.4 RELATING FUSION PERFORMANCE TO MILITARY EFFECTIVENESS

Because sensors and fusion are contributors to improved information accuracy, timeliness, and content, a major objective of many fusion analyses is to determine the effect of these contributions to military effectiveness. This effectiveness must be quantified, and numerous quantifiable measures of merit can be envisioned: engagement outcomes, exchange ratios (the ratio of blue-red targets killed), total targets serviced, and so on. The ability to relate data fusion performance to military effectiveness is difficult because of the many factors that relate improved information to improved combat effectiveness and the uncertainty in modeling them. These factors include

- Cumulative effects of measurement errors that result in targeting errors.
- Relations between marginal improvements in data and improvements in human decision making.
- Effects of improved threat assessment on survivability of own forces.

These factors and the hierarchy of relationships between data fusion performance and military effectiveness must be properly understood to develop measures and models that relate them. The Military Operations Research Society [14] has recommended a hierarchy of measures that relate performance characteristics of C^3 systems (including fusion) to military effectiveness (Table 11.1).

Dimensional parameters are the typical properties or characteristics that directly define the elements of the data fusion system elements (sensors, processors, communication channels, *et cetera*). They directly describe the behavior or structure of the system and should be considered to be typical measureable specification values (bandwidths, bit-error rates, physical dimensions, *et cetera*).

Measures of performance (MOPs) are measures that describe the important behavioral attributes of the system. MOPs are often functions of several dimensional parameters to quantify in a single variable a significant measure of operational performance. Intercept and detection probabilities, for example, are important MOPs that are functions of several dimensional parameters of both the data fusion system and the targets being detected.

Measures of force effectiveness (MOFEs) are the highest level measures that quantify the ability of the total military force (including the data fusion system) to complete its mission. Typical MOFEs include rates and ratios of attrition, outcome of engagements, and functions of these variables. In the overall mission definition, factors other than outcome of the conflict (e.g., cost, size of force, composition of force) may also be included in the MOFE.

Figure 11.7 depicts the relationship between these measures for a two-sensor system, showing the typical functions that relate lower-level dimensional parameters upward to higher level measures. In this example, sensor coverages (spatial and

Table 11.1
Four Categories of Measures of Merit

Measure	Definition	Typical Examples
Measures of Force Effectiveness (MOFE)	Measure of how a C^3 system and the force (sensors, weapons, C^3 system) of which it is a part perform military missions	Outcome of battle Cost of system Survivability Attrition rates Exchange ratio Weapons on targets
Measures of Effectiveness (MOE)	Measure of how a C^3 system performs its functions within an operational environment	Target nomination rate Timeliness of information Accuracy of information Warning time Target leakage Countermeasure immunity Communications survivability
Measures of Performance (MOP)	Measures closely related to dimensional parameters (both physical and structural) but measure attributes of system behavior	Detection probability False alarm rate Location estimate accuracy Identification probability Identification range Time from detect to transmission Communication time delay Sensor spatial coverage Target classification accuracy
Dimensional Parameters	The properties or characteristics inherent in the physical entities whose values determine system behavior and the structure under question, even when not operating	Signal-to-noise ratio Bandwidth, frequency Operations per second Aperture dimensions Bit error rates Resolution Sample rates Antijamming margins Cost

frequency), received signal-to-noise ratios, and detection thresholds define detection and false alarm rate MOPs. Sensor coverages and measurement accuracies of each parameter sensed are entered into a geometric model to derive target location accuracy MOPs. Probability of detection *versus* target range is another MOP that quantifies the warning time provided one's own forces prior to coming within lethal range of hostile forces. The improvements accrued by fusing the two sensors' data are then modeled to provide three MOPs:

- Detection probability and false alarm probability as a function of target location,

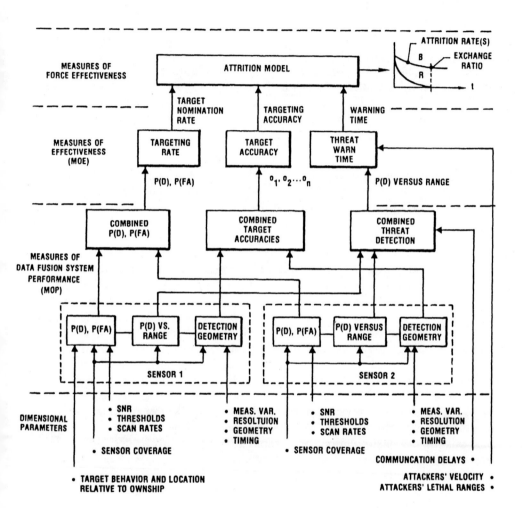

Figure 11.7 Relations among measures of merit.

- Location accuracy as a function of target location,
- Detection probability as a function of range of hostile targets to one's own forces.

These MOPs are then used to derive three MOEs that quantify the speed and accuracy with which targets can be engaged and the time that one's own forces have to evade or engage hostile forces. These MOEs then influence the ultimate attrition model that provides attrition rates and exchange ratio (when one force reaches zero) as MOFEs.

11.5 DATA FUSION SYSTEM MODELING CONSIDERATIONS

The objective of any model is to accurately represent the system under study to better understand its behavior, evaluate various operating strategies, and quantify certain measures that characterize its performance.

There are a number of options for the analyst to use when mathematically modeling the data fusion process (or the entire C^3 and combat process). Figure 11.8 provides the hierarchy of quantitative analysis techniques that may be applied.

Direct analytic solutions are applied where the equations of state can be fully represented as closed-form equations and system variables can be solved directly. Lanchester's equations are an example of such analytic solutions, where complex processes are represented in compact form to provide insight at a very high level of abstraction. As more detail (model fidelity) is added to reduce the level of abstraction, the set of equations soon becomes too large for direct solution.

In complex models, a simulation approach is often taken in which the system is functionally partitioned into elements (mathematical functions of time) that are interconnected and initialized with starting conditions. Time is represented by continuous operation of the network for analog computation or discrete time increments for digital computation. The discrete time increments can be constant time intervals (time-step simulation) or variable time increments so that computations occur only when changes of variables produce process changes of interest (event-step simulation). The simulation time scale is usually independent of real time: it may be synchronized to real time when used in conjunction with real-world operations (e.g., for man-in-the-loop testing).

Variables within the simulation may be represented deterministically or stochastically. Deterministic simulations are repeatable for a given set of inputs, because each incremental state is completely determined by the previous state. When the effects of randomness (due to noise, uncertain behavior, system errors, *et cetera*) are included as statistical models, the state-state transitions are dependent on probabilistic computations that are not repeatable from run to run. These stochastic simulations (also called *Monte Carlo simulations*) model the compound effects of multiple random variables within the system. Because they are not deter-

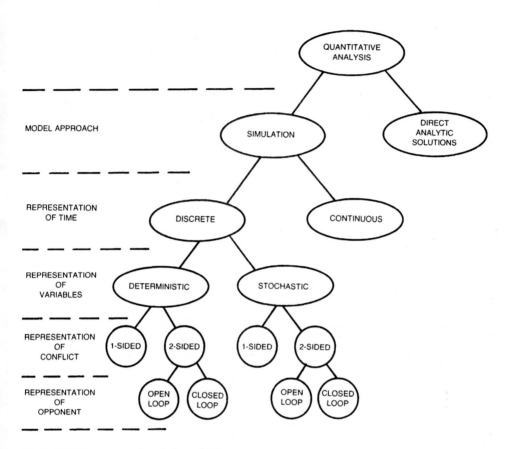

Figure 11.8 Taxonomy of analysis methods.

ministic, a large set of simulation runs is required to achieve a statistically significant set of results that can be analyzed to determine if variations between runs are due to experiment variables (e.g., different sensors, data fusion techniques) or simply due to natural variation from randomness in the model and selection of scenarios. The statistical method of analysis of variance, previously described in section 11.3.5, is one such method of determining the significance of results.

The interaction between forces may be represented by holding one side to a fixed, open-loop behavior (one-sided model) or by allowing both sides to react to each other's behavior in a closed-loop fashion (two-sided model). Two-sided models include feedback control that allows each force to independently sense the opponent, decide, and act. Figure 11.9 compares one- and two-sided simulations that follow the architecture of Lawson. At each time increment, both sides of the two-sided model estimate the location of opposing forces and act in accordance with a

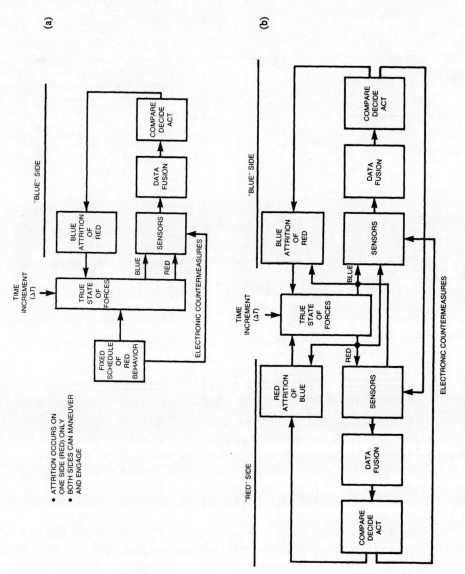

Figure 11.9 Comparison of one- (a) and two-sided (b) models.

battle doctrine. The attrition of each force is concurrently determined based on the actions of each time interval.

In addition to the selection of an appropriate analysis method, several other considerations are critical to a meaningful analysis. These are described in the following paragraphs.

11.5.1 Scenario Definition

The scenario describes the spatial and electronic environment under which the data fusion system must operate and the behaviors and capabilities of all participants in the military operation. The scenario must be representative of anticipated conditions (probable real futures) and quantitatively defined to describe the range of values that can be taken by the many variables that describe the scenario. Many of the scenario variables that characterize typical military models are summarized in Table 11.2. The number of variables can become very large for even a small scenario with less than a dozen targets, and careful attention must be paid to the selection of variables that influence the outcome and those that do not. Variation of the geometry and target behavior alone can spawn a virtually infinite set of scenarios for even a small number of targets. Operational experts (e.g., pilots, tacticians, field commanders) are usually called upon to prepare a manageable set of representative scenarios that span the range of expected futures.

11.5.2 Model Fidelity

Each element of a simulation must be mathematically modeled to some level of detail that represents all inputs, internal processes, and variations with time. This level of detail, or model fidelity, must be appropriate for the overall analysis objective: too little detail may ignore important behavioral characteristics that influence the outcome; too much detail may waste computational resources and development efforts.

Figure 11.10 illustrates three levels of fidelity that may be employed to model an airborne radar. The selection of model fidelity must be based upon sound judgment that considers the availability of data upon which the model will be built, the ability to validate the model with real measurements, and the required level of detail to meet analysis objectives.

11.5.3 Sensor Modeling

The sensor models must provide data to the fusion process that represents the actual sensor measurements in the real world. This must include random as well

Table 11.2
Major Military Scenario Variables

Variable Category	Typical Scenario Variables

Composition of Forces (order of battle)	• Initial location of all forces (targets located in x, y, z)
	• Initial capabilities of all weapons
	   Stores
	   Weapons capabilities
	   Countermeasure capabilities
	   Communications (interconnectivity)
	• Logistics, reload capabilities

Force Doctrine	• Offensive-defensive objectives • Time sequence of attack • Rules of engagement (conditions) • Preplanned trajectories for all mobile platforms (open-loop) or initial heading-speed (closed loop) • Group behavior doctrine

Electronic Order of Battle (including command, control, and communications)	• Countermeasure capabilities for all platforms
	   Soft kill (denial, deception)
	   Hard kill (e.g., home on jam)
	• Countermeasure tactics (behavior)
	• Sensor performance capabilities
	   Coverage volumes
	   Detection, track ranges
	   Target identification ranges
	   Performance accuracy
	• Interconnectivity between platforms
	   Voice communication links
	   Data links
	• Data fusion capabilities (speed, accuracy)
	• Presumed intelligence data access (prior knowledge)

Scenario Control	• Rules and restrictions
	• Time duration of simulation run or termination criteria
	• Environment
	   Weather (including effects sensor, maneuverability)
	   Background noise levels
	   Terrain (including effects on sensor)

as predictable effects that degrade performance and introduce errors in measurements. An example is provided in Figure 11.11 to illustrate the functions provided in a typical time-step simulation sensor model. The figure depicts an electronic support measure sensor (typical airborne radar warning receiver) that is represented by five submodels:

1. An intercept model computes the location of all targets relative to one's own ship each time-step to determine if the aspect of any target aircraft (with emit-

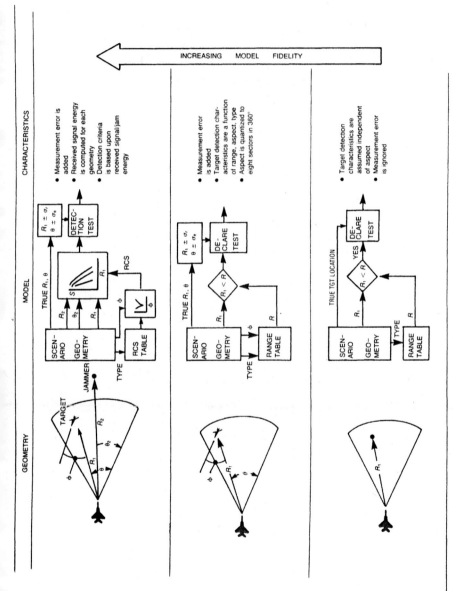

Figure 11.10 Three levels of sensor model fidelity.

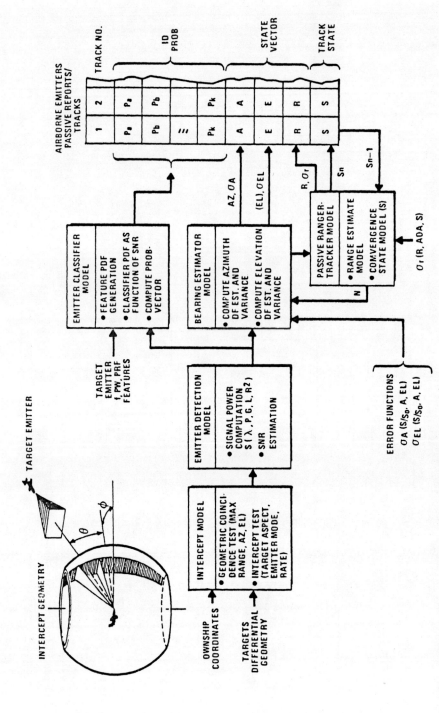

Figure 11.11 Airborne electronic support measures sensor model.

ter turned on) presents the emitter's radiation pattern to the sensor coverage (in this case, a sphere with polar areas clipped). A coincidence test is made and, if the target's emitter is within the geometry conditions for intercept, the target is determined to be a candidate for the sensor to detect the emitter.
2. A detection equation is computed, estimating received signal power as a function of range, effective radiated power, receiver sensitivity, antenna gain, and other loss factors. Detection probability is computed and, in the case of a hard-decision sensor, a threshold test is performed to determine if the emitter can be detected (intercepted). If detected, three models are initiated to model processing of the detected signal.
3. An emitter classifier model uses geometry and electronic noise data to compute the estimate of the emitter type and a measure of the confidence (or uncertainty) in the decision. This model may use a probability density function to adjust true emitter features and apply these data to the real signal classification algorithm used by the sensor.
4. A bearing estimator model computes estimated azimuth (and elevation, if modeling this capability) as a function of signal-to-noise ratio, computed in the detection model.
5. The emitter tracker is modeled by updating and improving detection-detection estimates of location. If a passive ranging capability is modeled, a range estimate and tracking state is computed as a function of signal-to-noise ratio and number of sequential measurements.

The identification of the emitter, its estimated location (geometric state vector), and its track number are then passed to the track file data base of the data fusion algorithm.

11.5.4 Fusion Process Modeling

The fusion process is often an algorithm that is not modeled, *per se,* because it can be implemented in a simulation in its complete form to evaluate its operation (timing, data handling, capacity, *et cetera*). In cases where human actions are involved in the fusion process, algorithms modeling human behavior are added or real-time "man-in-the-loop" simulations are employed to represent the effects of human assessment and decision making.

11.5.5 Simulation Architecture

Simulation architectures generally follow the structure of models presented earlier, in Section 11.1, and are partitioned into functional modules that compute variables at time or event intervals, depending on the simulation or module type. The major categories of functional modules are described for the representative, two-sided time-step simulation architecture provided in Figure 11.12.

Scenario generator accepts all input locations, headings or velocities, tactical objectives, and orders of battle criteria for all participating platforms in the simulation. Once the simulation is started, it computes new locations of all moving platforms, displaying platform coordinates for each time-step interval. In this example, red target tactics are all computed in this module based on blue target positions relative to red. Blue tactics are determined in the blue battle algorithm and fed back to the scenario generator as next-step changes. All kinematic computations (e.g., limiting target turn accelerations, velocities to realistic values) and coordinate transformations are usually performed here to provide common coordinate data for all targets. The generator also controls all activities or events programmed to occur at certain time or event criteria (e.g., the change in electronic or natural environment).

Sensor models enter target locations, relative geometries, signal environments, and LOS data (where applicable to determine obscuration or blockage) to compute sensor detections, tracks, and classifications. These models may operate at each time step, but output is timed to represent actual sensor timing, scanning, or output data rates.

Sensor controls emulate the management of sensors including mode selection, pointing, and on-off control to reduce emissions. These controls provide inputs (parameters and flags) to the sensor models.

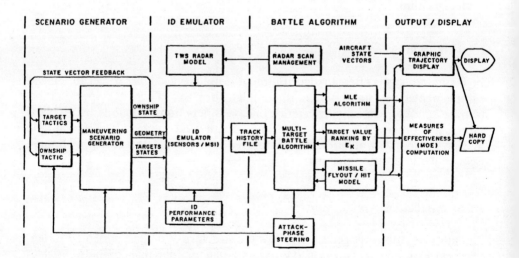

Figure 11.12 Typical simulation architecture.

Data fusion algorithm is the candidate algorithm being evaluated. Interfaces are carefully matched to input (sensors) and output (battle algorithm and display) models to be identical in data rate, accuracy, resolution, and timing to the physical C^3 system in which the fusion algorithm will be installed.

Situation data base is produced by the data fusion algorithm, including the set of all detections, tracks, identifications, and assessments of entities and events in the combat environment. It also includes all *a priori* and intelligence data, as well as data known about one's own forces.

Blue battle algorithm accepts the fusion data base as the best perception of the combat environment and makes tactical decisions on that basis. The decision criteria are based on own and hostile forces locations and behavior. Decisions include the control of own forces causing changes in kinematic behavior (steering) to be fed back to the scenario generator and the engagement of hostile forces by own-force weapons.

Weapons models are called when the battle algorithm causes a target to be engaged. The blue weapon system and red target geometries, velocities, *et cetera* are entered into the weapon model to compute a kill probability or, if the probability is thresholded, a hit-miss declaration. In cases where high weapon fidelity is required and weapons are slow relative to target velocities (e.g., some missiles), the weapon model may be a time-step model (e.g., a missile fly-out model) that is initiated to run in parallel with the overall simulation through the period of the weapons engagement. This allows the targets as well as weapons to interact during the engagement interval (sensor detections, target reactions, countermeasures, *et cetera*).

Merit models compute the measures of merit (MOPs, MOEs, MOFEs) and accumulate statistics throughout the simulation run. Outputs generally include event sequence lists (time annotated lists of major events and the data associated with them), graphic plots of trajectories, statistical summaries, and merit calculations.

11.5.6 Hierarchical Models

The development of large-scale military operational models that include system-level models of data fusion can become extremely large and complex if a uniform level of detail is maintained throughout. Even "simple" multiple-target simulations that include detailed sensor-communication models and data fusion algorithms can become very large (in the sense of development time and computational resources required) due to the detail required in all elements of the scenario, target characteristic models, sensors, and weapons effects. One solution is to structure the analysis as a hierarchy of models, along the lines of the hierarchy of measures described in Section 11.4. At lower levels, the model evaluates detailed performance on many simple scenarios with high-fidelity models. The results of these low-level simulations are then aggregated (normalized into representative values at a lower level of resolution) to provide inputs to the higher-level simulations. These simulations include lower-fidelity system models, appropriate for the level of fidelity of weapons and attrition models at the higher levels. The higher-level simulations consider larger, more complex scenarios at a lower level of fidelity.

Figure 11.13 illustrates a two-model approach in which a high-fidelity "system level" (S-level) model details one-on-one encounters between a weapon platform with multiple sensors (R, S2, S2, ...) and a data fusion system. The MOPs measured by this model are relative identification performance (ID accuracy, range at ID, time-to-ID after target detection, *et cetera*). The S-level model is used to evaluate a wide range of one-on-one target geometries with various system configurations of sensor suites (cases I, II, III, IV, V, ...). These results are then statistically aggregated to describe the identification improvement as a function of system configuration (Figure 11.13). The MOPs are the input variables to a higher-level "engagement" (E-level) simulation that evaluates the effect of varying identification performance on combat effectiveness.

The identification MOPs are inputs to the E-level simulation that include an emulation of the S-level data fusion identification performance. A battle algorithm models the conflict and results from fire control and target identification, providing a measurement of C^2 effectiveness as output MOEs. The functions relating MOPs to MOEs (Figure 11.13) can then be related, in this example, to determine the influence of various system configurations on identification, and then, in turn, on military effectiveness.

11.6 TESTBEDS AND SIMULATIONS

A range of modeling techniques is available to analyze the functions and performance of data fusion systems, each with its associated cost, realism, and ease of use. In addition to the analytic solutions and performance-level simulation methods already described, several more costly and realistic alternatives are available to increase the fidelity and scope of data fusion models.

Large-Scale Simulations

The scope of digital simulations may be increased well beyond the data-fusion system level models or engagement-level models described in the previous section. These simulations expand the scope of the simulation in terms of (1) number of participants, (2) fidelity of network (communication) modeling, (3) fidelity of coordinated behavior of participants, and (4) complexity of supporting models (battle strategies, logistics models, reload, topography, geography, *et cetera*) to sustain real-

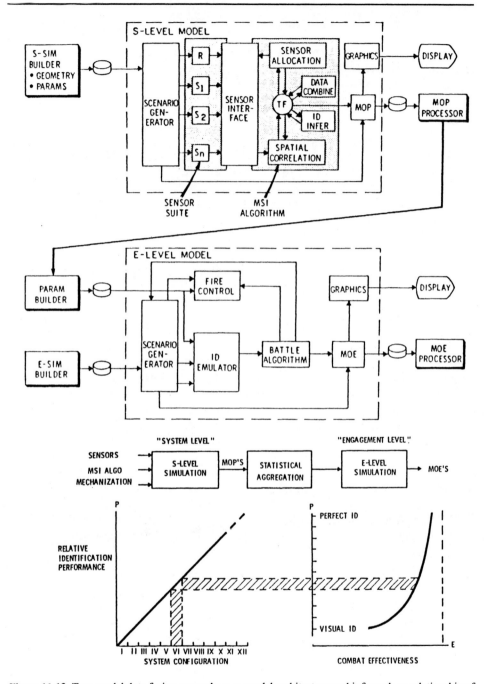

Figure 11.13 Two-model data fusion example: a = model architecture and infaces; b = relationship of measures of merit.

istic large-scale military operations. In these large-scale simulations, the effects of terrain (affecting LOS sensor and mobility), countermeasures, and logistics become important factors to sustain model realism.

Testbeds

Real-time digital simulations are combined with hardware and humans in testbeds to evaluate the human-machine interfaces in systems. "Man-in-the-loop" testbeds are real time simulations that provide displays to human operators and accept operator commands as an element of the simulation. The people (pilots, weapons operators, *et cetera*) introduce realistic gaming and response characteristics to the simulation that cannot be fully modeled digitally. Testbeds can also include real-time hardware (processors, displays, controls) to evaluate these items prior to commitment to production or employment. Such "hardware-in-the-loop" testbeds, combined with humans in the loop provide the highest degree of realism for assessing data fusion systems in which the real-time human-machine interface is an important element of the total system.

A much larger testbed reported by Darwin [15] has extensively tested theater-level data fusion performance to evaluate the method used in NATO C^2 systems to identify aircraft. In this testbed, tri-service operations were simulated by the interconnection of manned and digital aircraft and ground-based units through a common scenario data base that allowed all participants to conduct simulated combat operations in real time. Manned fighter simulators, air defense units, and airborne and ground surveillance units were all connected to the common simulation with digitally generated targets to create a large-scale air war and air defense simulation. The manned simulations were located throughout the continental United States and connected to the simulation testbed at Kirtland AFB, NM, by satellite data links to permit real-time operation of all units in a common, simulated geographic arena.

Simulations and testbeds offer several significant benefits for the test and evaluation of data fusion and sensor concepts:

- Costs of these methods are significantly lower than military exercises, flight test programs, and complete operational evaluations.
- Security is more easily maintained in the laboratory environment where emitters and tactics or doctrines are not revealed in environment that can be under surveillance.
- Ability to perform some tests is made possible only by simulation. The unavailability of foreign weapons, the inability to test some weapons (e.g., nuclear arms), and the inability to conduct exercises on a large scale are but some of the reasons why simulations and testbeds are the only methods of evaluating some systems.

Military Exercises

The highest level of realism is provided by military exercises, but also at the highest expense. Furthermore, exercises have the least flexibility to conduct sensitivity analyses by changing test variables or running large numbers of scenarios. Exercises do permit validation of data fusion and sensor performance characteristics that have been previously modeled in testbeds, simulations, or analyses. Exercises are carefully planned to evaluate top-level operational measures as well as critical MOPs in a realistic (and stressful) operational environment.

The normal course of development of new data fusion concepts will typically move from analytic studies and simulations (advanced development), to testbeds (engineering development, demonstration-validation phases), and then to *operational evaluation* (OPEVAL) following full-scale engineering development.

11.7 EVALUATING MILITARY WORTH

The previous sections have described the methods to develop data fusion system models, specify measures of merit for system evaluation, and conduct meaningful tests to quantify the merit of various systems. The results of these evaluations may provide accurate and representative data, but still pose a problem to the decision maker in choosing the "best" alternative. This is because multiple measures of merit and multiple conditions can provide very complex combinations of results.

Consider the earlier example presented in Section 11.3, in which three sensor-fusion system candidates were to be evaluated over three combat scenarios, with each scenario evaluated with and without electronic jamming (benign). This created 24 factors (or test bins) in the test matrix (Figure 11.6). Presume that two performance measures (MOP1 and MOP2) and a single effectiveness measure (MOE) are defined for evaluating the alternatives. The results of simulations can be displayed in the test matrix, as shown in Figure 11.14. Notice several features of the numerical results:

- Candidate A performs very well (high scores in all measures) in scenarios except number 4. In that case, it is the lowest performer of all.
- Candidate B appears to be most robust, with similar performance over all scenarios, both with and without jamming.
- Candidate C has very poor performance and effectiveness only in scenario 3; it performs very well in all other scenarios independent of electronic environment.

This example illustrates that more than a simple review of the test matrix may be required to select among complex alternatives. In this case, it is difficult to determine which of the alternatives is, indeed, "best."

scenario	1		2		3		4	
candidate \ benign/jam	B	J	B	J	B	J	B	J
A	.75 .81 .65	.65 .74 .61	.78 .91 .71	.71 .73 .68	.79 .88 .83	.73 .65 .74	.81 .79 .67	.15 .09 .21
B	.61 .74 .54	.60 .73 .53	.63 .75 .61	.60 .75 .60	.64 .80 .73	.61 .79 .51	.71 .61 .51	.68 .62 .50
C	.91 .94 .89	.87 .90 .83	.90 .83 .76	.87 .82 .70	.21 .34 .19	.11 .19 .09	.81 .83 .76	.79 .64 .61

format of values in each test bin:
```
MOP1
MOP2
MOE
```

Figure 11.14 Sample test matrix results for three measures per test.

Methods of decision analysis have been applied to such problems to evaluate quantitatively among candidates to determine their military worth (expressed as a value, called *utility*). *Multiple attribute utility* (MAU) analysis [16] is one method that can be directly applied to the evaluation of our example.

This technique develops a hierarchical structure that allows the decision maker to quantify the utility (or importance) of each factor in the evaluation, as well as each measure of merit. This amounts to a quantification of the criteria to define "best." The five steps used to apply MAU analysis follow.

Step 1. Factors (alternatives and test conditions) are decomposed into a hierarchy to provide a structure that relates measures (at the bottom) to a single total utility measure (at the top). Figure 11.15 depicts the MAU hierarchy for the previous example. The levels in the figure are

- Level 1 represents the single utility parameter that measures total military worth. This single value is used to compare the three system alternatives.
- Level 2 corresponds to three evaluation factors to be used: military effectiveness, cost, and interoperability. The military effectiveness branch value is provided by a performance-effectiveness simulation, the other values by separate analyses.

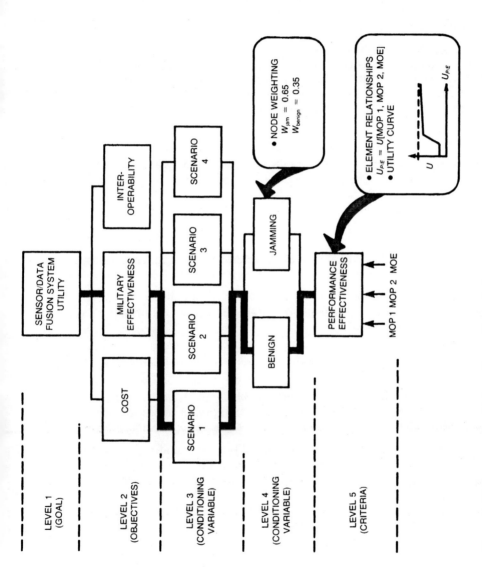

Figure 11.15 Structure of multiple attribute utility analysis.

- Level 3 represents the four scenarios under which each candidate must be evaluated.
- Level 4 represents the two electronic environments (benign and jamming).
- Level 5, finally, represents the individual measures of merit provided by the simulation runs of the test matrix.

The heavy line and shaded boxes in the figure depict one of many possible threads through this hierarchy by which a measure of merit can be related to total military worth. The establishment of such a hierarchy must carefully distinguish between conditioning variables (those contexts in which the alternatives will be evaluated, e.g., scenarios), the alternatives, and the measureable variables (MOPs, MOEs, *et cetera*).

Step 2. Element relationships in the structure are quantified by defining the mathematical equations that aggregate multiple input variables at a single node. In the example, a relationship between the three simulation measures is required to derive a common level 5 utility value:

$$U = U[\text{MOP1, MOP2, MOE}]$$

This function may be a simple weighted sum or a more complex function that relates the aggregate utility of the system for the set of measures.

Step 3. Parameter boundaries must be established to define limits of parameters, above or below which the change in parameter has no significant effect on the utility.

Step 4. Utility curves that relate utility at the bottom of each path to the performance parameters of each alternative must be defined. These curves (see Figure 11.15) can be adjusted by the decision maker to limit utility increases beyond boundaries (saturation) or define the rate of change of value as a function of input. The curves define the basis for placing value on individual performance parameters for each candidate.

Step 5. Weights that are multiplied by the utility at each node are defined to express the importance to the decision maker of that node's contribution to the total utility. The weights are selected to range from 0.0 to 1.0 and, across any level, usually sum to 1.0 (For the two electronic environments in our example, the weighting might be W [jamming] = 0.65, W [benign] = 0.35. This weighting would give credit for higher utility to those systems that perform better under jamming conditions: the decision maker has quantified the value of antijamming performance).

The quantification processes in steps 2, 4, and 5 are critical to the MAU process, because these specify the importance of each element and condition that contribute to overall value. The propagation of measures of merit, upward through utility functions and weights, provides a single utility value at the top level for each of the candidates. When properly structured, the resulting scores can be ranked to compare the alternatives.

These methods have been used in numerous military programs [17-19] to evaluate competing alternatives where a large number of system attributes had to be considered for evaluation.

REFERENCES

1. Engel, J.H., "A verification of Lanchester's Law," *Oper. Res.*, Vol. 2, No. 1, February 1954, pp. 163-171.
2. Weiss, H.K., "Combat Models and Historical Data: The U.S. Civil War," *Oper. Res.*, Vol. 14, No. 5, September-October 1966, pp. 759-790.
3. Lanchester, F.W., *Aircraft in Warfare: The Dawn of the Fourth Arm,* Constable, London, 1916.
4. Taylor, J.G., "Force on Force Attrition Modeling," Operations Research Society of America, Arlington, VA.
5. Everett, R.R., "Lanchester and C3," *Proc. of Symp. on Measures of Merit for Command, Control and Communications,* MITRE Corp., Bedford, MA, November 1972, pp. 29-50.
6. Dockery, J.T., "Lanchester Revisited: Progress in Modeling C2 in Combat," *Signal,* July 1988, pp. 41-48.
7. Lawson, J.S., Jr., "The Art and Science of Military Decisionmaking," *Phalanx,* Vol. 15, No. 4, Military Research Society, December 1982, pp. 1-13.
8. Rona, T.P., "Survey of C3 Assessment Concepts and Issues," *Proc. for Quantitative Assessment of the Utility of Command and Control Systems,* Washington, DC, January 1980, pp. 25-36.
9. Wohl, J.G., "Force Management Decision Requirements for Air Force Tactical Command and Control," *IEEE Trans. Systems, Man and Cybernetics,* Vol. SMC-11, No. 9, September 1983, pp. 618-639.
10. Zracket, C.A., "Issues in Command and Control R&D Evaluation," *Proc. for Quantitative Assessment of the Utility of Command and Control Systems,* Washington, DC, January 1980, pp. 49-62.
11. Goodman, I.R., "A General Theory for the Fusion of Data," *Proc. 1987 Data Fusion Symp.,* Vol. 1, June 1987, ONR/JDL Data Fusion Subpanel, pp. 254-270.
12. Goodman, I.R., and H.T. Nguyen, *Uncertainty Models for Knowledge-Based Systems,* North-Holland Press, Amsterdam, 1985.
13. Sweet, R., "Command and Control Evaluation Workshop," MORS C2 MOE Workshop, Military Operations Research Society, January 1985.
14. Johnson, R., and G. Bhattacharyya, *Statistics, Principles and Methods,* John Wiley and Sons, New York, 1985, Chapter 15, pp. 465-488.
15. Darwin, Capt. J.R., "Testing C3 Contributions to Combat Identification," IFFN Joint Test Force, Kirtland AFB NM, 14 January 1983.
16. Hayes, M.L., and M.F. O'Connor, "Relating a Promised Performance to Military Worth: An Evaluating Mechanism," *Defense Management J.,* October 1977, pp. 36-46.
17. Kemple, Maj. W.G., "Evaluating Concepts for C3 Systems," *Military Electronics/Countermeasures,* September 1980, pp. 56-66.
18. Geesey, R.A., "Combat Task Importance: The Multi-Attribute Utility of Combat Tasks in Theatre Operations," *Proc. NAECON,* IEEE, 1982, p. 1378.
19. Van Orden, R. Adm. M.D., "Management by Decision," *Signal,* September 1978, pp. 35-39.

Chapter 12
THE EMERGING ROLE OF ARTIFICIAL INTELLIGENCE TECHNIQUES

Most people would agree that artificial intelligence is a technological domain that encompasses a wide range of disciplines and techniques, and thereby finds application across a diversity of problems. In a sense, the data fusion domain has similar qualities, and the breadth of both fields makes their intersection, AI as applied to data fusion, equally wide. With the definition and characterization of AI chosen here, we will show that certain of its techniques find application to Level 1 fusion operations, but in general AI methods are appropriate to applications involving Level 2 and Level 3 fusion processes. However, this "mapping" of AI to fusion, as we will see, is not very precise or neatly bounded.

The AI technology component that has strongly dominated data fusion applications is the knowledge-based system component. Because our view in this chapter is generally experiential, the focus similarly is dominated by discussions related to KBS. Given the dominance of KBS, it is important to discuss the numerous technical issues applicable to the exploitation of KBS technologies in the face of the type of solution constraints imposed by fusion problems. For this reason, the chapter also includes sections on design issues (e.g., achieving real-time performance) and system test and evaluation issues for KBS applications to data fusion.

What is the role of AI in data fusion? In Chapter 1, we have said that inference and reasoning techniques are generally necessary to combine more abstract data (i.e., "symbolic" rather than numeric data) based on knowledge of relationships between such data sets. In turn, inference and reasoning capabilities are achieved with various AI techniques through an ability for what is called *symbolic processing*. Although one definition of AI [1] describes it as "the study of mental faculties through the use of computational models," the real distinction has to do with this symbolic rather than numeric processing. In their Turing Award lecture of 1975 [2], Newell and Simon introduce the concept of physical symbol systems as the basis for intelligent action. They defined a symbol as a physical pattern that can occur as a component of a *symbol structure* composed of a number of symbols related in some physical way. The codification of knowledge composed of facts, beliefs, and heuristics through symbolic representation and processing is a central and distinctive aspect of AI [3]. It is sufficient to think of symbols as strings of characters and symbol structures as list structures containing symbols. Predicate calculus is one formal language used to manipulate such symbol structures and develop inferences through such manipulations. Thus, formalisms have been developed that permit flexible symbolic representation and manipulation of knowledge to emulate, within limits, human logic, reasoning, and inference in computers.

Thus, AI comprises a set of *application areas* that build upon various *techniques* that in turn build upon this distinctive *symbolic processing* capability. The visible manifestations of AI technology in a given problem domain (as solutions) result primarily from the application capabilities. Thus, if we examine the application areas in the context of fusion problems, we can (based on our survey of AI applications to data fusion) make the following general comments:

- The primary AI applications used in solving data fusion problems are (1) expert or KBS, (2) *natural language processing* (NLP), (3) planning (i.e., plan recognition), (4) learning (in a very few advanced prototypes [see Chapter 9]), (5) intelligent assistance (all implementations could be called this).
- Similarly, the primary techniques used in data fusion are some type of pattern matching, the specifics of which frequently constitute an important processing kernel of the fusion system. The other techniques are important to varying degrees in given applications. Inference generation techniques are usually subsumed in the KBS application. The selection of techniques for searching (i.e., searching for viable (*satisfying* solutions), and for knowledge representation constitutes a crucial system design issue, but such techniques usually are not present as a solution strategy in a particular application. So, of the supportive techniques of AI, only pattern matching techniques, *per se,* are overtly visible as a type of solution for fusion problems; the other techniques are *part of the solution developed within* (and transparent within) an application. Because of this, only the pattern-matching technique is treated here as part of a family of solution types.

Another crucial aspect of the AI-data fusion relationship must be understood. As described earlier, this relationship is a result of and is characterized by the ability to apply inferential and reasoning techniques (via symbolic processing) to data and knowledge involved in a data fusion problem.

This broad definition makes it very difficult (perhaps impossible) to be highly *prescriptive* in the approach to describing the relationship and in describing how each application method is applied to data fusion problems. At Level 1, the processing goals are fairly precise (e.g., obtaining an optimum position estimate), and these precise goals permit the employment of relatively precise (and mostly numerical) solution methods such as optimization methods with specific objective functions. At Levels 2 and 3, the processing goals are less precise and have a much larger and variable scope. At these levels we seek estimates of "situations" and "threats" that have many possible elements (see Chapters 2 and 8) and context-dependent interpretations; therefore, it is difficult to prescribe a "usual" and "correct" approach. Reviews of work in this area substantiate this interpretation; solution strategies are far-ranging, creative, and disparate. Moreover, as just noted, the "ground truth" or "air truth" or "sea truth" that describes an actual situation or threat state is generally not available in other than laboratory experiments, so the *provably "correct"* approach, that is, the correct inferencing and reasoning strategy, is never really known. Because of this, and because inferencing and reasoning strategies that attempt to employ exhaustive assessments of typically combinatorially complex problems (tactical situation and threat states) are still unfeasible even with modern-day computers, knowledge-based approaches are characterized by solutions that are *satisfying* and knowingly suboptimal.

Nevertheless, some additional detail on a couple of AI application concepts (viz. planning and KBSs) and two brief descriptions of fusion-related projects will be included so that representative solution strategies can be appreciated (see Section 12.2). Coupled with the generic descriptions of AI application areas and techniques provided, we should be able to recognize the breadth of possible solution strategies.

In summary, in data fusion processing, AI forms a subset of the application concepts and techniques that provides a family of flexible solution strategies most often applied to portions of Level 2 and Level 3 data fusion problems, which provide the computational means for achieving reasoning power. Occasionally, such AI techniques have been used in combination with each other, and they are frequently used in combination with numerical methods in data fusion experiments and systems because reasoning strategies typically depend on the quantitative values of various parameters.

12.1   BROAD BENEFITS OF AI TECHNOLOGY

The components of AI technology applicable to data fusion problems just described are summarized in Table 12.1. The broad benefits these technologies bring to solving data fusion problems are summarized in Table 12.2. Each new application or technique of this table will be discussed.

Expert or KBS

At Level 1, numerical methods (see Chapters 6 and 7) are limited in their ability to deal with the tracking of, for example, highly maneuverable targets and in sensing or anticipating the onset of a maneuver. To overcome this and other tracking (and identification) algorithm deficiencies, KBSs that manage the employment of

Table 12.1
Employment of AI in Fusion Processing

AI COMPONENT	FUSION LEVEL		
	1	2	3
• **Application Area**			
Expert or KBS	X	X	X
Natural Language Processing		X	X
Plan Recognition		X	X
Learning		X	X
Intelligence Assistance	X	X	X
• **Technique**			
Pattern Matching	X	X	X

Table 12.2
Broad Benefits of AI in Fusion

AI COMPONENT	FUSION LEVEL		
	1	2	3
• **Application Area**			
Expert or KBS	Adaptability (e.g., sensor control and algorithm selection)	Multi-level inferencing; communicating and cooperative expertise	
NLP	–	Sensor collection management via auto msg. routing; fusion pre-processor; man-machine interface support	
Plan Recognition	–	Exploitation of doctrinal and/or exercise-based knowledge of hostile behavior. Plan provides fusion framework.	
Learning	–	Adaptation to extreme behaviors or parametric variance beyond elastic constraints	
Intelligence Assistance	←——— Decision Aiding ———→ ←——— Alerting (I&W) ———→ ←——— Attention-Focusing ———→		
• **Technique**			
Pattern Matching	Parametric approach to tracking and unit/group ID.	Potential for classical methods to discern various battlefield patterns of units or events.	

multiple algorithms (i.e., as algorithm selectors) have been researched. In such applications, the KBS applies contextual knowledge and knowledge of algorithm performance to select and invoke the "best" algorithm for the estimated current problem. In applications where Level 1 fusion is implemented as a feedback control process, KBSs have been used as means to apply expert knowledge for sensor control and employment strategies. Numerical methods for optimizing sensor control usually employ utility-based or scheduling theory-based approaches that, again, have limited awareness of problem context—KBSs can improve adaptability of such strategies by applying expert reasoning in combination with the numerical methods. As noted in Table 12.2, the key benefit that AI brings to Level 1 processing, via KBS methods, is adaptability in the overall solution.

For Levels 2 and 3, KBS application methods permit a broad range of reasoning and inference strategies as described in Section 12.5. At Levels 2 and 3, hierarchical strategies are used to overcome the combinatoric aspects of the problems (e.g., how to aggregate the large number of targets to represent an organized and describable threat). Solutions of this type typically require a corresponding multilevel inferencing strategy as noted in the table; that is, an ability to reason upward and downward in the hypothesized threat hierarchy. Further, situation and threat estimating performance can be improved in distributed or headquarters-type fusion centers (with fusion "subcenters") by applying decision and analysis support with communicating or cooperating KBSs, as noted in Table 12.2.

Natural Language Processing

Natural language processing methods can be powerful fusion support tools for message-based systems; that is, systems whose primary inputs are textual messages. NLP methods are basically used to "read" such messages. By doing so, these methods can be used for a variety of purposes, for example,

- Aiding in achieving optimum sensor tasking and management by providing the capability to automatically route messages to sensor systems or analysis functions;
- Provision of alerting functions by searching for and identifying key words, phrases, or "events" by analyzing and fusing the text from several messages; in general, this capability serves as a preprocessing function for other fusion processes such as enabled by KBSs;
- Provision of automated or semiautomated human-machine interface capability in fusion systems;
- NLP systems also frequently serve as interfaces to data management systems of various types, such as spatially oriented database systems; again, in this role, they serve a support function to fusion systems.

NLP systems can also be used to construct messages and thereby serve as output generators.

By and large, all such applications are appropriate to Levels 2 and 3. It is not clear how NLP functions can be applied at Level 1 unless unit positions and identity-related parameters are entered in the form of messages, in which case NLP again serves as a preprocessor to extract but usually not to fuse such data.

Generically, NLP capability fulfills a system support function in a fusion system. Figure 12.1 shows that the means by which such support is provided is through three primary operations: message processing, advanced *human-machine interfacing* (MMIF) methods supported by NLP, and support and interfacing improvements to data management operations via NLP.

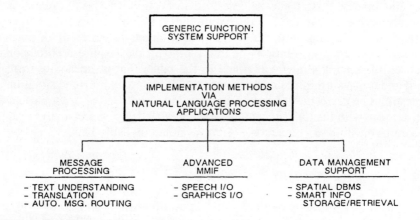

Figure 12.1 NLP support to data fusion systems.

Plan Recognition

Military operations, even those that are covert or involve surprise, are typically guided by a plan or set of plans, because such operations are complex, involve multiple resources and goals, and require significant coordination. If this assertion is true and if the general doctrines that guide red force actions are known to a blue force (at least in part) then the blue force can hypothesize the use of certain red force plans and, based on incoming multisensor data, assert the possible existence of particular red plans and use such assertions for decision making and action. This "plan recognition" process is used in certain data fusion applications (e.g., [4–7]), and based on formal theories of planning that are part of the AI milieu (see, e.g., [8, 9]).

Certain aspects of the plans involved with military operations, however, require special consideration in applying such theories in military data fusion problems. For example, standard AI models of planning have typically employed what might be termed a *predictability assumption;* that is, we assume that the planner's model of the world as well as of the effect of its actions on that world is complete and correct (e.g., STRIPS [10] and NOAH [11]). This assumption is violated if either an action fails to achieve its intended effect or plan relevant events lie outside the bounds of the planner's ability to control or predict.

Even assuming the planner's ability to predict relevant aspects of the world, the planner may still lack the effectiveness required to achieve the goal. There may be no sequence of the planner's primitive actions that achieve the goal state from the initial state. In such a case, classical planning models are typically designed to fail; that is, they provide no plan at all. This condition on planning success is often called the *effectiveness condition.*

Many types of military problems require some form of planning to achieve an overall goal where either the predictability assumption or the effectiveness condition are violated. The problem of how to plan in contexts where at least one of these is violated is frequently referred to as the *reactive planning problem* [12]. Note that when predictability or effectiveness is suspect, then plan execution must be monitored to detect failure and trigger some form of replanning. Such replanning is now interleaved with execution and in many contexts this places real-time constraints on the time available for replanning. Consequently, any approach to reactive planning must be sensitive to the issue of how to bound the time required for replanning. This is a particularly difficult problem because there is no guarantee that the failed plan can be revised or even that there is a solution to the problem from the state entered on plan failure. Several approaches have examined this problem (e.g., [13–16]) but no preferred solution has developed from these studies.

Thus, the red force planner is involved in a reactive planning process; at the same time, the blue force fusion analyst is faced with the same problem: to try to fuse the multisensor data into the best estimate of a red force plan, knowing that such plans are being reactively developed. Such estimates of red force plans form plan templates used in the fusion analysis. The fusion analysis essentially contructs the formation of a "best fit" of the dynamic, multisensor data into a set of dynamic, hypothesized red force plans considered possible at any given time. To do this usually requires a hierarchical set of plan constructs and the application of reasoning processes via KBSs to reason through the optimization process of judging how to guess the most likely red force plans (see Section 12.2 for an example).

Although we could argue that plan recognition methods could be used at Level 1 to detect the unit-level plans of single platforms, Table 12.2 shows them applied primarily in support of or as a means to assess situations and threat conditions at Levels 2 and 3.

Learning

In Chapter 9, we asserted that essentially all fusion systems rely on the red forces exhibiting some type of *expectation-based behavior* in order to make either estimates of such behavior or predictions of future behavior. However, our ability to anticipate, which forms the basis for developing a wide variety of fusion methods that rely on this ability, is limited for such complex processes and behaviors as can occur in combat or even in peace time. More robust fusion systems ideally would be able to learn to adapt to extreme (i.e., unanticipated) behaviors, as noted in Table 12.2. Although most expectation-based fusion systems attempt to build in some level of adaptability (e.g., by both numeric and reasoning procedures that can accommodate a *range* of values of various measured or derived parameters by having "elastic constraints" in these procedures), any red force behavior outside the permitted dynamic range causes at least partial failure by the system. Some advanced concept studies for fusion systems (e.g., [17]) are examining the use of various paradigms of machine learning (see [18]) for the design of highly adaptive data fusion systems (see Section 9.3).

As shown in Table 12.2, the relatively few applications of machine learning concepts have been to Levels 2 and 3 although, as for plan recognition procedures, it could be argued that machine learning concepts could also be effectively employed for the Level 1 type of problem. Most books and papers on machine learning (e.g., [19, 20]) postulate that learning is a very complex process and at least in part dependent on so-called deep knowledge. Specifying the details of this knowledge and inserting it in computers remains a topic of basic research and much progress will have to be made before feasible and then reliable machine learning capabilities are expected to be used in data fusion systems.

Intelligent Assistance

Intelligent assistance is a term that usually denotes a broad genre of computerized decision aids for human operators. These decision aids support the conduct of both analyses and operations and, in data fusion systems, are frequently focused on an alerting function, sometimes called an *indication and warning* (I&W) function. These alerting-type functions are the means to direct an operator to important or possibly critical observations. Modern human-computer interfacing techniques used in the design of intelligent assistance strategies perform similar "attention focusing" functions by sensing the problem context and the flow of objects that have drawn or are drawing operator attention in analysis activities. Methods for intelligent assistance involve various of the applications and techniques of AI. This application area is distinct from the others only in that special considerations for interacting with a human and improving human performance are included in the implementation approach. A particular focus of such applications is the implementation of a human-computer "dialogue," typically enabled through speech synthesis-recognition systems or natural language processing systems, such as CUBRICON (see, e.g., [21]), as well as KBS and possibly other elements of AI.

Other qualities peculiar to intelligent assistance methods are those that result from directed application of the principles and methods of cognitive psychology. For example, such systems sometimes maintain a "cognitive style profile" of its user group so that the system can optimally adjust to each user's problem-solving technique [22].

With this rather broad characterization, Table 12.2 shows intelligent assistance as generally applicable across all fusion levels. The only caveat is that the application of intelligent assistance cannot usually be done for time-critical problems where response time requirements are so short as to prevent human involvement; this usually means limited application for Level 1 processing.

Pattern Matching

Pattern matching techniques are widely applicable to data fusion problems. Chapter 7 discusses many of the aspects of such applications, mostly in the context of Level 1 object identification strategies. This class of techniques has also been used for target tracking applications because such techniques reduce the requirement for and effect of modeling of target dynamics and observational processes that Kalman-type approaches require (see Chapter 6 on tracking).

However, such techniques are also applied at Levels 2 and 3, for which event and object-network templating, essentially a pattern recognition technique, is perhaps the most popular solution method (e.g., [23–25] and see also Chapter 9 on situation and threat assessment). Moreover, there is at least the potential for the use of classical pattern-matching techniques at Levels 2 and 3 because such problems are rich with patterns of interest to the fusion analyst (e.g., object-class patterns, communication or logistics networks, sensor networks). Finally, our definition of pattern-matching techniques includes artificial neural network concepts, and such techniques have been researched for data fusion applications, even for Level 1 tracking problems [26, 27].

Thus, many of the elements of AI technology have great potential for solving part or all of many data fusion problems of interest. Most of the methods have been or are in the process of being applied to fusion problems with measurable success. References [28, 29, 30] provide some overviews on the application of AI technology to data fusion and C^3I problems.

12.2 REPRESENTATIVE PRESCRIPTIVE SOLUTIONS

As mentioned in the opening remarks to this chapter, and as perhaps can be better appreciated from the discussion in Section 12.1, no "standardized" prescriptive solution strategies govern the application of AI technology to data fusion problems. We have seen that there is not only a broad overlap between AI and data fusion but that each application concept or technique is itself highly flexible in permitting a broad range of particular solutions of that given class. However, to give a flavor for the type of solution strategies that can be applied with AI techniques, two representative "prescriptions" will be described: one is based on the use of planning theory and one on knowledge-based techniques. Each subsection has some background and overview information on the general methodology. Readers interested in this general topic should also see Chapter 9 and various sections of Chapter 2.

12.2.1 Applying Planning Theory

In everyday terms, *planning* means deciding on a course of action before acting. A plan is, thus, a representation of a course of action. It can be an unordered list of goals, but usually a plan has an implicit order to its goals. Most plans have a rich subplan structure; each goal can be replaced by a more detailed subplan to achieve it. Although a finished plan is a *linear* or *partial* ordering of problem-solving operators, the goals achieved by the operators often have a hierarchical structure.

Approaches to Planning

Four distinct approaches to planning are frequently described in the literature. They are *nonhierarchical planning, hierarchical planning, script-based planning* and *opportunistic planning*.

The use of these planning concepts finds its way into C^3I and data fusion applications in two ways: as part of the development of *blue* force plans (e.g., [31]) or as the core of a strategy for "plan recognition" wherein *red* force plans are hypothesized and, in essence, a "reverse engineering" process is applied, exploiting planning theories, to test whether such hypothesized red force plans are being invoked (e.g., [4–7]). Next, we will discuss one example of the latter type of application.

Multiagent Plan Recognition

The focus of the work by Azarewicz *et al.* [4] is to use a strategy called *multiagent plan recognition* to try to interpret the intentions of many potentially hostile air and ship platforms threatening a blue force naval battle group. The red platforms represent multiple "agents" operating either individually or in combination to carry out shared goals associated with the attack. The blue force decision maker's interpretation of the agents' behavior (essentially a tactical situation assessment problem) can be cast as a form of plan recognition. Azarewicz *et al.* assert that the blue decision maker interprets the red activity by hypothesizing red goals and inferring their plans, and the architecture and details of a *plan recognition model* (PRM) form the basis of the study.

The high-level design of the PRM derives from an analogy to human cognitive processing [32], which employs *long-term memory* (LTM), *procedural memory* (PM), and *short-term memory* (STM); the architecture is shown in Figure 12.2. This approach could be called *opportunistic*.

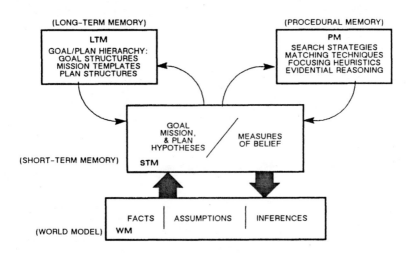

Figure 12.2 Plan recognition model architecture (after [4]).

In this architecture, the LTM contains blue's knowledge of the goals and plans that red agents pursue in specific tactical situations. PM is a component that employs a KBS approach to provide the reasoning power regarding multiagent activities. STM provides a "blackboard" data storage area for data referenced and manipulated in the plan recognition process. PM essentially compares the information coming in from the multiple blue sensors with candidate red plan templates that it draws from LTM and performs the best fusion possible of the data with the feasible plan templates. The type of goal-plan hierarchy contained in LTM is shown in Figure 12.3. Note that stored subplans, that is, scripts in effect, are also employed in this approach, and so the approach has a hybrid character.

In addition, PM employs both top-down and bottom-up reasoning to formulate multiagent mission hypotheses regarding current and future activities of the red platforms. The top-down process generates expectations of future behavior by hypothesizing high-level goals and subgoals and missions to accomplish the

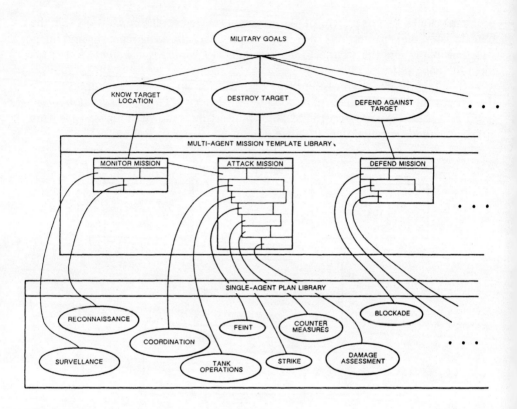

Figure 12.3 Concept of goal-plan hierarchy (after [4]).

subgoals. The goals are hypothesized primarily on the basis of the current assessment of problem *contextual* factors (i.e., nonbehavioral or dynamic factors); based on that contextual assessment, a candidate goal-plan template (script) is drawn from LTM and placed in STM. Then, the current data is assessed to see if it substantiates the candidate plan as viable.

In contrast, the bottom-up reasoning in PM is based upon an "opportunistic" approach that activates plan candidates on the basis of explicit platform (agent) *behavior*. Thus, the observation of an attack-type platform could generate a candidate attack plan even though, from a contextual viewpoint, indicators of an attack may be absent.

The plausible multiagent red plans are derived from this combined top-down and bottom-up approach. Fusion is achieved when input multisensor data populates the goal-plan hierarchy, as a result of the inference processes performed by the multiagent planners.

12.2.2 Applying Knowledge-Based Approaches

Knowledge-based systems employ human knowledge to solve problems that ordinarily require human intelligence. Most computers today perform tasks according to the decision-making logic of conventional programs, but these programs do not readily accommodate significant amounts of knowledge. Conventional programs consist of two distinct parts, algorithms and data. Algorithms determine how to solve specific kinds of problems, and data characterize parameters in the particular problem at hand. Human knowledge does not fit this model, however. Because much human knowledge consists of elementary fragments of knowhow, applying a significant amount of knowledge requires new ways to organize decision-making fragments into useful entities.

Knowledge systems collect these fragments in a knowledge base, then use the knowledge base to reason about specific problems. As a consequence, knowledge systems differ from conventional programs in the way they are organized, the way they incorporate knowledge, the way they execute, and the impression they create through their interactions. Knowledge systems attempt to emulate the inference processes of human specialists or experts. The human-machine interface can take on a variety of forms including graphics, menus, NLP, *et cetera*.

The primary building blocks of a knowledge system consists of those techniques that underlie many AI applications-symbolic programming: predicate calculus, search, and heuristics.

Knowledge systems frequently reason to solve problems by generating candidate solutions and evaluating them. Usually, solutions involve applying heuristic rules to given data to deduce logical or probable consequences and prove that these consequences satisfy the goal. These actions correspond to the basic propositional calculus mechanisms of inference and proof. Although most knowledge systems today do not actually employ formal logic programs, they achieve the same effects. Propositional calculus provides a formal foundation for their generally more limited inference and proof capabilities.

The most frequently used forms of knowledge representation involve constraints, assertions, rules, and certainty factors. A knowledge system incorporates constraints to express restrictions on allowable states, values, or conclusions. In fact, some knowledge systems derive their value primarily through an ability to recognize and satisfy complex, symbolic constraints sets. In this way, knowledge systems extend the class of constraint-satisfaction problems amenable to computation. Different from computer systems that focus primarily on linear constraints, knowledge systems address arbitrary symbolic constraints such as requirements on spatial, temporal, or logical relationships.

Assertional data bases provide means for storing and retrieving propositions. An assertion corresponds to a true proposition, a fact. Many simple forms of assertions lend themselves to relational database implementations, but more complicated patterns do not. In general, most knowledge systems today incorporate their own specialized assertional database subsystems.

Rules represent declarative or imperative knowledge of particular forms. The rules tell a knowledge system how to behave. Declarative rules, in general, describe the way things work in the world. On the other hand, imperative rules prescribe heuristic methods that the knowledge system should employ in its own operations.

Certainty factors designate the level of confidence or validity a knowledge system should associate with its data, or conclusions. These certainty factors may reflect any of a variety of different schemes for dealing with error and uncertainty. Subsection 12.3.5 discusses some of the detailed issues and approaches dealing with representation and calculation of uncertainty.

Rule-based systems (RBSs) appear to constitute the best currently available means for codifying the problem-solving knowhow of human experts; certainly they are the most frequently used method. Experts tend to express most of their problem-solving techniques in terms of a set of situation-action rules, and this suggests that RBSs should be the method of choice for building knowledge-intensive expert systems. Although many different techniques have emerged for organizing collections of rules into automated experts, all RBSs share certain key properties [33]:

1. They incorporate practical human knowledge in conditional if-then rules,
2. Their skill increases at a rate proportional to the enlargement of their knowledge bases,
3. They can solve a wide range of possibly complex problems by selecting relevant rules and then combining the results in appropriate ways,
4. They adaptively determine the best sequence of rules to execute,
5. They explain their conclusions by retracing their actual lines of reasoning and translating the logic of each rule employed into natural language.

RBSs address a need for capturing, representing, storing, distributing, reasoning about, and applying human knowledge electronically. They provide a practical means of building automated experts in application areas where job excellence requires consistent reasoning and practical experience.

Roughly speaking, an RBS consists of a knowledge base and an inference engine (see Figure 12.4). The knowledge base contains rules and facts. Rules always express a conditional, with an antecedent and a consequent component. The interpretation of a rule is that if the antecedent (left-hand side of the rule) can be satisfied, then the consequence (right-hand side of the rule) should be effected or enacted; that is, if evidence satisfies the rule antecedent, then the rule should be executed. The rule consequence defines an action, such as updating the dynamic memory with new knowledge or evidence, requesting added data, or executing a subprogram.

Because the behavior of all RBSs derives from this simple regimen, their rules will always specify the actual behavior of the system when particular problem-solv-

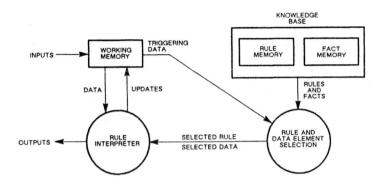

Figure 12.4 Rule-based system architecture (after [33]).

ing data are entered. In so doing, rules perform a variety of distinctive functions [33]:

1. They define a parallel decomposition of state transition behavior, thereby inducing a parallel decomposition of the overall system state that simplifies auditing and explanation. Every result can thus be traced to its antecedent data and intermediate rule-based inferences.
2. They can simulate deduction and reasoning by expressing logical relationships (conditionals) and definitional equivalences.
3. They can simulate subjective perception by relating signal data to higher-level pattern classes.
4. They can simulate subjective decision making by using conditional rules to express heuristics.

According to Hayes-Roth [33], from which much of the RBS description is derived, several key techniques for organizing RBSs have emerged. Rules can be used to express deductive knowledge, such as logical relationships, and thereby to support inference, verification, or evaluation tasks. Conversely, rules can be used to express goal-oriented knowledge that an RBS can apply in seeking problem solutions and cite in justifying its own goal-seeking behavior. Finally, rules can be used to express causal relationships, that an RBS can use to answer "what if" questions or to determine possible causes for specified events.

An RBS can solve problems only if it incorporates rules that use symbolic descriptions to characterize relevant situations and corresponding actions. The language employed for these descriptions imposes a conceptual framework on the problem and its solution. The rules may be precise or gross; the intermediate partial solutions abstract or detailed. Efforts to solve the problem may proceed top-down, outside-in, bottom-up, or in some other way. The meaning, importance, and contribution of each rule depend on its effectiveness as a contributor within the entire set of rules available for solving a problem.

Facts, the other kind of data in a knowledge base, express assertions about properties, relations, propositions, *et cetera*. In contrast to rules, which the RBS interprets as imperatives, facts are usually static and inactive—implicitly, a fact is silent regarding the pragmatic value and dynamic utilization of its knowledge. Thus, although in many contexts, facts and rules are logically interchangeable, in the context of RBSs, they are quite distinct.

In addition to its static memory for facts and rules, an RBS uses a working memory to store temporary assertions. These assertions record earlier rule-based inferences. We can describe the contents of working memory as problem-solving state information. Ordinarily, the data in working memory adhere to the syntactic conventions of facts. Temporary assertions thus correspond to dynamic facts.

The computing environment for rule interpretation consists of current facts and the inference engine itself. Together, these provide a context for interpreting the current state, understanding what the rules mean, and applying relevant rules appropriately. RBSs cannot obviate all the concerns of conventional computer programming (e.g., state sequences and variable scoping) because someone must still ensure that an RBS applies rules appropriately and in meaningful contexts.

The basic function of an RBS is to produce results. The primary output may be a problem solution, the answer to a question, or an analysis of some data. Whatever the case, an RBS employs several key processes in determining its overall activity. A "world" manager maintains information in working memory, and a built-in control procedure defines the basic high-level loop; if the built-in control provides for programmable specialized control, an additional process manages branches to and returns from special control blocks.

12.3 TECHNICAL ISSUES AND DESIGN FACTORS IN USING KBS FOR DATA FUSION

Whereas the potential and real applications of AI technology to data fusion described in Sections 12.1 and 12.2 are broad, many issues related to *implementation* of the technology must be dealt with. If we consider the elements of AI technology to be those shown in Figure 12.5, most application areas are generally immature as regards the breadth of understanding required for proper implementation across a broad range of problem types.

We emphasize that, of the AI components in Figure 12.5, those that have been applied with some reasonable frequency in the fusion problem domain include expert or knowledge-based methods, techniques from natural language processing, and methods of pattern recognition. In what follows, various issues that affect the application of such components to data fusion problems will be discussed.

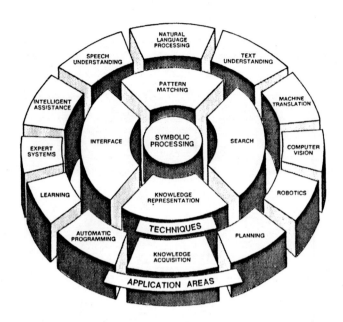

Figure 12.5 Elements of AI technology.

12.3.1 Difficulties in the Application of AI Components to Data Fusion

Even in relatively bounded applications for the lower echelons of the military, Level 2 and Level 3 assessments and the related decision-making processes will generally be difficult; that is, data fusion at these levels, even for problems of limited scope, is an intricate, complicated process. To solve this class of problem generally demands a fairly broad knowledge base; Figure 12.6 [34] shows the kinds of knowledge that can be required for a KBS. But such knowledge has other attributes that can greatly affect the solution approach. The design of a computational system, whether it be a KBS, PR, or NLP-type system depends on the size of the problem, data error characteristics, and reliability-uncertainty of the applied knowledge, each of which can influence the way in which a solution is derived. One of the most variable characteristics of KBSs is the way they *search* for solutions; as a consequence, searching techniques are one of the most studied topics in AI. Figure 12.7 shows how such problem requirements influence solution prescriptions (the bottom portion of each box); in [35], each case in the diagram is discussed at some length. In essence, the problem-solution interactions shown on Figure 12.7 and in [35] result from the effects of these problem characteristics on the *search* strategies employed in the solution approach.

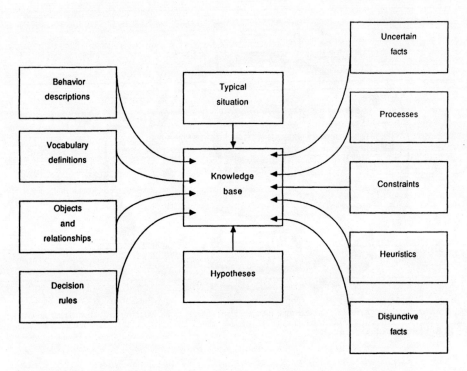

Figure 12.6 The kinds of knowledge that can go into a knowledge base (from [34]).

Although such knowledge organization and solution search strategy issues are crucial, still other factors of data fusion problems bear on the viability and nature of AI-based solutions. Among these are

- Real-time processing requirements,
- Time-varying data and solutions (temporal variance),
- Combined symbolic and numeric functions,
- Large data-knowledge requirements,
- Significant levels of uncertainty in data and knowledge,
- Human operator foibles and ideosyncracies.

Discussing the technical and system design implications of these characteristics is worthy of a book itself, and many fine texts abound on AI system design (e.g., [36–38]). However, we will briefly cover these issues because AI (especially KBS) components are such indivisible elements of many modern-day fusion system prototypes.

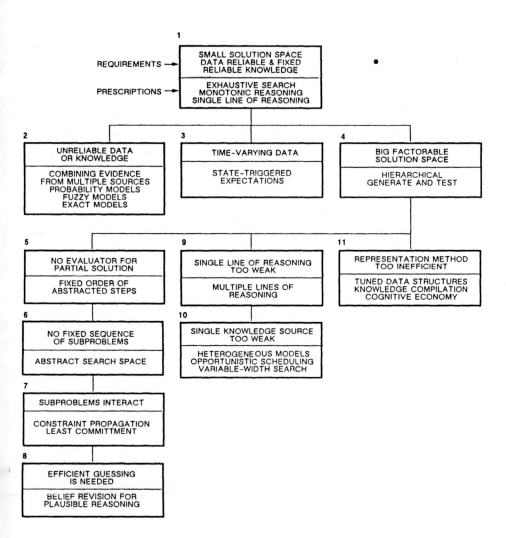

Figure 12.7 KBS design "prescriptions" (from [35]).

12.3.2 Real-Time Processing Requirements and Temporal Variance

A multisensor system that is usually observing a time-varying "world" state in an asynchronous manner must confront the general issue and design problem of selecting the system time granularity. That is, we must choose a time interval and time epochs over and at which to fuse and interpret the data. Such granularity determines the fundamental rate at which data is aggregated although cumulative, longer-term, and more broadly viewed inferencing of course can be applied as well. For example, we may have three levels of time granularity corresponding to the three levels of fusion. The shortest of these however determines the most time-critical nature of the process. There is, of course, an interplay between the rate at which the world state is expected to change (its temporal variance), the sampling rate designed into the sensor systems, and the minimal fusion interval that provides the needed estimate of the world.

Various methods have been used to deal with these issues; in general, they all employ some type of "expectation template" (see Chapter 9) and temporally and spatially elastic constraints. The elastic constraints are, in effect, confidence intervals of a sort that the knowledge engineer installs to represent the possible or probable variances in observables, events, knowledge, and data; Fall [23] and Noble [24] utilize such methods, for example.

Real-time performance of KBS systems has been achieved in a variety of problem settings through the use of several techniques, for example,

- Using a fast compiled language (rather than interpreted LISP) eliminates "garbage collection" and establishes a predictable search problem, but there is a loss of adaptability by not being able to dynamically create objects at run time; viability is domain dependent.
- Close linking, in both a logical and physical sense, can take place between frame slots or rule conditions and the sensor data paths.
- Priority-based metarules (high-level rules that control other rule systems) can be used as a dynamic control strategy.
- Parallel processing can be implemented (see Chapter 2 for details and examples).
- Time-based "value" functions can estimate the contribution to a goal of pursuing ("expanding") a search node in the knowledge base—that is, methods to set priorities and process the most "influential" knowledge first (see [39])—sometimes called *time-urgent searching*.
- Intelligent partitioning of the knowledge base and the use of intermediate goals states and "minisimulations" (fast exploration of intermediate goals to estimate optimum direction of the overall solution or major goal) can be used.

Other creative solution strategies continue to evolve. In operational military settings over the near-term (approximately five years), fusion solutions to real-time problems may also be constrained to those that can function (somehow) on five or ten year old computing equipment.

12.3.3 Combined Symbolic and Numeric Processing

In Chapter 2, we emphasized that fusion subsystems must maintain an interface with "real-world" system architectures; usually, this implies an interface between numeric and symbolic processing functions. Further, within the fusion subsystems themselves there is frequently a need to perform various numeric calculations. We can expect therefore that AI-based processing (symbolic processing) will have to be integrated with numeric processes in operationally useful data fusion systems.

Because each processing type usually requires different programming languages, there is typically the immediate issue of interlanguage functionality. In some cases this is feasible—that is, one language can call the other—but with current software this is not always possible. Another approach (see, e.g., [40]) is to let each process and language execute separately and to use a shared-memory base as a means of communication. In [40], event flags are used as a shared-memory control strategy. Figure 12.8 shows the multiprocess approach selected in an airborne surveillance data fusion application. Note that there is some "hybridization" because the FORTRAN data transfer routines have direct interface with the PROLOG routines, because this particular multiprocess approach resulted from solving memory management conflicts for those routines that required system library

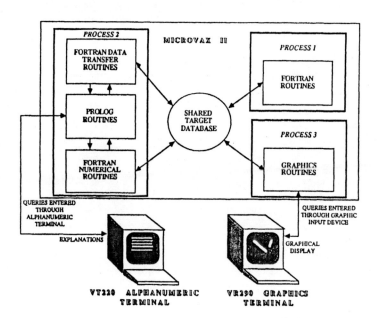

Figure 12.8 Integration of symbolic and numeric processing for air surveillance application (from [40]).

access—the FORTRAN routines in "Process 2" do not. This multiprocess (shared memory) approach to symbolic-numeric integration has three advantages:

1. Separation of the code and therefore the development and modification processes,
2. Separation of fast and slow input-output functions,
3. Provision of the groundwork for an actual parallel processing approach.

Note, of course, that this approach presumes that there is some reason why the symbolic processing would not be coded in a conventional language; this is yet another issue. Some good summaries of recent directions in multilanguage interfacing are included in [41], and [42, 43] discuss the particular issues associated with ADA and AI. Yet another software technology issue has to do with optimization of the AI code to improve execution speed, and various techniques are being experimented with to achieve some level of optimization [44–46].

12.3.4 Large Data-Knowledge Requirements

As in all problems and all software systems, in general size is a driver for complexity and, in turn, many other aspects of the subject process and its implementation. For AI systems we want to make a few special points. First, let us examine the size of systems that appear to be performing useful military functions. In [47] we find the following:

Application	1	2	3	4
Function	Threat Assessment	Force Management	Logistics	Aircraft Loading
Knowledge Base Size	>2400 rules	5000–8000 rules (projected)	833 Rules	300 Rules

Application Name
1 = Combat Action Team (CAT)
2 = Force Reallocation ES (FRESH)
3 = Inventory Manager's Asst (IMA)
4 = Automated Air Load Planning System (AALPS)

Those KBSs oriented to dynamic combat problems (1 and 2 in the table) appear to have knowledge bases in the range of an order of magnitude more than "managerial" systems (systems 3 and 4). This is not unexpected but the numbers lend credence to the scaling. Systems of several thousand rules will clearly require careful design depending on search space complexity (recall Figure 12.7).

An implementation complexity that such sizes introduce is that of "scaling." By *scaling* we mean the problem of estimating both the viability or feasibility and implementation parameters (lines of code, knowledge base size, cost, schedule, *et cetera*) for projected operational systems from the laboratory prototypes. Because of potential geometric scaling with size of the knowledge base, the projection of solution viability even from a laboratory version to a field version is a risky process.

12.3.5 Uncertainty in Data and Knowledge

The broad expansion of research and development in AI technology over the past several years has promulgated extensive work in the area of uncertainty representation and management in AI-based systems. Again, much of this work has centered on KBS applications, where (especially for military problems) various sources of uncertainty can affect the design and implementation of the KBS or RBS.

Sources of Uncertainty (derived from [48])

The presence of uncertain information in an expert or knowledge-based or rule-based system can be associated with at least four causes [49]. The first type is related to the reliability of information: uncertainty can be present in the factual knowledge (i.e., the set of assertions or facts) as a result of ill-defined concepts in the observations or inaccuracy and poor reliability of the instruments used to make the observations. Uncertainty of this type can also occur in the knowledge base (i.e., the rule set) as a result of weak implications. This occurs when the system builder is unable to establish a strong correlation between premise and conclusion.

The *aggregation* of this type of uncertainty is a recurrent need in expert system applications. Facts must be aggregated to determine whether the premise of a given rule has been satisfied, to verify the extent to which external constraints have been met, to propagate the amount of uncertainty through the triggering of a given rule, to summarize the findings provided by various rules or knowledge sources or experts, to detect possible inconsistencies among the various sources, and to rank different alternatives or different goals.

The second type of uncertainty is caused by the inherent imprecision of the rule representation language. If rules are not expressed in a formal language, their meaning cannot be interpreted exactly. Thus, a "lexical" matching is no longer adequate to compare subsets of facts with the premise. Instead, this requires a "semantic" matching comparing the approximate meaning of facts and premise.

Approximate reasoning is implemented using a "generalized modus ponens" [50]. In classical (two-valued) logics, modus ponens allows (Y is B) to be derived from the assertion of the crisp statements:

$(X$ is $A)$ and $[(X$ is $A) \rightarrow (Y$ is $B)]$

However, the inference can be made only if the unconditional assertion (X is A) is identical to the premise of the conditional assertion (rule). Therefore, to cover all possible situations, we need as many rules as the number of different values X can take. On the other hand, if A and B are described by fuzzy values, we have to define a fuzzy production rule (fuzzy implications) $[(X$ is $A) \rightarrow (Y$ is $B)]$ and we need a mechanism to match the meaning of any given situation with the premise of the rule. The generalized modus ponens proposes a definition for such a mechanism. The interpolating capabilities provided by the generalized modus ponens allow the model builder to describe only a few relevant situations using elastic constraints (fuzzy values). This scheme of inference still allows conclusions to be derived from statements describing situations that are close in meaning to the premise of some fuzzy production rule.

The third type of uncertainty occurs when inference is based on incomplete information. In this case, we need to partially match facts and premise; that is, we wish to allow for the value "unknown" during the evaluation of the degree of uncertainty of the premise. Furthermore, we need to be able to distinguish between necessary evidence *versus* possible (optimal) evidence and treat them appropriately in the partial matching process. Approaches to deal with this kind of uncertainty have ranged from default reasoning with consistent assumptions [51] to analogical reasoning [52].

The fourth type of uncertainty arises from the aggregation of rules from different knowledge sources or different experts. Four possible errors can occur in knowledge represented as production rules [53]: conflicting, redundant, subsuming, and missing rules. Conflicting rules, which succeed under the same circumstances but come to contradictory conclusions, increase the level of uncertainty by creating inconsistencies. Redundant rules, under which the same circumstances make the same conclusion, may create an overinflated assessment of the uncertainty of the conclusion. A subsumed rule, in which the premise of the first rule is a subset of the second, can create an overestimate of the certainty of the common conclusion. Missing rules, which fail to provide a needed conclusion under the right circumstances, create uncertainty of the third type, in which inference is based on incomplete information. In each of these cases, it is essential to detect a constraint violation and any resulting inconsistency. Attempts to reduce this kind of uncertainty have led to the compilation of the rule set into a network. By analyzing the resulting network, it is possible, at compilation time, to detect unreachable nodes (unused and missing information) and common substructures shared by two or more structures (potential subsumption or redundancy). At running time, warnings of potentially reachable, conflicting states (contradictions) could also be issued.

Representing Uncertainty

Representations of uncertainty in the expert system literature have been classified by Bonissone [48] into three categories:

- Numerical characterizations,
- Reasoned assumptions,
- Qualitative endorsements.

Bonissone [48] also formed a desiderata list for uncertainty representation and propagation techniques, and evaluated these methods in the context of these desirable qualities. Table 12.3 shows the desiderata list and Table 12.4 shows the evaluation.

Table 12.3
Desirable Requirements for Representing Uncertainty (from [48])

1. Combination rules not based on independence of evidence.
2. Combination rules not based on exhaustivity or exclusiveness.
3. Explicit representation of amount of evidence supporting or refuting a hypothesis.
4. Explicit representation of reasons for supporting or refuting hypothesis.
5. Representation technique should accommodate varying levels of detail in uncertainty.
6. Explicit representation of consistency, detection of conflicts.
7. Explicit representation of ignorance.
8. Clear distinction between ignorance and violation of consistency (conflict).
9. Second-order measure of uncertainty; uncertainty of the measure itself.
10. Representation natural and understandable.
11. Representation system in a closed form.
12. Pairwise comparisons of uncertain knowledge feasible.
13. Traceability through the reasoning or combination process.

Based on these characterizations, Figure 12.9 shows a taxonomy of uncertainty representation and propagation methods, viewed from a slightly different perspective. In Figure 12.9, the numeric methods are divided into one-, two-, and fuzzy-valued techniques and the symbolic methods into the formal and heuristic approaches. As shown, the numeric methods permit some type of "calculus" for their propagation, whereas symbolic methods use logic and rules for propagation.

Summary Comments

The uncertainty of some types of evidence or facts is a complex subject and it is unlikely that a single, uniform representation or propagation method will satisfy all practical implementation constraints. However, the data fusion system designer cannot avoid taking a stand on this issue; nearly all data fusion problems have uncertain elements of the type described earlier, and this will be especially true when countermeasures, concealment and deception techniques, and low observable technology are factors in the operational problem. However, trying to be "accurate" in estimating, inferring, and calculating uncertainty may not necessarily lead

Table 12.4
Comparison of Selected Uncertainty Representations (from [48])

REPRESENTATIONS	REQUIREMENT NUMBER													Y/N RATIO
	1	2	3	4	5	6	7	8	9	10	11	12	13	
NUMERIC														
Modified Bayesian	N	N	N	N	N	N	N	N	N	Y	Y	Y	N	3/13
Confirmation	N	Y	Y/N	N	N	N	Y	N	N	N	Y	Y	N	4/13
Upper & Lower Probabilities	N	N	Y	N	Y	Y	Y	Y	Y	Y	Y	Y	N	9/13
Evidential Reasoning	N	N	Y	N	Y	Y	Y	Y	Y	Y	Y	Y	N	9/13
Probability Bounds	Y	Y	Y	N	Y	Y	Y	Y	Y	Y	Y	Y	N	11/13
Fuzzy Necessity & Possibility	Y	Y	Y	N	Y	Y	Y	Y	Y	Y	Y	Y	N	11/13
Evidence Space	Y	Y	Y	N	Y	Y	Y	Y	Y	Y	Y	Y	N	10/13
SYMBOLIC														
Reasoned Assumptions	Y	Y	N	Y	N	Y	N	N	Y	Y	Y	N	Y	7/13
Endorsements	Y	Y	N	Y	N	Y	N	N	N	Y	Y	N	Y	7/13

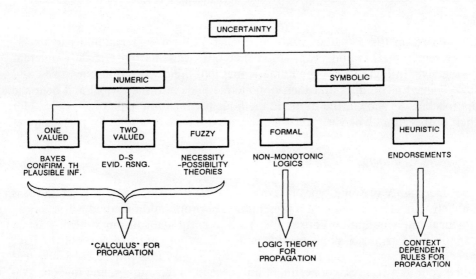

Figure 12.9 Relationships between uncertainty representation and uncertainty propagation.

to a provably "accurate" result nor may the effort required be cost effective. Judging exactly how to deal with uncertainty in KBS-RBS applications is simply part of the overall system engineering problem.

12.3.6 Human Factors

The advent of computer-based systems that exhibit intelligent behavior has had profound effects both on the development process for such systems and on the way such systems interact with humans. Lack of understanding or respect for the human engineering aspects of such systems can lead not only to suboptimal performance but, in the worst case, to rejection of the system by potential users. Compared to those first KBS-based consultation systems that achieved some success, military applications (real-time advisory systems) demand still further extensions of AI technology.

Generally, the development of optimal intelligent systems requires a much greater sensitivity to the fact that an intelligent system is being built. This may sound straightforward but such a requirement truly has a "cultural" impact on system design in formalizing procedures for incorporating human factors in system design and in requiring special, highly qualified human factors personnel in the design and development process. In designing an intelligent system, such human factors experts would focus attention on the following steps:

- Problem specification,
- Task analysis,
- Task allocation (human-machine),
- Determination of human informational needs,
- Application of human factors principles to interface design.

What must be considered in all this is the overall human-machine interaction at a *cognitive* level. This, in turn, is a result of the nature of the *communication* between human and machine. In intelligent systems there is much more overt communication with the operator; this happens in various ways; for example,

- Asking questions,
- Presenting a number of action choices,
- Providing support information,
- Asking for information,
- Recommending decision strategies,
- Asserting conclusions,
- Providing explanations for reasoning.

Lehner [54] argues that in fact most of the distinctive aspects of KBS-human communication result from the explanation requirement. Because average users generally are not domain experts, justification of recommendations or conclusions must account for a level of ignorance of knowledge-base contents. Additionally, average users generally applying problem-solving methods that may be incompatible with the normal flow of the inference system; thus, explanations must account for differences in solution perspectives. Finally, all this has to be done in small amounts of time, requiring "at a glance" rationales and indicators of what basic data the inference process is keying on.

Intelligent systems are also creating more "human *out*-of-the-loop" problems. When humans are relegated to monitoring functions, their performance often degrades compared to when they are involved. Further, users are often faced with a choice of deciding when to interrupt system processing to provide their own recommendations or decisions. Because this most often occurs at the boundaries of the system's knowledge, frequent application to difficult problems (which is what they are designed for in the first place) can lead to distrust or disappointment in the system.

Suffice it to say that the human-machine interface in intelligent systems requires the application of new design principles and both close contact with and training of potential users. In fact, the interface itself should be intelligent. Research in this area is occurring; Neal and Shapiro [55] have been working on an "intelligent, multimedia interface" concept that extends the concept of "natural language" to representations in different media forms. Projects such as these are also studying how to aid the user cognitively, with features that aid in maintaining the user's focus of attention on solution goals. Because users will not be prone to adapting to system needs, systems such as these will greatly improve system flexibility and adaptability.

Future systems will improve on the overall human-machine coupling. Important features of such systems will include ability to adapt to user expertise and learn by expanding their own knowledge bases from experience.

12.4 SPECIAL ASPECTS OF TESTING AND EVALUATING KBS

As the technology of knowledge-based systems grows past its infancy, there will be ever increasing pressures from the customer base, especially those in the Department of Defense, to somehow increase the level of confidence in the output of such systems. As for the *testing and evaluation* (T&E) of "conventional" software systems (i.e., non-KBS software), complexities associated with exhaustive T&E are basically driven by the size of the software system although, in general, many other factors affect the complexity of T&E. If basic "black-box" or "white-box" testing concepts are employed [56] then, correspondingly, either exhaustive input-output or exhaustive path tesing is required to fully examine program behavior. For most military systems with program sizes measured in tens or hundreds of thousands of lines of code, exhaustive testing of any type is not an economically viable option, and most often statistical testing techniques are employed to study program behavior. Thus, even in the case of "conventional" software development, there are strategies that must be developed to define an approach to T&E, given the *a priori* realization that not all the software can be examined.

This situation is no better in the case of KBS software. In fact, it is worse, basically because the technology is younger and the necessary standards, models, and processes for T&E have not yet been established. KBSs are, after all, software systems and their inherent nature is such that the design, development, and T&E processes are more difficult than for conventional applications. Both the complexities of KBS design and those of running time operation prevent the software engineer from fully understanding all potential behaviors of the system. Whereas this is true to a degree for conventional systems as well, it is judged more severe for KBSs because

- KBSs employ heuristic techniques in their control structures (as the problems are otherwise not computable),
- They often have convoluted, intricate program control,
- Their structures are irregular and reflect individual programmer styles,
- The scaling of performance with size is poorly understood,
- The sensitivity of results to changes in the knowledge base can be difficult to predict,

among other issues. Hence, the so-called state transition patterns of KBS software systems can be expected to be extremely difficult to fully comprehend. As a result, developing effective T&E processes for KBS software can be expected to be at least as difficult as for conventional software systems.

Strangely, in spite of a rush of funding in this area over the past few years, and the corresponding development of numerous demonstration and prototype systems, little energy has been devoted to serious assessments of the T&E process. A few research papers have been published (e.g., [57–60]), and some of the recent textbooks speak to the issues (e.g., [3, 57]) but, by and large, no serious guidelines have emerged from the AI-KBS community for software that must behave and perform "like a human." This perspective is certainly among the issues that must be dealt with in establishing T&E guidelines. If KBS software is to come out of the laboratory into operational use, serious dialogs on the issues must begin.

There are conceptual and philosophical issues regarding what constitutes "goodness" for software of any type that can, more of less, alter the complexity of the T&E issue. For example, there is the issue of reliability versus trustworthiness. Testing oriented toward measuring reliability is often "classless"; that is, it occurs without distinction of the type of failure encountered. Thus, reliability testing often derives an unweighted likelihood of failure, without defining the class or, perhaps more important, a measure of the severity of the failure. This perspective derives from a philosophy oriented to the unweighted conformance of the software with the software specifications, a common practice among the DoD and its contractors.

According to Parnas [58], however, trustworthiness of software is a more desirable goal to achieve via the T&E processes. Parnas defines trustworthiness as a measure of the software's likelihood of failing catastrophically. Thus, the trustworthiness characteristics can be described by a function that yields the probability of occurrence for all significant levels of severe failures. This probabilistic function provides the basis for estimating a confidence interval for trustworthiness. The system designer-developer (or customer) can thus have a basis for assuring that the level of failures will not, within specified probabilistic limits, exceed certain levels of severity.

For KBSs employed in critical applications (e.g., fusion applications) there is the additional, crucial measure of the accuracy of the system; that is, the static accuracy of the embedded knowledge, and the dynamic accuracy of the advice, recommendations, or conclusions the system provides. Static accuracy is determined from static evaluations in which, typically, various domain experts compare the system's knowledge base with their own, looking for consistency and completeness. Dynamic accuracy is derived by domain experts analyzing the system's line of reasoning and its ultimate output and comparing it with their own equivalents.

In sum, it is important to realize that KBSs *are* different type software systems that will require some different techniques for T&E. However, in the rush to demonstrate "successful" prototypical behavior, many efforts in the AI community have produced shallow results. By returning to good system engineering principles, by using many existent T&E principles for conventional software, and by working on those special aspects of T&E for KBSs, the applications should be able to demonstrate a maturation process that takes them from the defense laboratories to the operational commands; the technology base is substantive—the development process needs improvement.

12.5 THE RANGE OF APPLICATIONS OF AI TO FUSION PROBLEMS

Even a cursory literature review of the work in data fusion will reveal a large number of applications of AI to data fusion problems; thus, a serious review of such work would produce a sizeable bibliography, beyond the scope of this book. However, we will attempt to characterize the *range* of applications at each level of fusion in this brief section to show the breadth of research activity.

Based on research surveys conducted at the Tri-Service Data Fusion Symposiums [61, 62], it seems that the R&D community is heavily focused on Level 1 processing. However, this level is dominated by numeric processing applications and the extent of AI applications is somewhat limited. Relatively little work has been done on the use of AI in tracking, although AI-enhanced classification processes (largely based on contextual augmentation of attribute-based classification processing) has received serious study.

In the tracking domain, AI has been used for data association by employing pattern recognition techniques (e.g., [63]). It has also been used for some sensor management applications; in these problems, KBS methods have typically been used to aid a person in managing sensors, either in centralized or distributed architectures (e.g., [64–66]). Grimshaw and Amburn [67] have applied a KBS for the selection of an optimal filter in a multiple-model adaptive filter architecture; this is one of what seem to be only a very few papers on the use of AI methods to augment the detailed processing steps of trackers. Kuczewski [68] has presented a paper that explores the use of neural network techniques for multitarget tracking. AI technology has also been applied to the relocatable target search problem, with sensor management based on maximizing the chance of finding a moved target [69]. Kadar and Lodaya [70] examine a KBS-based decision aid concept for interactive tracking and identification. Hartless [71] applied KBS methods for an adaptive multisensor control system for low-observable target (cruise missile threat) tracking problems.

Various components of AI have been used in the classification domain. Applications have employed pattern recognition methods, KBS methods, and neural networks. For example, Farhat *et al.* [72] have used *artificial neural systems* (ANS) for noncooperative recognition of airborne targets with microwave-type sensors. ANS methods have also been used for hull-emitter classification problems based on ELINT data [73]. Bowman [74] has provided a comprehensive overview of the use of ANS techniques for target identification. Spence *et al.* [75] have examined the biophysical aspects of ANS applied to target localization and identification. Garvey [76] has done a survey of AI methods applied to information integration problems oriented to classification. Spiesbach and Gilmore [77] described the general approach to using AI for contextual analysis in target recognition applications. Several papers on AI for classification are included in the SPIE Proceedings of the 1988 Sensor Fusion Conference [78]. KBS methods have been actively applied to the correlation of multisensor data for ship identification (e.g., [79, 80]).

At levels 2 and 3, AI methods have been broadly applied to the problems of situation and threat assessment; Section 12.2 of this chapter and the bulk of Chapter 9 discuss several of the solution strategies used for such applications. This breadth of application is only natural in that such problems are more amenable to symbolic processing-based solution techniques. We provide here reference to a few such applications (e.g., [81–85]), but note that many such studies in the fusion literature that examined the SA-TA problems of each of the services have employed several types of solution techniques (although these have been dominated by KBS methods) and have been rather creative. Integrated situation assessment and planning and the general topic of data fusion form two complete chapters of [86], which is a good overview of the role of AI in C^2. However, it is significant to note the exceptions and omissions, in particular the absence of classical pattern recognition techniques and, as yet, the use of ANS techniques at these levels. Another significant observation is that relatively few of these methods have been tested in any serious or thorough fashion, nor have they found their way into operational settings. Fox and Arkin [87] describe some of the theoretical and practical aspects of building and deploying large-scale data fusion systems, based largely on their development experience with the Navy's OSIS Baseline Upgrade Program; they cite many useful lessons learned and recommendations for dealing with real data, supporting software and data bases, and security control and system recovery issues. As noted in Chapter 10, the building of real fusion systems entails many aspects beyond the specifics of numeric or symbolic fusion algorithms.

REFERENCES

1. Charniak, E., C.K. Riesbeck, and D.V. McDermott, *Artificial Intelligence Programming*, Lawrence Erlbaum Associates, Hillsdale, NJ, 1979.
2. Newell, A., and H.A. Simon, "Computer Science as Empirical Enquiry: Symbols and Search" (the 1975 Turing Award Lecture), *Communications of the ACM*, Vol. 19, No. 3, 1976.
3. Hayes-Roth, F., D.A. Waterman, and D.B. Lenat, *Building Expert Systems*, Addison-Wesley, Reading, MA, 1983.
4. Azarewicz, J., et al., "Template-Based Multi-Agent Plan Recognition for Tactical Situation Assessment," *Proc. 5th Conf. on Artificial Intelligence Applications*, Miami, March 1989.
5. Carbonell, J.G., "Counterplanning: A Strategy Based Model of Adversarial Planning in Real-World Situations," *Artificial Intelligence*, Vol. 16, 1981.
6. Hayslip, I.C., and J.P. Rosenking, "Adaptive Planning for Threat Response," in *Applications of Artificial Intelligence VII*, Proc. of the SPIE, Vol. 1095, March 1989.
7. Azarewicz, J., et al., "Plan Recognition for Airborne Tactical Decision Making," *Proc. 5th Nat. Conf. on Artificial Intelligence*, Philadelphia, August 1986.
8. Gevarter, W.B., *Intelligent Machines: An Introductory Perspective to Artificial Intelligence and Robotics*, Prentice-Hall, Englewood Cliffs, NJ.
9. Willensky, R. *Planning and Understanding—A Computational Approach to Human Reasoning*, Addison-Wesley, Reading, MA, 1983.
10. Fikes, R.E., et al., "STRIPS: A New Approach to the Application of Theorem Proving to Problem Solving," *Artificial Intelligence*, Vol. 2, 1971.
11. Sacerdoti, E.D., *A Structure for Plans and Behavior*, Elsevier, New York, 1977.
12. Schmidt, C.F., et al., "Reactive Planning Using a Situation Space," *Proc. Annual AI Systems in Government Conf.*, Washington, DC, March 1989.
13. Schoppers, M.J., "Universal Plans for Reactive Robots in Unpredictable Environments," *Proc. 8th IJCAI*, Milan, Italy, August 1987.
14. Hendler, J.A., and J.C. Sanborn, "A Model of Reaction for Planning in Dynamic Environments," *Proc. Knowledge-Based Planning Workshop*, Austin, TX, December 1987.
15. Firby, R.J., "An Investigation into Reactive Planning in Complex Domains," *Proc. Nat. Conf. on Artificial Intelligence*, Seattle, August 1987.
16. Agre, P.E., and D. Chapman, "Pengi: An Implementation of a Theory of Activity," *Proc. Nat. Conf. on Artificial Ingellicence*, Seattle, August 1987.
17. Kadar, I., "Data Fusion by Perceptual Reasoning and Prediction," *Proc. 1987 Tri-Service Data Fusion Symp.*, Johns Hopkins University, Baltimore, June 1987.

18. Michalski, R.S., et al., *Machine Learning*, Vols. 1 (1983 and 2 (1986), Tioga Press.
19. Kohonen, T., *Self Organization and Associative Memory*, Springer-Verlag, New York, 1984.
20. Jackson, A.H., "Machine Learning," *Expert Systems*, Vol. 5, No. 2, May 1988.
21. Neal, J.G., and S.C. Shapiro, "Intelligent Integrated Interface Technology," *Proc. 1987 Tri-Service Data Fusion Symp.*, Johns Hopkins University, Baltimore, June 1987.
22. Halpin, S.M. and F.L. Moses, "Improving Human Performance through the Application of Intelligent Systems," *Signal*, Vol. 40, No. 10, June 1986.
23. Fall, T.C., "Evidential Reasoning with Temporal Aspects," *Proc. AAAI 5th Nat. Conf. on Artificial Intelligence*, Philadelphia, August 1986.
24. Noble, D., "Managing Temporal Uncertainty in Situation Assessment," *Proc. 1988 Tri-Service Data Fusion Symp.*, Johns Hopkins University, Baltimore, May 1988.
25. Hall, D.L., and R.J. Linn, "Comments on the Use of Templating Techniques for Multisensor Data Fusion," *Proc. 1989 Tri-Service Data Fusion Symp.*, Johns Hopkins University, Baltimore, May 1989.
26. Bowman, C., "Artificial Neural Network Adaptive Systems Applied to Multisensor ID," *Proc. 1988 Tri-Service Data Fusion Symp.*, Johns Hopkins University, Baltimore, May 1988.
27. Gorman, R.P. and T.J. Sejnowski, "Analysis of Hidden Units in Layered Network Trained to Classify Sonar Targets," *Neural Network J.*, Vol. 1, November 1987.
28. Harris, C.J. (ed.), *Application of Artificial Intelligence to Command and Control Systems*, Peter Peregrinus Ltd., London, 1988.
29. Shumaker, R.P., and J. Franklin, "Artificial Intelligence in Military Applications," in *Principles of Command and Control*, J.L. Boyes and S.J. Andriole (eds.), AFCEA International Press, Washington, DC, 1987.
30. Harris, C.J. and I. White (eds.), *Advances in Command, Control and Communication Systems*, Peter Peregrinus Ltd., London, 1987.
31. Meng, A.C., "AMPES: Adaptive Mission Planning Expert System for Air Mission Tasks," *Proc. NAECON*, Dayton, OH, 1988.
32. Anderson, J., *The Architecture of Cognition*, Harvard University Press, Cambridge, MA, 1983.
33. Hayes-Roth, F., "Rule Based Systems," *Communications of the ACM*, Vol. 28, No. 9, September 1985.
34. Fikes, R., and T. Kehler, "The Role of Frame-Based Representation in Reasoning," *Communications of the ACM*, Vol. 28, No. 9, September 1985.
35. Stefik, M., et al., "The Organization of Expert Systems: A Tutorial," *Artificial Intelligence*, Vol. 18, 1982.
36. Cohen, P.R., and E.A. Feigenbaum, *The Handbook of Artificial Intelligence*, Vol. 3, Heuris Tech. Press, Stanford, CA.
37. Barr, A., and E.A. Feigenbaum (eds.), *The Handbook of Artificial Intelligence*, Vols. 1 and 2, William Kaufman, Los Altos, CA, 1981 and 1982.
38. Buchanan, B.G., and E.H. Shortliffe, *Rule-Based Expert Systems: The MYCIN Experiments of the Heuristic Programming Project*, Addison-Wesley, Reading, MA, 1983.
39. Slagle, J., et al., "BATTLE: An Expert System for Fire Support Command and Control," Naval Res. Lab. Memo Rpt. 4847, July 1982.
40. Kurien, T., and A. Caromicoli, "Combined Numerical and Symbolic Processing for Airborne Surveillance," *Proc. AIAA/IEEE 8th Digital Avionics Systems Conf.*, San Jose, CA, October 1988.
41. Falk, H., "AI Techniques Enter the Realm of Conventional Languages," *Computer Design*, October 15, 1988.
42. Schwartz, R.L., et al., "On The Suitability of ADA For Artificial Intelligence Applications," SRI Rep. on Contract DAAG29-79-C-0126, (NTIS AD A090790), July 1980.
43. Naedel, D., "ADA and AI," *Defense Electronics*, April 1986.

44. Bennett, M.E., "Real-Time Continuous AI Systems," *IEEE Proc.* (U.K.), Vol. 134, Part D, No. 4, July 1987.
45. Deering, M.F., "Hardware and-Software Architectures for Efficient AI," *Proc. IJCAI-81*, Vancouver, BC, Canada, August 1981.
46. Ulug, M.E., "A Real-Time AI System for Military Communications," *Proc. 3rd Conf. on Artificial Intelligence Applications*, Kissimmee, FL, February 1987.
47. Simpson, R.L., "DoD Applications of Artificial Intelligence: Successes and Prospects," *Proc. SPIE*, Vol. 937, Applications of Artificial Intelligence, 1988.
48. Bonissone, P.P., and R.M. Tong, "Editorial: Reasoning with Uncertainty in Expert Situations," *Int. J. Man-Machine Studies*, Vol. 22, 1985, p. 241–50.
49. Tong, R.M., "The Representation of Uncertainty in DISCIPLE," AIDS Techn. Note TN1018-1, 1982.
50. Zadeh, L.A., "A Theory of Approximate Reasoning," Electronics Research Laboratory Memo. No. UCB/ERL M77/58, University of California, Berkeley, 1977. Also in Hayes, J., D. Michie, and L. Mikulich (eds.), *Machine Intelligence*, Vol. 9, pp. 149–194.
51. Reiger, R., "A Logic for Default Reasoning," *Artificial Intelligence*, Vol. 13, 1980, pp. 81–132.
52. Winston, P.H., "Learning and Reasoning by Analogy," *Communications of the ACM*, Vol. 23, No. 12, 1980, pp. 689–703.
53. Suwa, M., A.C. Scott, and E.H. Shortliffe, "An Approach to Verifying Completeness and Consistency in a Rule-Based Expert System," HPP-81-5, Heuristic Programming Project, Stanford University, Stanford, CA, 1981.
54. Lehner, P.E., "Man/Machine Interface Issues in the Application of Expert System Technology to Tactical Fusion/Correlation," *Proc. AFCEA Seminar on Artificial Intelligence Applications to the Battlefield*, Ft. Monmouth, NJ, May 1985.
55. Neal, J.G., and S.C. Shapiro, "Intelligent Multi-Media Interface Technology," in *Architectures for Intelligent Interfaces: Elements and Prototypes*, J. Sullivan and S. Tyler (eds.), Addison-Wesley, Reading, MA, 1989.
56. Myers, G.J., *The Art of Software Testing*, John Wiley and Sons, Wiley-Interscience, New York, 1979.
57. Buchanan, B.G., and E. H. Shortliffe, *Rule-Based Expert Systems*, Addison-Wesley, Reading, MA, 1985.
58. Suydam, W., "Approaches to Software Testing Embroiled in Debate," *Computer Design*, November 15, 1986.
59. Llinas, J., and S. Rizzi, "The Test and Evaluation Process for Knowledge-Based Systems," RADC-TR-87-181, Rome Air Development Center, November 1987.
60. Bjerregaard, B.S., *et al.*, "Computer Aided Diagnosis of the Acute Abdomen: A System from Leeds Used on Copenhagen Patients," (in Reference 18).
61. F.E. White (Chairman, Data Fusion Subpanel), "Technical Proceedings of the 1987 Tri-Service Data Fusion Symposium," Johns Hopkins University, Laurel, MD, June 1987.
62. F.E. White (Chairman, Data Fusion Subpanel), "Technical Proceedings of the 1988 Tri-Service Data Fusion Symposium," Johns Hopkins University, Laurel, MD, May 1988.
63. Graham, M.L., *et al.*, "A Hierarchical Approach to Data Fusion for Tracking in an Underwater Environment," *Proc. 1987 Tri-Service Data Fusion Symp.*, Johns Hopkins University, Laurel, MD, June 1987.
64. Carney, E.R., and J.J. Bourguin, "Smart Operator Interface for Sensor Allocation of Multisensor Target Acquisition System," *Proc. 1987 Tri-Service Data Fusion Symp.*, Johns Hopkins University, Laurel, MD, June 1987.
65. Cowan, R.A., "Improved Tracking and Data Fusion through Sensor Management and Control," *Proc. 1987 Tri-Service Data Fusion Symp.*, Johns Hopkins University, Laurel, MD, June 1987.
66. Addison, E.R., and B.D. Leon, "A Blackboard Architecture for Cooperating Expert System to

Manage a Distributed Sensor Network," *Proc. 1987 Tri-Service Data Fusion Symp.*, Johns Hopkins University, Laurel, MD, June 1987.
67. Grimshaw, J., and P. Amburn, "Kalman Filter Residual Expert System," *1988 NAECON Proc.*, Dayton, OH, May 1988.
68. Kuczewski, R.M., "Neural Network Approaches to Multi-Target Tracking," *1st Int. Conf. on Neural Networks*, San Diego, CA, June 1987.
69. Cofer, R.H., "Data Fusion as the Guiding Principle in the Broad Area Search Problem," *Proc. 1988 Tri-Service Data Fusion Symp.*, Johns Hopkins University, Laurel, MD, May 1988.
70. Kadar, I., and M. Lodaya, "Decision Aiding: An Interactive Tracking and Identification System," *Proc. 1988 Tri-Service Data Fusion Symp.*, Johns Hopkins University, Laurel, MD, May 1988.
71. Hartless, M.L., "Multisensor Adaptive Control for the ADI Engagement Process," *Proc. 1988 Tri-Service Data Fusion Symp.*, Johns Hopkins University, Laurel, MD, May 1988.
72. Fahat, N.H., *et al.*, "Non-Cooperative Identification of Airborne Targets Using Microwave Diversity Imaging and Artificial Neural Networks."
73. Priebe, C., and D. Marchette, "An Adaptive Hull-to-Emitter Correlator," *Proc. 1988 Tri-Service Data Fusion Symp.*, Johns Hopkins University, Laurel, MD, May 1988.
74. Bowman, C., "Artificial Neural Network Adaptive Systems Applied to Multisensor ID," *Proc. 1988 Tri-Service Data Fusion Symp.*, Johns Hopkins University, Laurel, MD, May 1988.
75. Spence, C.D., *et al.*, "A Neural Network Computational Map Approach to Sensory Fusion," *Proc. 1988 Tri-Service Data Fusion Symp.*, Johns Hopkins University, Laurel, MD, May 1988.
76. Garvey, T.D., "A Survey of AI Approaches to the Integration of Information," *Proc. SPIE*, Vol. 782, *Infrared Sensors and Sensor Fusion*, May 1987.
77. Spiessbach, A.J., and J.F. Gilmore, "AI Context Analysis for Automatic Target Recognition," *Proc. Army Conf. on Application of AI to Battlefield Information Management*, U.S. Navy Surface Weapons Ctr., White Oak, MD, April 1983.
78. Weaver, C.B. (ed.), "Sensor Fusion," *Proc. SPIE*, Vol. 931, Orlando, FL, April 1988.
79. Xue-min, C., *et al.*, "A Knowledge-Based Reasoning System for Ship Identification," *1988 NAECON Proc.*, Dayton, OH, May 1988.
80. Davis, L.C., and J.R. Aldrich, "A Knowledge-Based Approach to Naval Multisensor Integration," *Proc. IEEE Conf. on Automating Intelligent Behavior*, Nat. Bureau of Standards, Gaithersburg, MD, May 1983.
81. Atwater, F.M., "An Expert System Approach to Situation Assessment Using Sensor Fusion," *Proc. AFCEA Seminar on AI Applications to the Battlefield*, Ft. Monmouth, NJ, May 1985.
82. Spain, D.S., "Application of Artificial Intelligence to Tactical Situation Assessment," *Proc. IEEE 16th EASCON*, Washington, DC, September 1983.
83. Gibbons, G.D., *et al.*, "ASAP: AI-Based Situation Assessment and Planning," *Proc. 1988 NAECON*, Dayton, OH, May 1988.
84. Chubb, D.W.J., "An Automated Tactical Situation Assessment Methodology: Its Limitations and Impact upon Tactical Command and Control," *Proc. 1988 Tri-Service Data Fusion Symp.*, Johns Hopkins University, Laurel, MD, May 1988.
85. Babcock, D.F., "An Architecture for the Application of AI techniques to Threat Warning," *Proc. Army Conf. on Application of AI to Battlefield Information Management*, U.S. Navy Surface Weapons Ctr., White Oak, MD, April 1983.
86. Harris, C.J., *Application of Artificial Intelligence to Command and Control Systems*, Peter Peregrinus Ltd., London, 1988.
87. Fox, G., and S. Arkin, "From Laboratory to the Field: Practical Considerations in Large Scale Data Fusion System Development," *Proc. 1988 Tri-Service Data Fusion Symp.*, Johns Hopkins University, Laurel, MD, May 1988.

INDEX

A priori data, 78
ACDS, 373
Adaptive neural networks, 228-230
ALLIES, 340-341
AMUID (automated multi-unit indentification), 334-337
Analysis of variance (ANOVA), 402
Angle of arrival (AOA), 156, 173
Antisubmarine warfare (ASW), 51-54
Architecture,
    connectionist, 370-373
    functional, 17-18
    generic level 1, 20-30
    integrated, 376
    military data fusion, 60-67
    parallel, 40-41
    parallel processing, 363
    real-world, 23-28
    tracker-correlator, 20-23
Artificial intelligence (AI),
    benefits, 427-433
    description, 425
    technology, 440-441
ASE (advanced sensor exploitation), 335
Assignment,
    report-to-track, 177-178
    sensors-to-targets, 143-147
Attentional allocation, 273
Attribute classification, 7
Attribute fusion, 63-66
Automatic target recognition (ATR), 23, 97

Battlefield warfare, 57-58
Bayesian theory,
    a posteriori probability, 242
    a priori probability, 242, 243
    axioms, 240-241
    historical background, 9
    likelihood, 242
    mathematics, 239-240
    probability theory, 239-245
Bayes' criteria, 83-84
Belief networks, 238-239
BETA (battlefield exploitation and target recognition), 317-318
Blackboard architectures, 33-40

Canonical model of C^3, 396
Classification (security), 75
Classification,
    nonparametric, 90
    nonstatistical, 100
    parametric, 90
    statistical, 100
    syntactic, 100
    unsupervised, 100
Cluster analysis, 224
COMINT (communications intelligence), 116, 121
Conditional probability matrix (CPM), 90
Connectionist architectures, 370-373
Contextual analysis, 299
Countermeasures, 74

Data association,
    batch processes, 188
    Bayesian methods, 192
    definition, 159
    description, 167
    dynamic, 163-165

nearest neighbor, 193
non-Bayesian methods, 192
recursive processes, 188
report-to-track, 187–193
static, 168
track-to-track, 193–198
Data fusion,
centralized architectures, 373–376
data base management requirements, 358–360
definition, 1–2
functional model, 5–8
integrated architectures, 376–382
levels of, 15
military applications, 3
modeling, 406–416
motivations, 2–5
parallelism, 363–368
physical architectures, 349–350
process model, 15–20
processing requirements, 357
research sources, 12
role of artificial intelligence, 425–427, 441–442
Database management system (DMS), 360
DECADE (decision and development and evaluation), 328
Decision support, 60–67
Decision theory, 8–9
Decisions,
hard, 69
soft, 69
Dempster-Shafer,
axioms, 240–241
evidence theory, 245–252
frame of discernment, 247–249
mathematics, 239–240
method, 222–224
rule of combination, 247–249
versus Bayesian, 237–260
Detection,
centralized, 86
definition, 80–81
distributed, 86
Direction of arrival (DOA), 173
Distance measures, 168–169
spatial, 168
statistical, 169

ELINT (electronic intelligence), 58, 73, 308–310
Electronic support measure (ESM) sensors, 53, 58
Emission control (EMCON), 134
Entropy methods, 230–232
ESIAS (expert system for intelligence analysis support), 328–334
Estimation, definition, 81
Estimation theory, 9
Event detection, 273
Event-activity profiling, 308–316

Features, 98
Figure of merit (FOM), 316–327
Flowdown of requirements, 350, 351
Ford-Fulkerson algorithm, 147
Fused information, characteristics, 67–73
Fusion,
common aperture, 21
measurement, 20
track file, 20
Fuzzy set theory, 236–237

Gating test, 169–170
Geometric dilution of precision (GDOP), 173

Hierarchical simulation models, 415–416
HUMINT (human intelligence), 121
Hypothesis
generation, 276
maintenance, 192
management, 180
scoring, 177–178, 192

Identity estimation, 214
Indirect data, 78
Inference,
Bayesian, 220–222
classical, 216–219
Information security (INFOSEC), 73–74
Integrated avionics, 378–382
Integrated CNI, 377
Integrated sensors, 376, 377–378
Intelligence preparation of battlefield (IPB), 301–308
Intelligent assistant, 432–433

Joint Directors of Laboratories (JDL) data fusion subpanel (DFS), 10, 11, 15

Lanchester models, 390–393
Learning, 432
Likelihood ratio, 84
Line of position (LOP), 173
Links,
    definition, 78
    design considerations for data fusion, 123–124
    information exchanged, 122–123
    tactical data, 123
Logical templating, 232–234

Measures of effectiveness (MOE), 404
Measures of force effectiveness (MOFE), 403
Measures of performance (MOP), 403
Military exercises, 419
Military problem solving, 265–277
Military worth, 419–423
Multiagent recognition, 434–436
Multiple attribute utility theory (MAUT), 333, 420
Multiple hypothesis attribute combination, 367–368
Multiple hypothesis tracking (MHT), 368
Munkres algorithm, 147

Natural language processing (NLP), 429–430
Neyman-Pierson criteria, 84
Noncooperative target recognition (NCTR), 97, 151

Parallelism,
    decomposition of algorithms, 366–367
    logic programming, 44
    processing architectures, 363–366
    rule-based systems, 41–42
    semantic networks, 43–44
Pattern matching, 433
Pave pillar, 376
PENDRAGON, 334–337
Perceptual reasoning machine (PRM), 341
Plan position indicator (PPI), 69
Plan recognition, 430–431
Prioritization (of targets), 136–137
Probabilities,
    classical, 243
    empirical, 242
    subjective, 243
Problem solving,
    communicative, 33
    cooperative, 33
    opportunistic, 33
Pulse recognition model (PRM), 435
Pulse repetition interval (PRI), 218–219

Quantitative assessment, 389–390

REACT (rapid expert assessment to counter threats), 38–39
Receiver operating characteristic (ROC), 85
Remote sensing, 77
Rule-based systems (RBS), 438–440

Satisfying solutions, 426
Search problems, 441
Segmentation, 102–103
Sensors,
    acoustic, 118–120
    cueing, 147, 149–151
    definition, 77–78
    electronic support measures (ESM), 114–118
    electro-optical, 113
    general model, 78–80
    hand-off, 147
    hard-decision, 88, 92
    identification friend or foe (IFF), 113–114
    infrared search and track, 110–113
    interfaces, 135–136
    modeling, 409–413
    prediction, 133
    radar, 106–110
    requirements definition, 353–354
    soft-decision, 88, 92, 93, 97–103
SHOR (stimulus-hypothesis-option-response) model, 169–177, 269–277
SIGINT (signal intelligent), 116, 308–310
Signal-to-noise ratio (SNR), 187
Simulation of data fusion, 406
Situation and threat assessment (STA), 293–294
Situation assessment,
    dealing with CC&D, 278–283
    functional requirements, 28–33
Situation displays, 70
Source data, 121–122
Sources, definition, 78
Spatial alignment, 175

Stansfield analysis, 173
STARS/PRM, 341–343
State estimation,
    definition, 160
    description, 167
    general problem, 198–200
    least squares, 200–201
    maximum likelihood, 201–203
    minimum variance, 203
    multiple sensor, 208–210
    weighted least squares, 201
State transition model, 265
State vector model of C^2, 396–397
Surface of position (SOP), 172
Systems analysis, 399–402
Systems engineering process, 350–358

Tactical air warfare (TACAIR), 54–57
Tactical assessment expert system (TAES), 341
Target localization, 160–161, 163
Templates,
    decision support, 307–308
    doctrinal, 303–304
    event, 306–307
    expectation, 294, 300–301
    situation, 305
    terrain analysis, 304–305
Templating, 232–234
Temporal alignment, 175
Terrain analysis, 304–305
Testbeds, 418
Testing and evaluation (T&E), 452–454
Thermodynamic model of C^3, 394–396
Threat assessment,
    expectation template, 290
    force capability, 286–287
    functional requirements, 28–33
    hostile intent, 287–288
    shifting perspectives, 288–289
    threat elements, 286–289
Threat evaluation, 301–302
Threat integration, 302
Time difference of arrival (TDOA), 173
Time of arrival (TOA), 173
Tracker-correlator,
    autonomous, 21
    centralized, 21
    functional architectures, 20–23
    hybrid, 21
Truth maintenance system (TMS), 40

Uncertainty,
    representation, 448–449
    sources of, 447–448

Voting methods, 230

THE AUTHORS

Edward L. Waltz earned his BS in Electrical Engineering at the Case Institute of Technology and an MS in Computer, Information and Control Engineering at the University of Michigan. At the Bendix Aerospace Systems Division, he was responsible for digital signal processing systems design on numeous imaging and non-imaging sensor systems, including the LANDSAT remote sensing satellite, airborne remote sensing systems, ocean surveillance projects, and Space Shuttle sensor experiments. Now with the Bendix Communications Division of the Allied-Signal Aerospace Company, he has been responsible for data fusion research and development as well as managing both R&D and full-scale development data fusion programs. He is currently the Systems Engineering Manager for the U.S. Tri-Service Mark XV IFF program.

James Llinas earned his BS in Aerospace Engineering at the Polytechnic Institute of New York, and an MS in Engineering Science and PhD in Operations Research and Statistics at the State University of New York at Buffalo. He served as Assistant Professor at the State University of New York at Albany for over two years, and was an Adjunct Professor at George Washington University, Washington, D.C. for two years. He was a co-author and lecturer of 2 and 3 day seminars in "Data Fusion and Multi-Sensor Correlation," presented numerous times in the U.S., Europe, and Australia. For nearly the last 5 years he has been a technical advisor to the DoD/Joint Directors of Laboratories Data Fusion Corporation.

The Artech House Radar Library

David K. Barton, *Series Editor*

Modern Radar System Analysis by David K. Barton

Introduction to Electronic Warfare by D. Curtis Schleher

High Resolution Radar by Donald R. Wehner

Electronic Intelligence: The Analysis of Radar Signals by Richard G. Wiley

Electronic Intelligence: The Interception of Radar Signals by Richard G. Wiley

Pulse Train Analysis Using Personal Computers by Richard G. Wiley and Michael B. Szymanski

RGCALC: Radar Ramage Detection Software and User's Manual by John E. Fielding and Gary D. Reynolds

Over-The-Horizon Radar by A.A. Kolosov, *et al.*

Principles and Applications of Millimeter-Wave Radar, Charles E. Brown and Nicholas C. Currie, eds.

Mulitple-Target Tracking with Radar Applications by Samuel S. Blackman

Solid-State Radar Transmitters by Edward D. Ostroff, *et al.*

Logarithmic Amplification by Richard Smith Hughes

Radar Propagation at Low Altitudes by M.L. Meeks

Radar Cross Section by Eugene F. Knott, *et al.*

Radar Anti-Jamming Techniques by M.V. Maksimov, *et al.*

Radar System Design and Analsis by S.A. Hovanessian

Aspects of Radar Signal Processing by Bernard Lewis, Frank Kretschmer, and Wesley Shelton

Monopulse Principles and Techniques by Samuel M. Sherman

Monopulse Radar by A.I. Leonov and K.I. Fomichev

Receiving Systems Design by Stephen J. Erst

High Resolution Radar Imaging by Dean L. Mensa

Radar Detection by J.V. DiFranco and W.L. Rubin

Handbook of Radar Measurement by David K. Barton and Harold R. Ward

Statistical Theory of Extended Radar Targets by R.V. Ostrovityanov and F.A. Basalov

Radar Technology, Eli Brookner, ed.

The Scattering of Electromagnetic Waves from Rough Surfaces by Petr Beckmann and Andre Spizzichino

Radar Range-Performance Analysis by Lamont V. Blake

Interference Suppression Techniques for Microwave Antennas and Transmitters by Ernest R. Freeman

Signal Theory and Random Processes by Harry Urkowitz

Techniques of Radar Reflectivity Measurement by Nicolas C. Currie

SIGCLUT: Surface and Volumetric Clutter-to-Noise, Jammer and Target Signal-to-Noise Radar Calculation Software and User's Manual by William A. Skillman

Radar Reflectivity of Land Sea by Maurice W. Long

Aspects of Modern Radar, by Eli Brookner, *et al.*

Analog Automatic Control Loops in Radar and EW by Richard S. Hughes

Introduction to Sensor Systems by S.A. Hovanessian

VCCALC: Vertical Coverage Plotting Software and User's Manual by John E. Fielding and Gary D. Reynolds

Electronic Homing Systems by M.V. Maksimov and G.I. Gorgonov

Principles of Modern Radar Systems by Michel H. Carpentier

Secondary Surveillance Radar by Michael C. Stevens

Printed in the United States
3274